Hiromi Nishida • Taku Oshima
Editors

DNA Traffic in the Environment

Springer

Editors
Hiromi Nishida
Biotechnology Research Center
& Department of Biotechnology
Toyama Prefectural University
Toyama
Japan

Taku Oshima
Biotechnology Research Center
& Department of Biotechnology
Toyama Prefectural University
Toyama
Japan

ISBN 978-981-13-3410-8 ISBN 978-981-13-3411-5 (eBook)
https://doi.org/10.1007/978-981-13-3411-5

Library of Congress Control Number: 2019932127

This Springer imprint is published by the registered company Springer Nature Singapore Pte Ltd.
The registered company address is: 152 Beach Road, #21-01/04 Gateway East, Singapore 189721, Singapore

Preface

Recent advances in microbial genomics technologies, environmental genomics, molecular microbiology, and information and theoretical sciences have made it possible to show that horizontal transfer of genetic elements has been occurring during the evolution of archaea and bacteria and is ongoing in the environment, which could be a major driving force for the evolution of these prokaryotes. Evidence suggests that they are acquiring abilities to carry out new actions through horizontal transfer of genetic elements, which may have both good and bad outcomes for human beings: for instance, bioremediation on the one hand and antibiotic resistance on the other. Thus, we believe that the horizontal transfer of genetic elements will continue to be an important topic for agricultural, medical, microbial, and other sciences.

This book consists of ten chapters written by experts in the field, covering the recent progress of studies into multiple aspects of DNA traffic in the environment: DNA carriers (phages, plasmids, transposons, and vector particles, in Chaps. 4–8), effects of DNA transfer (antibiotic resistance and bioremediation, in Chaps. 9 and 10), and regulation to support DNA transfer (gene silencing, in Chaps. 1 and 2). Also described are the exchange of DNA resources by horizontal transfer between archaea and bacteria (Chap. 2) and similarity of defense systems to restrict the invasion of foreign DNA (Chap. 3)—for example, clustered, regularly interspaced, short palindromic repeats (CRISPR)—and how mobile DNA can overcome those barriers (Chap. 1). The chapters show how these organisms have used DNA resources in the environment and affected the environment itself.

We express our deep appreciation to all authors for providing their cooperation and contributions for the publication of this book.

Toyama, Japan Hiromi Nishida
Toyama, Japan Taku Oshima

Contents

Xenogeneic Silencing and Horizontal Gene Transfer

1

Chiho Suzuki-Minakuchi and William Wiley Navarre

Abstract

The genomes of free-living eubacteria and archaea are constantly shaped by recombination and genetic exchange; a fact that allows prokaryotes to evolve at rates far faster than if they had to evolve new traits via mutation. Microbes contain many defenses against foreign DNA, and the competing forces of evolvability and genomic security determine the extent to which a microbe can take advantage of genetic material obtained from other cells. Over the past decade, we have learned that many bacterial species employ silencing proteins to specifically regulate genes obtained via horizontal gene transfer, including many involved in virulence and drug resistance. The "xenogeneic silencing" paradigm implies that bacteria, by silencing expression of genes that have %GC-content lower than the genome average, are able to acquire genes that otherwise would be toxic if left unregulated. This paradigm also implies that bacteria maintain bias in their genomes to distinguish self-DNA from nonself-DNA and that the nonself-DNA is subject to different evolutionary and regulatory forces than the genes in the ancestral genome. In this chapter, we discuss the forces that shape bacterial genomes and how silencing enables the spread of genetic elements. We will discuss the properties of different silencers and speculate on the role of silencing proteins encoded on many mobile genetic elements including plasmids.

C. Suzuki-Minakuchi
Biotechnology Research Center, The University of Tokyo, Tokyo, Japan
e-mail: ucsmina@mail.ecc.u-tokyo.ac.jp

W. W. Navarre (✉)
Department of Molecular Genetics, University of Toronto, Toronto, ON, Canada
e-mail: william.navarre@utoronto.ca

H. Nishida, T. Oshima (eds.), *DNA Traffic in the Environment*,
https://doi.org/10.1007/978-981-13-3411-5_1

Keywords
Xenogeneic silencing · H-NS · MvaT · Lsr2 · Horizontal gene transfer · Lateral gene transfer · Bacterial evolution · Bacterial genomes · Bacterial plasmids

1.1 Prokaryotic Evolution: Harnessing Power of Mobile DNA

Prokaryotes (archaea and bacteria) and eukaryotes have adopted distinct modes of evolution. Eukaryotes prioritize genomic security and generally shun genetic exchange with organisms outside of their own species. To generate novelty and diversity, eukaryotes make extensive use of sexual reproduction, gene duplication, and domain swapping (modularity) to engineer and refine novel proteins from their preexisting complement of genes [1–5]. This is reflected in the structure of the eukaryotic proteins, which often comprise arrays of modular domain types (e.g., SH2, SH3, WW, and CARD) assembled in different arrangements [6, 7].

Prokaryotes, on the other hand, make extensive use of horizontal (lateral) gene transfer (HGT)—defined as the transfer of genes between strains, species, and even phyla [8–11] (see also Chap. 2). This is reflected in the structures of prokaryotic genomes which show remarkable differences in gene content even within members of the same species. Nowhere is the role of HGT more visible than in the evolution of new pathogens and in the rapid emergence and spread of antibiotic resistance (see also Chaps. 6 and 10). Comparisons of thousands of prokaryotic genomes reveal that most free-living (i.e., not obligately symbiotic) species have "open genomes" that are shaped both by canonical mutational forces and by the acquisition and loss of large segments of DNA [12].

Whole genome sequencing of thousands of prokaryotic genomes over the past two decades has dramatically improved our understanding of bacterial genome content, evolution, and structure. While the genomes of all cells encode a largely universal set of "informational" genes (e.g., factors essential for transcription, DNA replication, protein synthesis, cell wall synthesis, cell division, core metabolism, nutrient transport, etc.), bacteria also encode additional functions that are more or less unique to their clade. Comparisons of *Escherichia coli* genomes found that while a typical isolate will encode approximately 5000 genes, only slightly less than 1500 of these are universally conserved among all strains of *E. coli*—and less that 1000 genes if *Shigella* sp. are included in the analysis [13, 14]. That means 70–80% of the genome of any given *E. coli* isolate comprises the noncore accessory genes. A recent analysis of *Salmonella*, for which there are more than 100,000 genome sequences available, reveals that whether a gene is called "core" or "accessory" is a matter of how stringent different criteria are used. If relatively relaxed criteria (a so-called soft-core genome) are used, the Salmonellae share a common core of 3002 genes. Here, the definition used was that a gene had to be present >98% or more, intact in >94%, and of "unexceptional diversity" in 3144 representative *Salmonella* genomes [15]. If an absolute presence/absence criteria is used, however, the number of core genes drops nearly tenfold to a mere 360 genes. This means that depending on the criteria used, the *Salmonella* core genome comprises either half or less than 8% of any given *Salmonella* isolate. It is likely, however, that the more relaxed inclusion criteria are ultimately more meaningful in terms of understanding the biology of the species as a whole.

The clade-specific core genes often give us clues as to the habitat and lifestyle of the species. One example, of many, is the *Salmonella* pathogenicity island 1 (or SPI-1), a cluster of several genes that encode a type 3 secretion system that is central to *Salmonella*'s pro-inflammatory lifestyle [16, 17]. All strains of *Salmonella enterica* and *Salmonella bongori* encode SPI-1, and the island is unique to the species *Salmonella*, although many bacteria encode similar type 3 secretion systems [15, 18, 19]. In addition to the SPI-1 system, most Salmonellae also encode metabolic genes that enable them to take advantage of the metabolic by-products of inflammation including peroxide, tetrathionate, and nitrate/nitrite [16, 17].

For bacterial species like *E. coli* and *Salmonella*, the "pan-genome," or the complete set of genes found within the total membership of the species, is far larger than the core genome in bacteria. A recent study estimates that the *Salmonella* pan-genome comprises approximately 25,000 genes, while *E. coli* was estimated to have larger pan-genome of at least 38,000 genes despite fewer strains being included in the analysis [19, 20]. These observations tell us that the majority of bacterial genes are drawn from a vast pool of sequences that flow between members of a given bacterial species and, less often, across species. This pool of mobile genes have been termed the "accessory" or "flexible" genome. The impact that accessory genes have on the ability of bacteria to rapidly adapt to novel and challenging environments is enormous. However, while the functions of most core genes in bacterial genomes are largely understood or can be inferred, the accessory genome is replete with genes of unknown function [21]. Most of these enigmatic accessory genes have no measurable impact on cellular fitness during standard laboratory conditions but probably are critical to survival in a specific set of conditions. It is likely that many of these genes play a role in enhancing competitive fitness by promoting interspecies cooperation or by protecting against an array of competing microbes, viruses, and potentially deleterious mobile genetic elements [22]. In such cases, the functions of these genes cannot be understood when microbes are grown in monoculture.

There is of course a balance that must be struck between genomic flexibility (i.e., the ability to acquire new DNA from foreign sources) and genomic security. To lessen the deleterious effects of HGT, bacteria encode a diverse array of defense systems including restriction systems, CRISPR systems, BREX systems, abortive infection and T/A systems, and nucleases like RecBCD that destroy free ends of incoming DNA [23–27]. Bacterial cells also constantly alter their cell surfaces to prevent phage infection and limit their susceptibility to rampant conjugation [26, 27] (see also Chap. 3).

Correspondingly, mobile genetic elements like transposons, temperate phages, and conjugal plasmids utilize complex, diverse, and mechanistically remarkable ways to ensure their survival and spread to new host genomes. Conjugal plasmids often encode factors that block restriction enzymes and blunt the induction of the host SOS response that is triggered when single-stranded plasmid DNA enters the cell during the conjugation process. Plasmids and phages have been shown to encode both anti-restriction factors and anti-CRISPR proteins [28–30]. In spite of their activity to subvert host defenses, most evolutionarily successful (widespread) mobile genetic elements will balance their negative impact by encoding factors that improve the fitness of their hosts [22, 31–34]. These include resistance to antibiotics, heavy metals, or by protecting the cell from other mobile elements [35]. For example, many temperate phages encode factors that prevent infection by other phages [23].

Therefore, there exists a complex interplay not only between a microbe and mobile genetic elements but also between different mobile genetic elements as they compete for territory in their varied microbial hosts [22, 31–34]. Our limited understanding of the varied forces that shape bacterial genomes is largely due to the fact that the majority of experimental research in this arena has focused on a few lab-adapted isolates of a small number of bacterial species.

For the rest of this chapter, we will focus on how some bacterial species employ gene silencing as a mechanism to both take advantage of and protect themselves from the potential dangers of HGT. Recent discoveries about bacterial gene silencing have given us tremendous insight into how many accessory genes are regulated and how bacterial regulatory networks can be resilient in the face of rampant gene exchange.

1.2 Bacterial Genomes Are Built to Prevent Parasitism

There are several sequence and structural features of bacterial genomes that allow it to distinguish its own DNA from foreign DNA. For example, most bacterial chromosomes are circular (i.e., no free ends), double stranded, negatively supercoiled, and methylated at specific sequences. In contrast, DNA entering a cell from phages, plasmids, and via competence usually enters the cell with a free end, relaxed, often single-stranded, and many times without the methylation patterns that would prevent it from being cleaved by restriction enzymes (see also Chap. 3). The free ends of newly introduced dsDNA are subject to degradation by exonucleases including the RecBCD complex, which not only can destroy foreign DNA directly but can also provide templates for CRISPR spacer acquisition [36].

This tells us that bacterial cells encode systems, like RecBCD/SOS, that constantly surveil for signatures of "nonself" nucleic acid that can indicate whether the cell is being invaded by a phage or plasmid. It also tells us that many fundamental and intrinsic features of the bacterial chromosome actually arose as a means to prevent genomic parasitism.

This brings up the question of what types of signatures can be used to tell self-DNA from nonself-DNA. Below, we will make the case that GC-content and codon bias are also features that bacteria use to discriminate genetic material that is foreign (see also Chap. 6).

1.3 Characteristics of Core Genes vs. Accessory (Noncore) Genes in the Genome

In 1952, Erwin Chargaff first observed that the genomic DNA of three bacterial species (*Haemophilus*, *Mycobacterium*, and *Serratia*) differed in the relative amount of their constituent nucleotides [37]. While *Serratia* and *Mycobacterium* were relatively rich in guanine and cytosine, *Haemophillus* was shown to be rich in adenine and thymine. These analyses were subsequently expanded to other microbes, revealing that each had a characteristic nucleotide composition [38, 39]. Indeed, this rather simple metric has become a reliable characteristic to describe individual species and is sometimes group different bacteria into particular phylogenetic clades. The Actinobacteria, for example, are often referred to as the "GC-rich Gram-positive bacteria," while the Firmicutes are known as the "AT-rich Gram-positive bacteria."

While bacterial GC-contents can range from less than 20% to slightly greater than 75%, the genomic GC-contents of any species tends to fall in a very narrow range. For example, almost every isolate of *E. coli* (10,680 out of 10,720 sequenced isolates) has a genomic %GC-content between 50 and 51.2. A scan of the NCBI Microbial Genomes Database (https://www.ncbi.nlm.nih.gov/genome/microbes/) reveals that similarly narrow ranges are also found among thousands of isolates of *S. enterica* (all between 51% and 52.5%), *Staphylococcus aureus* (32.4–33.5%), and *Pseudomonas aeruginosa* (65.1–66.8%).

The root causes and consequences of compositional bias are largely not understood and are controversial [40–45]. It is often assumed, for example, that prokaryotes with GC-rich genomes might be those that live at higher temperatures, but a careful analysis reveals that this is not universally true. The most GC-rich bacterial genome sequenced to date (%GC = 76) is from *Geodermatophilus telluris*, an Actinobacteria with an optimal growth temperature of approximately 30 degrees [46]. This contrasts with the thermophiles *Sulfolobus* (%GC = 35), *Chloroflexus aurantiacus* (%GC = 53), and *Thermophilus* %GC = 68. In fact, there is no clear extrinsic environmental factor (oxygen, temperature, pH, and salinity) that has been shown to directly dictate the genomic GC-content of a given microbe, although loose correlations have been identified [43, 47].

Another factor that influences GC-content may relate to the "social lifestyle" of the microbe. That is, microbes that live in complex communities that must compete with other microbes for limiting resources tend to have large genomes with higher %GC, although this correlation is not absolute [48, 49]. In the opposite direction, microbes that live in isolation or as symbionts of other microbes or animal hosts, such as endosymbionts, almost uniformly have small and functionally reduced genomes with very low GC-contents [50]. The most extreme example is the endosymbiont *Candidatus Zinderia insecticola* (%GC = 13) with a genome of only 200 genes [50]. However, very few, if any, free-living microbes have GC-contents below 25%.

Regardless of cause, however, these observations tell us that bacterial genomes adhere to a biased set of compositional codes that are unique to their species. Simply put, bacterial genomes have a "sense of self." The flip side of this insight is that compositional biases can be used to distinguish whether a given gene in a genome is ancestral or whether it was acquired in the past from a different bacterial species. Indeed, with the sequencing of the first bacterial genomes it was immediately apparent that the accessory and foreign genes often have different nucleotide compositions than those of the core genome. Specifically, they frequently differ in their %GC and they employ codons in ratios that deviate from the species norm.

In Fig. 1.1, we show the nucleotide compositions of bacterial genes with respect to whether they are core or noncore (accessory) genes. Here, the %GC of genes from *E. coli* and *S. aureus* are displayed, grouped by whether they are common to all members of the species (core genes), or only found in a subset of species members and hence either were acquired via horizontal gene transfer or were core genes that were discarded from a specific subset of genomes. Here, we make no distinction between the two possibilities and will simply call genes present in >97% of isolates "core" genes, and those that are present in fewer than 97% of species isolates "noncore." Using these criteria, we find that *E. coli* strains share approximately 2700 genes—similar to the number of "soft-core" genes found in *Salmonella*.

From Fig. 1.1, a few facts are apparent:

First: The Core/Housekeeping Genes Display the Narrowest Distribution of %GC In contrast, the noncore genes display a wider range of nucleotide compositions. This suggests that microbial core genes adopt a more uniform set of characteristics that can distinguish them from the noncore genes. That is, the "core" genes are biased toward a particular set of characteristics, while the foreign genes are more random in their compositions.

Second: The Noncore Genes Are Generally More AT Rich than the Core Genes, Even in AT-Rich Bacteria Remarkably, despite its low %GC the accessory genes in *S. aureus* are generally even more AT rich. This suggests that AT-rich bacteria do not accumulate GC-rich genes. The staph genome also is smaller and more "closed" (has fewer noncore genes as a percentage of the total genome) than that of *E. coli*. What is particularly surprising is that even noncore genes that are "almost core" (e.g., appear in 65 out of 70 strains) are dissimilar to the core genes in their nucleotide composition.

Third: Accessory Genes Do Not Necessarily Ameliorate to the GC-Contents of Their Host Some theories on genome evolution posit that horizontally acquired genes, that initially contain the biases of their previous host genome, will gradually acquire the compositional characteristics of their host genome over long periods of time because they become subject to the same mutational forces [51, 52]. It is notable, however, that many noncore genes remain more AT rich even when they are widely distributed within the species, suggesting that many accessory genes are under selective pressure to remain compositionally distinct from the core genome.

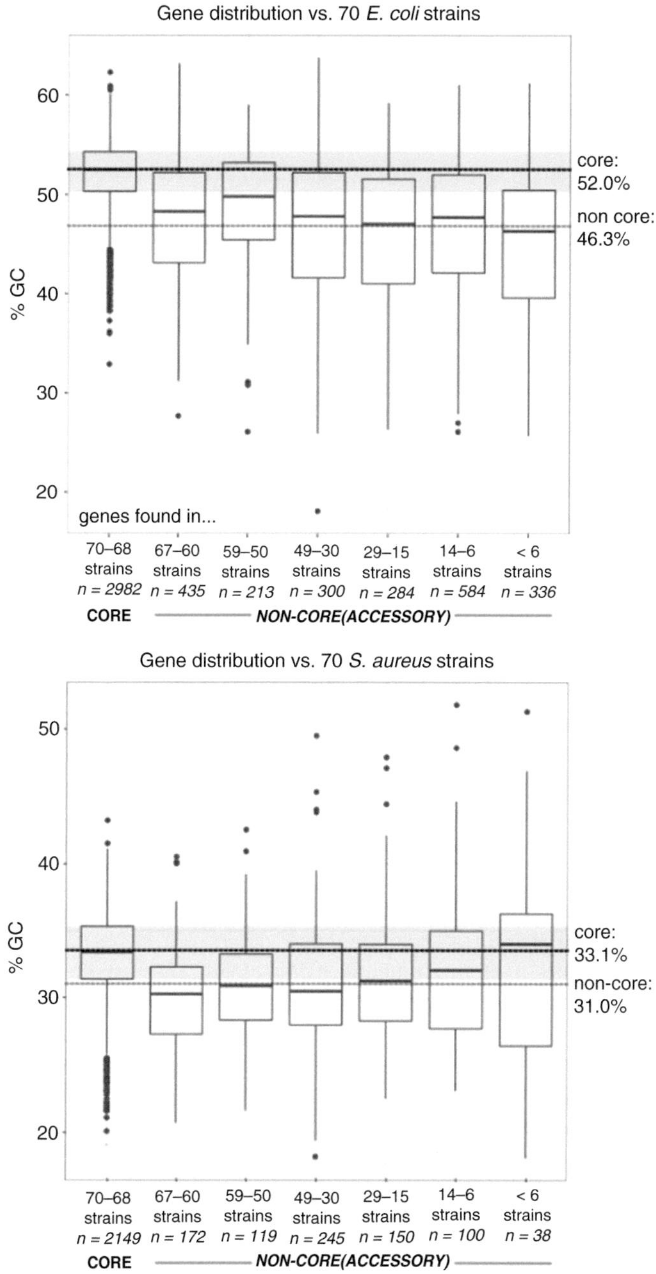

Fig. 1.1 A comparison of %GC-content of genes relative to their distributions in *Escherichia coli* and *Staphylococcus aureus*. The protein-coding genes of *E. coli* strain Xuhou21 (left panel) or *S. aureus* strain USA300 (right panel) were compared by GC-content relative to their distribution among 70 other diverse strain isolates from their species

From the perspective of a researcher, these species-specific compositional signatures can be useful to infer the origins and phylogeny of both bacteria and their individual genes. In addition to GC-contents, other subtle sequence characteristics are unique to different bacterial species. Different species will vary in their use of specific dinucleotide, trinucleotide, or higher-order "k-mer" motifs. The fingerprint of these motifs (often up to 18-mers), when calculated over a large area, is almost as robust as using 16S sequencing to identify a particular bacterial species. This is the basis of a new suite of epidemiological software tools including "Kmer Finder" [53] and "Kmer ID" [54] that identify species by comparing their k-mer fingerprint against a database of known isolates. Accordingly, finding sequences that deviate from the genome norm is a way to identify sequences that have been acquired by HGT, which is the basis of algorithms like "Alien Hunter" from the Wellcome Trust Sanger Institute [55–58].

From the perspective of the microbe, genomic compositional biases can serve as a way to distinguish self-DNA from nonself-DNA. However, that means microbes must employ specific systems to "read" the compositional signatures of their DNA. For the remainder of the chapter, we will discuss how %GC-bias is exploited by many bacterial species to specifically silence the expression of DNA acquired by HGT.

1.4 Xenogeneic Silencing: Repressing Foreign DNA As a Balance Between Evolvability and Security

In 2006, four laboratories, taking advantage of advances in microarray technology, independently discovered that the H-NS (heat-stable nucleoid structuring) proteins of *E. coli* and *Salmonella* specifically bind to and repress expression from AT-rich sequences in the genome that show evidence of being acquired by HGT [59–62]. All four groups utilized chromatin immunoprecipitation assays to determine the specific localization of H-NS (or epitope-tagged versions of H-NS) and analyzed the global gene expression of *hns* mutants using cDNA microarrays. These studies demonstrated that the H-NS protein specifically binds and represses transcription from regions that are significantly more AT rich than the *E. coli* and *Salmonella* chromosome and, with rare exception, these AT-rich regions showed evidence of having been acquired by HGT.

The transcriptional repression of foreign DNA on the basis of its atypical GC-content was termed "xenogeneic silencing." Xenogeneic silencers have since been identified in a diverse array of bacterial species where they regulate the majority of xenogeneic sequences (i.e., sequences derived from a foreign source). These silencers are grouped into distinct families based on conserved sequence signatures within their DNA-binding domains: the *H-NS*-like proteins found in several Gram-negative proteobacteria, the *MvaT*-like proteins of *Pseudomonas*, and the *Lsr2*-like proteins of *Actinobacteria* [63]. Bacteria frequently encode multiple paralogs of these proteins that are often encoded on genomic islands and plasmids acquired via LGT [64]. All strains of *E. coli* and *Salmonella* encode two H-NS paralogs (H-NS and StpA—53% identical to H-NS), but some strains encode homologues on islands or plasmids [65, 66]. The role of StpA remains mysterious as it has almost no phenotype when H-NS is present and it is largely ineffective as a silencer at most H-NS regulated genes even when expressed at similar levels in the cell [67]. This is despite the fact that chromatin immunoprecipitation studies find StpA binds virtually identical locations as H-NS [68]. Likewise, the plasmid-encoded H-NS-like Sfh protein appears to share some functional redundancy with H-NS [69]. In contrast, Ler, encoded on the *LEE* pathogenicity island of enteropathogenic *E. coli*, is an H-NS homologue that activates promoters within the *LEE* island by antagonizing H-NS [70, 71]. Therefore, we cannot assume that every H-NS, MvaT, or Lsr2 homologue identified bioinformatically is functionally similar. Another protein, Rok, from *Bacillus subtilis* may also serve as a xenogeneic silencer in that organism, but its role in silencing is less well studied. We will discuss the differences and similarities between these families in the sections below.

As a result of their activity, xenogeneic silencing proteins play a central role in the regulation of most virulence-associated genes in important pathogens including *E. coli*, *Shigella*, *Salmonella*, *Klebsiella*, *Yersinia*, *Vibrio*, *Bordetella*, *Pseudomonas*, and *Mycobacteria* [61, 72–77]. For this reason, there has been considerable interest toward understanding the role that xenogeneic silencing plays in the evolution and control of virulence. Over the past decade, we have gained a greater understanding of how silencers specifically target AT-rich DNA, how oligomerization is critical for silencing, how silencing impacts bacterial evolution and fitness, and how silencing can be alleviated via "countersilencing" to activate gene expression in response to specific environmental and physiological cues.

H-NS was first identified as a heat-stable factor that could stimulate *E. coli* in vitro transcription from phage templates at low concentrations but inhibit transcription when added at high concentrations [78, 79]. H-NS was again found during biochemical screens for bacterial proteins isolated under conditions used to isolate eukaryotic histones [80, 81]. Its name (heat-stable nucleoid structuring protein or H-NS) was given during early studies of bacterial "histone-like" proteins [82–85]. Subsequently, H-NS was repeatedly identified as responsible for regulating a number of disparate biological processes. As such, H-NS and its corresponding gene, *hns*, have been at some point referred to as H1 [78], H1a [86], 16 K [87], B1 [81], *bglY* [88], *osmZ* [89], *drdX* [90], *virR* [91], *cur* [92], and *pilG* [93]. The fact that H-NS was identified to be involved in so many different pathways ranging from virulence, pili expression, osmotic defense, and metabolism is due to its role as a master regulator of AT-rich accessory genes in the *E. coli* chromosome.

The fact that H-NS was originally identified in screens looking for bacterial analogs of eukaryotic histones has played a large part in how its function in the cell is viewed. The H-NS protein is typically considered a member of the "nucleoid-associated proteins" (NAPs), a poorly defined classification for a set of abundant DNA-binding proteins with loose sequence specificity believed to play roles in structuring the bacterial chromosome [94]. Other well-studied NAPs include Fis, IHF, and HU [95]. The NAP terminology, however, is fairly uninformative. It should be noted that the NAPs are not functionally equivalent, do not share a common evolutionary origin, and are not spread similarly or uniformly across the bacterial phylogenetic tree. While HU-like proteins are found across the bacterial domain, H-NS is found only in a subset of proteobacteria and the Fis protein is only found in close relatives of *E. coli*. In addition, many studies of "the" bacterial chromosome are derived from studies on *E. coli*, which for a variety of parameters is not representative of the majority of bacterial genomes. Therefore, it is not meaningful, and perhaps counterproductive, to place NAPs into the same descriptive category. There is no evidence that any NAPs, apart from those related to the known silencers (H-NS/StpA, MvaT, and Lsr2), play a role in defense against foreign DNA.

How can silencing of AT-rich DNA benefit both the bacterial host and the newly acquired gene? The impact of silencing on genome content is difficult to infer directly because we can only observe gene transfer events that are evolutionarily successful. HGT events that led to the demise of the recipient cell are erased from the evolutionary record and hence are invisible to us. However, as discussed below, there is evidence from both an analysis of sequenced genomes and experimental fitness studies that silencing protects cells from the unchecked expression of genes acquired by HGT. Simply put, silencing seems to improve the ability of AT-rich genes to be tolerated by their new host genome because they are not expressed at high levels. Indeed, studies of gene transfer using plasmids expressing recombinant proteins find that the most highly expressed genes are the ones that are most selected against [96]. This also suggests that the observation that AT-rich accessory genes have accumulated in the genomes of *E. coli* and *Salmonella* is because silencing by H-NS has lessened their negative impact on the fitness of the bacterial cell.

If this is true, we would expect to observe negative fitness consequences in the absence of silencing. While this is generally observed, the ablation of silencing reduces bacterial fitness to varying degrees in different species. For example, H-NS appears to be essential in *Yersinia* sp. [72, 97, 98], while it is largely dispensable in most strains of *E. coli* and at least one strain of *Klebsiella pneumoniae* [99]. *S. enterica* sv. Typhimurium strains lacking H-NS display strong fitness defects despite the fact that *Salmonella* is more closely related to *E. coli* than *Klebsiella* or *Yersinia* [60, 61]. The silencers MvaT and MvaU of *Pseudomonas* are also essential for the viability of *P. aeruginosa* strain PAO1 [100].

Several studies have explored the underlying causes of fitness defects in silencing deficient bacterial strains. The picture painted from these studies is that silencing can benefit the bacterial cell in both general and specific ways. For example, the essential role of MvaT and MvaU in *P. aeruginosa* strain PAO1 was explored by identifying single-gene deletion mutants for their ability to tolerate the loss of MvaT (via targeted degradation) in an MvaU-deficient background [100]. Strains harboring deletions in the Pf4 prophage and the Type-IV pilus that acts as its receptor remained viable in the absence of silencing, suggesting that silencing can be a beneficial protection against some phages both by silencing phage expression and by limiting the expression of phage receptors.

In another study, experimental evolution was used to identify second-site suppressor mutations that could restore fitness to *Salmonella hns* mutants [67]. Six independent *hns* mutant cultures were repeatedly passaged for ~300 generations, during which time the growth rates of all six lineages improved to that of wild-type *Salmonella* by the end of the experiment. Genome sequencing revealed that five of the evolved lineages independently incurred large deletions in the SPI-1 pathogenicity island—the first island acquired by *Salmonella* during its evolution, and central to its inflammatory lifestyle. SPI-1 misregulation was found to account for ~50% of fitness impact caused by the loss of H-NS. Additional mutations were identified in the oligomerization domain of the poorly understood H-NS paralog StpA. These StpA variants formed complexes on DNA with mobility similar to that of H-NS in electrophoretic mobility shift assays, suggesting that subtle changes to the StpA oligomerization are sufficient to alter its activity to be more like that of H-NS.

Together, these studies show how silencing enabled both *Pseudomonas* and *Salmonella* to acquire genetic elements (a prophage or SPI-1) that otherwise would have been toxic. H-NS, in fact, enabled the genesis of Salmonellae by buffering the fitness impact of the seminal and defining genetic locus of the species. However, these studies also show that there is no single reason why silencing can be beneficial to the bacterial cell and that the essentiality of silencing will often depend on the specific and aggregate properties of the particular genes that are present in any given bacterial genome.

In addition to protecting bacterial cells from the unmitigated expression of particular genes, it's also clear that silencing also prevents additional negative consequences that can occur from the acquisition of AT-rich DNA in more general fashion. Specifically, AT-rich sequences are far more likely to contain accidental promoter-like sequences that can titrate away the fairly limited amounts of RNA polymerase in the cell [101]. Transcription from these spurious promoters can drain resources, make unwanted antisense RNAs, cause frequent and unregulated conflicts with both the replication machinery and the transcription of nearby genes, and perhaps form potentially toxic R-loops [102–104].

1.5 The Mechanistic Basis for Xenogeneic Silencing

The H-NS, MvaT, and Lsr2 silencers are dissimilar in primary sequence and structure, but they share two common features. First is the ability to selectively bind AT-rich DNA, without recognizing a specific sequence, via their C-terminal DNA-binding domain [105]. Second is the ability to self-assemble into oligomers on DNA via interactions within the N-terminal part of the protein. These proteins initiate binding (nucleate) at higher-affinity sites on DNA before polymerizing cooperatively and laterally along AT-rich DNA [74, 106–109]. Remarkably, these unrelated proteins are able to functionally substitute for one another (e.g., Lsr2 from *Mycobacteria* can complement *E. coli hns* mutants, and vice versa), suggesting that they form fundamentally similar nucleoprotein complexes [110–113].

Studies of H-NS at several genes in *E. coli* and *Salmonella* indicate that silencing can occur by different mechanisms at different loci [114]. This mechanistic diversity may reflect the fact that silencers must downregulate alien sequences the cell has not previously encountered, and that they bind AT-rich DNA regardless of its location relative to the gene [59–61]. At some genes, H-NS simply prevents RNA polymerase (RNAP) from binding the promoter (occlusion). At others, H-NS binding appears to prevent RNAP from escaping the promoter after binding (promoter trapping). H-NS can also affect expression of some genes by binding downstream of promoters where it can trigger Rho-dependent termination [115]. Indeed, a large subset of transcripts terminated by Rho are in AT-rich genes bound by H-NS [116].

A number of studies have implicated Rho and H-NS as cooperative partners in preventing the transcription of foreign DNA. Rho is an essential hexameric ATPase/helicase that mediates ~20% of all transcription termination events in *E. coli* [117]. Rho binds to nascent mRNA emerging from RNAP during transcription at C-rich "*rut*" sequences. Rho progressively translocates 5′ to 3′ along the mRNA molecule and terminates transcription if it catches up to a stalled RNAP-elongation complex. Translating ribosomes on the mRNA can block Rho from reaching a stalled RNAP (translational/transcriptional coupling). Rho terminates many noncoding (i.e., not translated) intragenic and antisense transcripts made from "accidental" promoters that occur by chance in AT-rich sequences [104, 117]. One study suggests that H-NS facilitates the Rho-dependent termination of these "accidental" transcripts by inducing RNAP stalls while it transcribes through AT-rich regions [116].

Oligomerization is essential for function and mutations in the oligomerization domain, that have negligible effects on DNA binding both in vitro and in vivo, can have dramatic effects on silencing [118, 119]. In solution, H-NS dimers are held together by an N-terminal dimerization domain [120]. At high protein concentrations, or when bound to DNA, homotypic interactions via the central dimerization domain enable elongated nucleoprotein filaments. The structure of the H-NS oligomerization domain (residues 1–80—without the DNA-binding domain) has been solved by X-ray crystallography—showing that H-NS polymers are formed by two independent dimerization domains: N-terminal (1–46) and central (65–80), that assemble "head-to-head/tail-to-tail" to generate a corkscrew-shaped filament [121]. In this arrangement, DNA-binding domains are predicted to project on both sides of the filament.

H-NS:DNA complexes have been studied using atomic force microscopy, electron microscopy, and single-molecule approaches. These studies find that the H-NS nucleoprotein complex can adopt at least two different conformations; one termed "bridging" where H-NS facilitates compaction of DNA by crosslinking adjacent duplexes, and one called "stiffened" where the H-NS/DNA complex forms a rigid filament that is resistant to compaction or bending [122, 123]. The conformation that the protein:DNA complex adopts in vitro critically depends on the specific ionic conditions used. A key parameter was found to be the concentration of divalent ions like magnesium and calcium in the binding buffer [124]. Specifically, at concentrations of divalent cations above ~2 mM the H-NS:DNA complex condenses and shows bridging behavior, while at lower concentrations the complex adopts a stiffened state.

The underlying mechanism behind the stiffening/bridging transition remains unclear. Divalent cations could theoretically affect the nucleoprotein complex via effects on H-NS, the DNA, or both. It is known that DNA flexibility increases in response to Mg^{2+} and other ions [125]. Also, in vivo studies indicate that DNA inside of a cell is much more flexible than DNA in a low-ionic strength buffer, likely through a combination of cytoplasmic ions (potassium, magnesium, and spermidine), and DNA-binding proteins [126]. Therefore, the low-ionic strength conditions under which the stiffened H-NS nucleoprotein filaments form in vitro may not represent the physical state of the DNA polymer within the cell.

Whether the stiffened or bridged complexes of DNA are the biologically relevant modes for silencing has been addressed in a several recent studies. In one study, it was found that H-NS-mediated silencing could be alleviated by SsrB, an H-NS antagonist, under conditions that induced stiffening mode [127]. These findings that stiffening is the relevant mode of binding in vivo are also supported by recent work measuring the effects of various protein and magnesium concentrations of H-NS with specific targets in vitro [128]. A different study examined the effects of bridging or stiffening on the ability of H-NS to block progression of RNAP with respect to pause sites that trigger Rho-dependent termination [115]. This study found that conditions that favored bridging enabled longer pausing of RNA polymerase at a subset of pause sites. Moreover, it was the bridged complexes, and not stiffened, that promoted Rho-dependent termination. Previously, it has been shown that activity of RNAP can be stalled with the increase in DNA torque [129]. This suggests that bridged nucleoprotein structures can constrain DNA in plectonemes, causing accumulation of DNA torsional stress in front of the polymerase by preventing twist diffusion which, in turn, may lead to a stall that inhibits transcription [115, 130].

1.6 How Genes Are Regulated in the Face of Silencing

Xenogeneic silencing is a major revision to the "textbook" of bacterial gene regulation. The first novel concept that arose from our understanding of silencing is that the bacterial cell can partition the regulation of genes into those that are "self" and those that are nonself. For the most part, the ancestral genes are regulated by classical repression and activation whereby RNAP, in complex with a sequence-specific sigma factor, will recognize a short promoter sequence. Binding to the promoter can be modulated by sequence-specific transcription factors that bind in close proximity to the promoter to either block or recruit the RNAP holoenzyme.

Expression of AT-rich xenogeneic genes, however, requires both RNAP recruitment *and* the alleviation of silencing. Studies of several H-NS-regulated genes indicate that, like silencing, countersilencing happens differently at different genes. Bacteria appear to integrate new genes into their regulatory networks ad hoc, often repurposing ancestral DNA-binding proteins as countersilencer proteins [114, 131]. The protein's ancestral role does not correlate with its ability to countersilence—activators, repressors, and even plasmid-partitioning proteins serve as CSPs at various promoters. The transcription factor PhoP, common to most enteric bacteria, activates a few ancestral genes involved in Mg^{2+} transport by binding at the −35 promoter sequence to recruit RNAP [132]. During *Salmonella* evolution, however,

PhoP's role was expanded to regulate many virulence genes in response to Mg^{2+} limitation within the host environment [132]. The PhoP-binding sites in these AT-rich sequences map to a variety of locations and orientations relative to the promoter, some within the downstream coding regions [132, 133]. Noncanonical binding locations have also been identified for other CSPs that antagonize H-NS [131, 134–136].

It is unclear how proteins like PhoP, by binding a short 14-nucleotide DNA sequence tens or hundreds of bp upstream or downstream the promoter, counteract silencing by an H-NS/Hha nucleoprotein polymer that coats hundreds of bases. However, it is likely that it occurs not by direct interactions with RNAP but instead by disrupting the nucleoprotein structure to alleviate specific blocks to transcription. For example, it is possible that binding sites far downstream of the transcription start site serve to prevent RNAP stalling and Rho-dependent termination caused by H-NS rather than by allowing RNAP access to the promoter.

1.7 Diversity of Xenogeneic Silencing Proteins

H-NS was previously thought to exist only in enterobacteria and closely related bacteria [137]. The first functional homologue of H-NS found outside of this group is BpH3, which was isolated from *Bordetella pertussis* [75]. Since then, functional homologues of H-NS have been identified in many bacteria, and now xenogeneic silencing proteins are considered to be widely distributed in bacteria. In this section, we discuss what is known about the major families of non-H-NS silencers.

1.7.1 MvaT

MvaT was firstly found in *Pseudomonas mevalonii* as a transcriptional regulator of *mvaAB* operon, which encodes the enzymes that catalyze the initial reactions of mevalonate catabolism [138]. It was then identified as a functional homologue of H-NS due to the capability of complementing the serine susceptibility of the *E. coli hns* mutant, though they share less than 20% identity in amino acid sequence [112]. MvaT homologues are distributed only in gamma-Proteobacteria, mostly in pseudomonads, and all pseudomonads have at least one MvaT homologue [64, 139]. In *P. aeruginosa*, MvaT was found to be a global regulator which controls the expression of genes involved in quorum sensing, virulence, biofilm formation, and multidrug resistance [140–144]. Similar to H-NS and StpA in enterobacteria, *P. aeruginosa* also has an additional MvaT homologue named MvaU whose functions are majorly overlapped with MvaT [73, 100, 145].

Although MvaT homologues are known to form nucleoprotein complexes like those of H-NS [113, 123, 146], it had not been clarified whether they have the same mechanisms in oligomerization and DNA binding. Recently, the structures of the N-terminal and C-terminal regions of MvaT homologues were reported. The

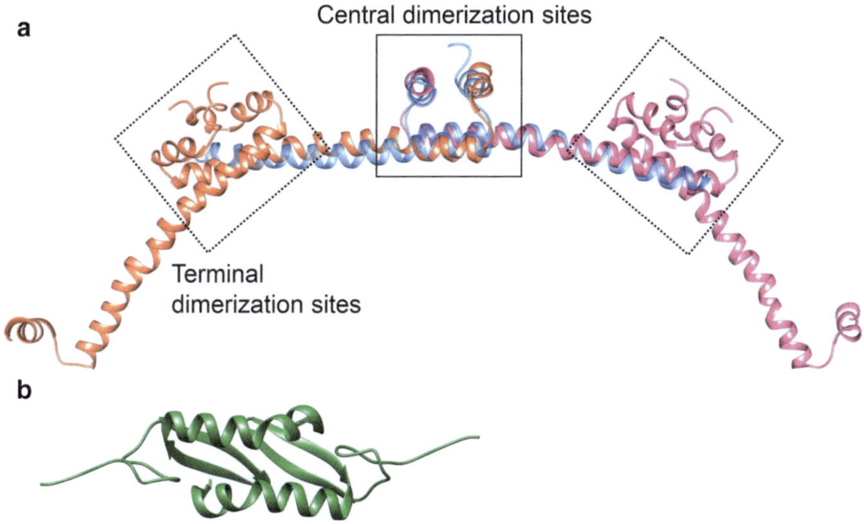

Fig. 1.2 Dimerization and oligomerization domains of the xenogeneic silencers. (**a**) Superimposed structure of the N-terminal dimerization and oligomerization domains of H-NS (two dimers are shown in orange and pink) and the MvaT homologue TurB (blue) [121, 151]. (**b**) Structure of the N-terminal region of Lsr2 [152]

N-terminal region, which is responsible for the dimerization and oligomerization ability, had been thought be similar to H-NS in structure [147]. The crystal structure of N-terminal region of TurB, which is an MvaT homologue in *Pseudomonas putida* KT2440 [148, 149], revealed that it has two dimerization sites: the central dimerization site near the linker region and the terminal dimerization site at the N-terminus [150] (Fig. 1.2a). The central dimerization site actually has almost the same structure with the corresponding region of H-NS in *S. enterica* [7]; both of them are stabilized by a hydrophobic core and salt bridges. On the other hand, the terminal dimerization site of TurB has only one helix, whereas the corresponding region of H-NS consists of three helices which are interlocked and stabilized by a hydrophobic core and salt bridges ([121, 150]) (Fig. 1.2a).

The C-terminal region of MvaT in *P. aeruginosa* was used for solution NMR to solve its DNA-binding mechanism [153]. The overall structure is similar to the C-terminal domain of H-NS [154, 155], but DNA-binding mechanism is different. As for H-NS, docking models indicated that a loop containing the QGR motif (Q112-G113-R114) is inserted into the minor groove of DNA, where the side chains of glutamine and arginine extend in opposite directions [154] (Fig. 1.3a). The structure of the C-terminal domain of MvaT and DNA complex revealed that MvaT does not have the residue corresponding to the glutamine in the QGR motif of H-NS; instead, the side chain of arginine on another loop (R80) holds the DNA with the loop which contains glycine and asparagine (G99 and N100) corresponding to "G" and "R" in the QGR motif [153] (Fig. 1.3b). This DNA-binding mode

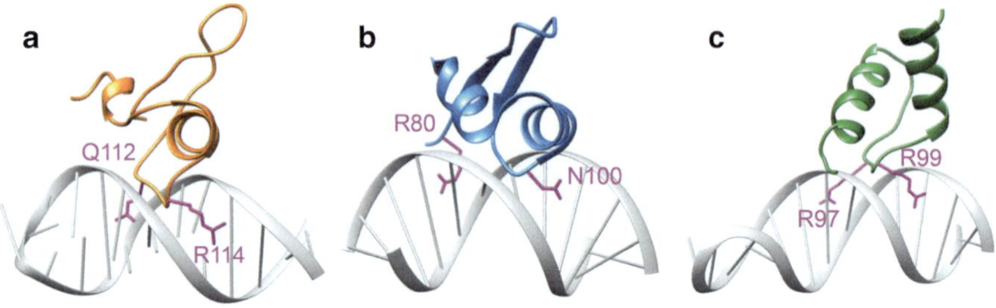

Fig. 1.3 DNA-binding mode of the xenogeneic silencers. C-terminal DNA-binding domains of H-NS (**a**), MvaT (**b**), and Lsr2 (**c**) in complex with DNA [74, 153, 154]. The residues comprising the "AT-hook" and "AT-pincer" motifs are highlighted in magenta. Note that docking models are shown in panels **a** and **c**, while the structure of MvaT and DNA complex solved by NMR is shown in the panel **b**

can cause the slight but clear difference between MvaT and H-NS in DNA-sequence preferences, where MvaT much prefers flexible DNA sequences with multiple TpA steps [153].

1.7.2 Lsr2

Lsr2 is the only H-NS family protein in Gram-positive bacteria whose function was experimentally confirmed to be equivalent to H-NS; it can complement β-glucoside utilization, mucoidy, motility, and hemolytic activity in an *E. coli hns* mutant [111]. It was firstly identified as an immunoreactive protein in *Mycobacterium leprae* [156, 157]. The distribution of Lsr2 is limited to Actinobacteria, and most of them have only one Lsr2 protein [64, 139]. In *Mycobacterium tuberculosis*, *lsr2* is an essential gene probably because Lsr2 physically protects DNA from degradation by reactive oxygen species [158]. It binds to AT-rich sequences, bridges DNA, and regulates transcription of many genes including those related with antibiotic resistance, biofilm formation, and colony morphology [110, 159, 160]. Both of the DNA-binding and oligomerization abilities are essential for the function of Lsr2 [110].

The N-terminal region of Lsr2 has a totally different structure from those of H-NS and MvaT. It is composed of an α-helix and β-sheets and forms a dimer to pack the helix and the sheets in an antiparallel manner [152] (Fig. 1.2b), whereas the corresponding regions of H-NS and MvaT are composed of only α-helices [7, 150] (Fig. 1.2a). The C-terminal DNA-binding domain of Lsr2 also shares no overall structural homology with those of H-NS and MvaT [74]. Intriguingly, protein-binding microarray analyses revealed that Lsr2 exhibits similar DNA-binding specificity to H-NS [154]. This can be explained by the DNA recognition mechanism of Lsr2; it inserts a loop containing the RGR motif (R97-G98-R99), which corresponds to the QGR motif in H-NS, into the minor groove of DNA [154] (Fig. 1.3c). Despite the low identity in amino acid sequences, both H-NS and Lsr2 use a "hook"-like structure to bind to the minor groove of DNA, while MvaT uses a "pincer"-like structure.

1.8 H-NS Family Proteins Encoded on Plasmids

Considering that the major role of H-NS family proteins is to suppress the expression of genes acquired through horizontal gene transfer, it is interesting that some of the H-NS family proteins are encoded on plasmids [64, 161, 162]. Since the newly acquired genetic elements often produce a fitness cost in the host cell [96], the expression of the genes on plasmids should be appropriately controlled. Shintani and colleagues reported that many bacterial hosts containing plasmids with H-NS family proteins also carry the same type of H-NS family proteins on their chromosomes, suggesting that the plasmid-borne H-NS family proteins function cooperatively with their homologues encoded on the host chromosomes [64]. If that is the case, do they have the same function or plasmid-borne H-NS family proteins have a special function? Here, we focused on the H-NS family proteins encoded on the plasmids of enterobacteria (H-NS homologues) and that of the genus *Pseudomonas* (an MvaT homologue).

1.8.1 Sfh and H-NS$_{R27}$

Sfh, which is encoded on the IncHI plasmid pSfR27 found in *Shigella flexneri*, is considered to have functions that reduce fitness cost in host cells after obtaining the plasmid [163]. When pSfR27 is transferred into a host cell, a large number of H-NS proteins expressed from the chromosome could bind to the plasmid DNA because it contains more than 100 DNA-binding sites for H-NS [69]. Expression of Sfh from pSfR27 can avoid a transient decline in the H-NS-to-DNA ratio and undesirable upregulation of the genes repressed by H-NS [163]. Indeed, Sfh was found to have similar functions with the chromosomally encoded H-NS proteins and it can form heteromeric complexes with them [164, 165]. The DNA-binding sites of Sfh are also overlapped with those of H-NS expressed from the chromosome, suggesting that Sfh functions as a molecular backup of H-NS in the *hns* mutant [69].

Similarly, H-NS$_{R27}$ expressed from the plasmid R27 (whose nucleotide sequence is 99.7% identical to pSfR27 [166]) has a common function with the chromosomally encoded H-NS; both of them affect thermoregulation of R27 transfer [167]. However, further studies revealed the different features between H-NS$_{R27}$ and chromosome-borne H-NS. Transcriptome analyses suggested that H-NS$_{R27}$ selectively regulates genes on horizontally acquired DNA regions, while chromosomally encoded H-NS targets both core genomes and horizontally acquired DNA regions [168]. It is possible that the flexibilities in the linker region cause the differences in regulatory preferences between plasmid-encoded H-NS and chromosomally encoded H-NS proteins [65].

1.8.2 Pmr

The IncP-7 catabolic plasmid pCAR1 [169] carries three genes encoding the following nucleoid-associated proteins: Pmr, an MvaT homologue, Pnd, an NdpA homologue, and Phu, an HU homologue [64, 162]. These proteins function synergistically to affect pCAR1 replication, maintenance, transfer, and biofilm morphology in a host *P. putida* KT2440, but these phenotypic changes clearly appeared only when *pmr* and *pnd* or *pmr* and *phu* were simultaneously disrupted [151, 170]. Pmr preferentially binds to the horizontally acquired DNA regions and forms homo- and hetero-oligomers with TurA and TurB, which are expressed from the chromosome of *P. putida* KT2440 [171–174]. Transcriptome analyses using the deletion mutants of *pmr*, *turA*, or *turB* showed that genes regulated by Pmr were less overlapped with those regulated by TurA or TurB [174]. This phenomenon can be partly explained by the different coupling ratio of the three proteins [171].

1.9 Summary

Xenogeneic silencers are essentially analogs of eukaryotic histones that assemble on DNA in a loosely specific manner to package DNA and protect the genome from parasites like viruses and transposons. Like histones form heterochromatin, silencers form a basic repressive nucleoprotein complex that sequence-specific transcription factors modulate to turn select genes on or off. Unlike eukaryotic chromatin, however, we still lack a basic understanding of bacterial chromatin structure and how factors can modulate it to control xenogeneic gene expression. Future work in this area will reveal whether H-NS-like proteins are all similar in function and how the structure of the silencing nucleoprotein complex is modulated to license or block transcription of AT-rich genes. Studies mentioned here raise the possibility that plasmid-encoded H-NS family proteins are not just a backup of the chromosomally encoded H-NS family proteins but have unique mechanisms to maintain fitness in the host cells.

Acknowledgment WWN acknowledges support from the Canadian Institutes for Health Research (CIHR) grant (MOP-86683).

References

1. Daubin V, Ochman H. Start-up entities in the origin of new genes. Curr Opin Genet Dev. 2004;14:616–9. https://doi.org/10.1016/j.gde.2004.09.004.
2. Gao Y, Zhao H, Jin Y, Xu X, Han GZ. Extent and evolution of gene duplication in DNA viruses. Virus Res. 2017;240:161–5. https://doi.org/10.1016/j.virusres.2017.08.005.
3. Light S, Basile W, Elofsson A. Orphans and new gene origination, a structural and evolutionary perspective. Curr Opin Struct Biol. 2014;26:73–83. https://doi.org/10.1016/j.sbi.2014.05.006.
4. Markov GV, Sommer RJ. The evolution of novelty in conserved gene families. Int J Evol Biol. 2012;2012:490894. https://doi.org/10.1155/2012/490894.

5. Zhaxybayeva O, Doolittle WF. Lateral gene transfer. Curr Biol. 2011;21:R242–6. https://doi.org/10.1016/j.cub.2011.01.045.
6. Moon HS, Bhak J, Lee KH, Lee D. Architecture of basic building blocks in protein and domain structural interaction networks. Bioinformatics. 2005;21:1479–86. https://doi.org/10.1093/bioinformatics/bti240.
7. Pawson T, Nash P. Assembly of cell regulatory systems through protein interaction domains. Science. 2003;300:445–52. https://doi.org/10.1126/science.1083653.
8. Hacker J, Kaper JB. Pathogenicity islands and the evolution of microbes. Annu Rev Microbiol. 2000;54:641–79.
9. Ochman H. Miles of isles. Trends Microbiol. 1997;5:222.
10. Ochman H, Davalos LM. The nature and dynamics of bacterial genomes. Science. 2006;311:1730–3.
11. Ochman H, Lawrence JG, Groisman EA. Lateral gene transfer and the nature of bacterial innovation. Nature. 2000;405:299–304.
12. Daubin V, Moran NA, Ochman H. Phylogenetics and the cohesion of bacterial genomes. Science. 2003;301:829–32.
13. Kaas RS, Friis C, Ussery DW, Aarestrup FM. Estimating variation within the genes and inferring the phylogeny of 186 sequenced diverse *Escherichia coli* genomes. BMC Genomics. 2012;13:577 Artn 577. https://doi.org/10.1186/1471-2164-13-577.
14. Lukjancenko O, Wassenaar TM, Ussery DW. Comparison of 61 sequenced *Escherichia coli* genomes. Microb Ecol. 2010;60:708–20. https://doi.org/10.1007/s00248-010-9717-3.
15. Alikhan NF, Zhou Z, Sergeant MJ, Achtman M. A genomic overview of the population structure of Salmonella. PLoS Genet. 2018;14:e1007261. https://doi.org/10.1371/journal.pgen.1007261.
16. Nuccio SP, Baumler AJ. Comparative analysis of Salmonella genomes identifies a metabolic network for escalating growth in the inflamed gut. MBio. 2014;5:e00929–14. https://doi.org/10.1128/mBio.00929-14.
17. Winter SE, Baumler AJ. Dysbiosis in the inflamed intestine: chance favors the prepared microbe. Gut Microbes. 2014;5:71–3. https://doi.org/10.4161/gmic.27129.
18. Groisman EA, Ochman H. How Salmonella became a pathogen. Trends Microbiol. 1997;5:343–9.
19. Laing CR, Whiteside MD, Gannon VPJ. Pan-genome analyses of the species Salmonella enterica, and identification of genomic markers predictive for species, subspecies, and serovar. Front Microbiol. 2017;8:1345. https://doi.org/10.3389/fmicb.2017.01345.
20. Laing C, et al. Pan-genome sequence analysis using Panseq: an online tool for the rapid analysis of core and accessory genomic regions. BMC Bioinformatics. 2010;11:461 Artn 461. https://doi.org/10.1186/1471-2105-11-461.
21. Price MN, et al. Mutant phenotypes for thousands of bacterial genes of unknown function. Nature. 2018;557:503–9. https://doi.org/10.1038/s41586-018-0124-0.
22. Harrison E, Truman J, Wright R, Spiers AJ, Paterson S, Brockhurst MA. Plasmid carriage can limit bacteria-phage coevolution. Biol Lett. 2015;11:20150361. https://doi.org/10.1098/rsbl.2015.0361.
23. Bondy-Denomy J, Qian J, Westra ER, Buckling A, Guttman DS, Davidson AR, Maxwell KL. Prophages mediate defense against phage infection through diverse mechanisms. ISME J. 2016;10:2854–66. https://doi.org/10.1038/ismej.2016.79.
24. Chaudhary K. BacteRiophage EXclusion (BREX): a novel anti-phage mechanism in the arsenal of bacterial defense system. J Cell Physiol. 2018;233:771–3. https://doi.org/10.1002/jcp.25973.
25. Doron S, et al. Systematic discovery of antiphage defense systems in the microbial pangenome. Science. 2018;359:eaar4120. https://doi.org/10.1126/science.aar4120.
26. Dy RL, Richter C, Salmond GP, Fineran PC. Remarkable mechanisms in microbes to resist phage infections. Annu Rev Virol. 2014;1:307–31. https://doi.org/10.1146/annurev-virology-031413-085500.

27. Seed KD. Battling phages: how bacteria defend against viral attack. PLoS Pathog. 2015;11:e1004847. https://doi.org/10.1371/journal.ppat.1004847.
28. Bondy-Denomy J, Pawluk A, Maxwell KL, Davidson AR. Bacteriophage genes that inactivate the CRISPR/Cas bacterial immune system. Nature. 2013;493:429–32. https://doi.org/10.1038/nature11723.
29. Liang W, Xie Y, Xiong W, Tang Y, Li G, Jiang X, Lu Y. Anti-restriction protein, KlcAHS, promotes dissemination of carbapenem resistance. Front Cell Infect Microbiol. 2017;7:150. https://doi.org/10.3389/fcimb.2017.00150.
30. Tock MR, Dryden DT. The biology of restriction and anti-restriction. Curr Opin Microbiol. 2005;8:466–72. https://doi.org/10.1016/j.mib.2005.06.003.
31. Carattoli A. Plasmids and the spread of resistance. Int J Med Microbiol. 2013;303:298–304. https://doi.org/10.1016/j.ijmm.2013.02.001.
32. Dennis JJ. The evolution of IncP catabolic plasmids. Curr Opin Biotechnol. 2005;16:291–8. https://doi.org/10.1016/j.copbio.2005.04.002.
33. Harrison E, et al. Bacteriophages limit the existence conditions for conjugative plasmids. MBio. 2015;6:e00586. https://doi.org/10.1128/mBio.00586-15.
34. Rotger R, Casadesus J. The virulence plasmids of Salmonella. Int Microbiol. 1999;2:177–84.
35. Bondy-Denomy J, Davidson AR. When a virus is not a parasite: the beneficial effects of prophages on bacterial fitness. J Microbiol. 2014;52:235–42. https://doi.org/10.1007/s12275-014-4083-3.
36. Levy A, et al. CRISPR adaptation biases explain preference for acquisition of foreign DNA. Nature. 2015;520:505–10. https://doi.org/10.1038/nature14302.
37. Zamenhof S, Brawerman G, Chargaff E. On the desoxypentose nucleic acids from several microorganisms. Biochim Biophys Acta. 1952;9:402–5. https://doi.org/10.1016/0006-3002(52)90184-4.
38. Gandelman B, Zamenhof S, Chargaff E. The desoxypentose nucleic acids of three strains of *Escherichia coli*. Biochim Biophys Acta. 1952;9:399–401.
39. Lee KY, Wahl R, Barbu E. Contenu en bases puriques et pyrimidiques des acides désoxyribonucléiques des bactéries. Ann Inst Pasteur. 1956;91:212–24.
40. Bohlin J, Eldholm V, Pettersson JH, Brynildsrud O, Snipen L. The nucleotide composition of microbial genomes indicates differential patterns of selection on core and accessory genomes. BMC Genomics. 2017;18:151. https://doi.org/10.1186/s12864-017-3543-7.
41. Musto H, Naya H, Zavala A, Romero H, Alvarez-Valin F, Bernardi G. Genomic GC level, optimal growth temperature, and genome size in prokaryotes. Biochem Biophys Res Commun. 2006;347:1–3. https://doi.org/10.1016/j.bbrc.2006.06.054.
42. Raghavan R, Kelkar YD, Ochman H. A selective force favoring increased G+C content in bacterial genes. Proc Natl Acad Sci U S A. 2012;109:14504–7. https://doi.org/10.1073/pnas.1205683109.
43. Reichenberger ER, Rosen G, Hershberg U, Hershberg R. Prokaryotic nucleotide composition is shaped by both phylogeny and the environment. Genome Biol Evol. 2015;7:1380–9. https://doi.org/10.1093/gbe/evv063.
44. Rocha EP, Feil EJ. Mutational patterns cannot explain genome composition: are there any neutral sites in the genomes of bacteria? PLoS Genet. 2010;6:e1001104. https://doi.org/10.1371/journal.pgen.1001104.
45. Wu H, Zhang Z, Hu SN, Yu J. On the molecular mechanism of GC content variation among eubacterial genomes. Biol Direct. 2012;7:2 Artn 2. https://doi.org/10.1186/1745-6150-7-2.
46. Montero-Calasanz Mdel C, et al. Geodermatophilus telluris sp. nov., an actinomycete isolated from Saharan desert sand. Int J Syst Evol Microbiol. 2013;63:2254–9. https://doi.org/10.1099/ijs.0.046888-0.
47. Moura GR, et al. Species-specific codon context rules unveil non-neutrality effects of synonymous mutations. PLoS One. 2011;6:e26817. https://doi.org/10.1371/journal.pone.0026817.
48. Konstantinidis KT, Tiedje JM. Trends between gene content and genome size in prokaryotic species with larger genomes. Proc Natl Acad Sci U S A. 2004;101:3160–5. https://doi.org/10.1073/pnas.0308653100.

49. Konstantinidis KT, Tiedje JM. Genomic insights that advance the species definition for prokaryotes. Proc Natl Acad Sci U S A. 2005;102:2567–72. https://doi.org/10.1073/pnas.0409727102.
50. McCutcheon JP, Moran NA. Functional convergence in reduced genomes of bacterial symbionts spanning 200 my of evolution. Genome Biol Evol. 2010;2:708–18. https://doi.org/10.1093/gbe/evq055.
51. Lawrence JG, Ochman H. Amelioration of bacterial genomes: rates of change and exchange. J Mol Evol. 1997;44:383–97.
52. Marri PR, Golding GB. Gene amelioration demonstrated: the journey of nascent genes in bacteria. Genome. 2008;51:164–8. https://doi.org/10.1139/g07-105.
53. Larsen MV, et al. Benchmarking of methods for genomic taxonomy. J Clin Microbiol. 2014;52:1529–39. https://doi.org/10.1128/JCM.02981-13.
54. Chattaway MA, Schaefer U, Tewolde R, Dallman TJ, Jenkins C. Identification of *Escherichia coli* and shigella species from whole-genome sequences. J Clin Microbiol. 2017;55:616–23. https://doi.org/10.1128/JCM.01790-16.
55. Oliveira de Carvalho M, Loreto ELS. Methods for detection of horizontal transfer of transposable elements in complete genomes. Genet Mol Biol. 2012;35:1078–84.
56. Vernikos GS, Parkhill J. Interpolated variable order motifs for identification of horizontally acquired DNA: revisiting the Salmonella pathogenicity islands. Bioinformatics. 2006;22:2196–203. https://doi.org/10.1093/bioinformatics/btl369.
57. Bernard G, Chan CX, Ragan MA. Alignment-free microbial phylogenomics under scenarios of sequence divergence, genome rearrangement and lateral genetic transfer. Sci Rep. 2016;6:28970. https://doi.org/10.1038/srep28970.
58. de Carvalho MO, Loreto EL. Methods for detection of horizontal transfer of transposable elements in complete genomes. Genet Mol Biol. 2012;35:1078–84.
59. Grainger DC, Hurd D, Goldberg MD, Busby SJ. Association of nucleoid proteins with coding and non-coding segments of the *Escherichia coli* genome. Nucleic Acids Res. 2006;34:4642–52.
60. Lucchini S, Rowley G, Goldberg MD, Hurd D, Harrison M, Hinton JC. H-NS mediates the silencing of laterally acquired genes in bacteria. PLoS Pathog. 2006;2:e81.
61. Navarre WW, Porwollik S, Wang Y, McClelland M, Rosen H, Libby SJ, Fang FC. Selective silencing of foreign DNA with low GC content by the H-NS protein in *Salmonella*. Science. 2006;313:236–8.
62. Oshima T, Ishikawa S, Kurokawa K, Aiba H, Ogasawara N. *Escherichia coli* histone-like protein H-NS preferentially binds to horizontally acquired DNA in association with RNA polymerase. DNA Res. 2006;13:141–53.
63. Ali SS, Xia B, Liu J, Navarre WW. Silencing of foreign DNA in bacteria. Curr Opin Microbiol. 2012;15:175–81. https://doi.org/10.1016/j.mib.2011.12.014.
64. Shintani M, Suzuki-Minakuchi C, Nojiri H. Nucleoid-associated proteins encoded on plasmids: occurrence and mode of function. Plasmid. 2015;80:32–44. https://doi.org/10.1016/j.plasmid.2015.04.008.
65. Fernández-de-Alba C, Berrow NS, Garcia-Castellanos R, García J, Pons M. On the origin of the selectivity of plasmidic H-NS towards horizontally acquired DNA: linking H-NS oligomerization and cooperative DNA binding. J Mol Biol. 2013;425:2347–58. https://doi.org/10.1016/j.jmb.2013.03.006.
66. Sun Z, Vasileva D, Suzuki-Minakuchi C, Okada K, Luo F, Igarashi Y, Nojiri H. Growth phase-dependent expression profiles of three vital H-NS family proteins encoded on the chromosome of Pseudomonas putida KT2440 and on the pCAR1 plasmid. BMC Microbiol. 2017;17:188. https://doi.org/10.1186/s12866-017-1091-6.
67. Ali SS, Soo J, Rao C, Leung AS, Ngai DH, Ensminger AW, Navarre WW. Silencing by H-NS potentiated the evolution of Salmonella. PLoS Pathog. 2014;10:e1004500. https://doi.org/10.1371/journal.ppat.1004500.
68. Ueda T, Takahashi H, Uyar E, Ishikawa S, Ogasawara N, Oshima T. Functions of the Hha and YdgT proteins in transcriptional silencing by the nucleoid proteins, H-NS and StpA, in *Escherichia coli*. DNA Res. 2013;20:263–71. https://doi.org/10.1093/dnares/dst008.

69. Dillon SC, Cameron AD, Hokamp K, Lucchini S, Hinton JC, Dorman CJ. Genome-wide analysis of the H-NS and Sfh regulatory networks in Salmonella Typhimurium identifies a plasmid-encoded transcription silencing mechanism. Mol Microbiol. 2010;76:1250–65. https://doi.org/10.1111/j.1365-2958.2010.07173.x.
70. Garcia J, Cordeiro TN, Prieto MJ, Pons M. Oligomerization and DNA binding of Ler, a master regulator of pathogenicity of enterohemorrhagic and enteropathogenic *Escherichia coli*. Nucleic Acids Res. 2012;40:10254–62. https://doi.org/10.1093/nar/gks846.
71. Winardhi RS, Gulvady R, Mellies JL, Yan J. Locus of enterocyte effacement-encoded regulator (Ler) of pathogenic *Escherichia coli* competes off histone-like nucleoid-structuring protein (H-NS) through noncooperative DNA binding. J Biol Chem. 2014;289:13739–50. https://doi.org/10.1074/jbc.M113.545954.
72. Baños RC, Pons JI, Madrid C, Juarez A. A global modulatory role for the Yersinia enterocolitica H-NS protein. Microbiology. 2008;154:1281–9.
73. Castang S, McManus HR, Turner KH, Dove SL. H-NS family members function coordinately in an opportunistic pathogen. Proc Natl Acad Sci U S A. 2008;105:18947–52.
74. Gordon BR, et al. Lsr2 is a nucleoid-associated protein that targets AT-rich sequences and virulence genes in Mycobacterium tuberculosis. Proc Natl Acad Sci U S A. 2010;107:5154–9. https://doi.org/10.1073/pnas.0913551107.
75. Goyard S, Bertin P. Characterization of BpH3, an H-NS-like protein in Bordetella pertussis. Mol Microbiol. 1997;24:815–23.
76. Picker MA, Wing HJ. H-NS, its family members and their regulation of virulence genes in shigella species. Genes (Basel). 2016;7:E112. https://doi.org/10.3390/genes7120112.
77. Russell R, Wang H, Benitez JA, Silva AJ. Deletion of gene encoding the nucleoid-associated protein H-NS unmasks hidden regulatory connections in El Tor biotype Vibrio cholerae. Microbiology. 2018;164(7):998–1003. https://doi.org/10.1099/mic.0.000672.
78. Cukier-Kahn R, Jacquet M, Gros F. Two heat-resistant, low molecular weight proteins from *Escherichia coli* that stimulate DNA-directed RNA synthesis. Proc Natl Acad Sci U S A. 1972;69:3643–7.
79. Jacquet M, Cukier-Kahn R, Pla J, Gros F. A thermostable protein factor acting on in vitro DNA transcription. Biochem Biophys Res Commun. 1971;45:1597–607.
80. Bakaev VV. Structure of chromosomal deoxyribonucleoproteins. XI. Organization of deoxyribonucleoprotein complex in bacterial cells. Mol Biol (Mosk). 1981;15:1350–63.
81. Varshavsky AJ, Nedospasov SA, Bakayev VV, Bakayeva TG, Georgiev GP. Histone-like proteins in the purified *Escherichia coli* deoxyribonucleoprotein. Nucleic Acids Res. 1977;4:2725–45.
82. Falconi M, Gualtieri MT, La Teana A, Losso MA, Pon CL. Proteins from the prokaryotic nucleoid: primary and quaternary structure of the 15-kD *Escherichia coli* DNA binding protein H-NS. Mol Microbiol. 1988;2:323–9.
83. Gualerzi CO, et al. Proteins from the prokaryotic nucleoid. Structural and functional characterization of the *Escherichia coli* DNA-binding proteins NS (HU) and H-NS. In: Gualerzi CO, Pon CL, editors. Bacterial chromatin. Berlin: Springer; 1986. p. 101–34.
84. Lammi M, et al. Proteins from the prokaryotic nucleoid: biochemical and 1H NMR studies on three bacterial histone-like proteins. In: Hubscher U, Spadari S, editors. Proteins involved in DNA replication, Advances in experimental medicine and biology, vol. 179. New York: Plenum Press; 1984. p. 467–77.
85. Paci M, Pon CL, Canonaco MA, Gualerzi CO. Proteins from the prokaryotic nucleoid. [1H]-NMR studies on *Escherichia coli* histone-like proteins NS and H-NS and their interactions with DNA. In: Gualerzi CO, Pon CL, editors. Bacterial chromatin. Berlin: Springer; 1986. p. 135–54.
86. Spassky A, Rimsky S, Garreau H, Buc H. H1a, an *E. coli* DNA-binding protein which accumulates in stationary phase, strongly compacts DNA in vitro. Nucleic Acids Res. 1984;12:5321–40.
87. Laine B, Sautiere P, Spassky A, Rimsky S. A DNA-binding protein from *E. coli* isolation, characterization and its relationship with proteins H1 and B1. Biochem Biophys Res Commun. 1984;119:1147–53.

88. Defez R, De Felice M. Cryptic operon for beta-glucoside metabolism in *Escherichia coli* K12: genetic evidence for a regulatory protein. Genetics. 1981;97:11–25.
89. May G, Dersch P, Haardt M, Middendorf A, Bremer E. The osmZ (bglY) gene encodes the DNA-binding protein H-NS (H1a), a component of the *Escherichia coli* K12 nucleoid. Mol Gen Genet. 1990;224:81–90.
90. Goransson M, Sonden B, Nilsson P, Dagberg B, Forsman K, Emanuelsson K, Uhlin BE. Transcriptional silencing and thermoregulation of gene expression in *Escherichia coli*. Nature. 1990;344:682–5.
91. Hromockyj AE, Tucker SC, Maurelli AT. Temperature regulation of Shigella virulence: identification of the repressor gene virR, an analogue of hns, and partial complementation by tyrosyl transfer RNA (tRNA1(Tyr)). Mol Microbiol. 1992;6:2113–24.
92. Diderichsen B. cur-1, a mutation affecting a phenotype of sup+ strains of *Escherichia coli*. Mol Gen Genet. 1980;180:425–8.
93. Spears PA, Schauer D, Orndorff PE. Metastable regulation of type 1 piliation in Escherichia coli and isolation and characterization of a phenotypically stable mutant. J Bacteriol. 1986;168:179–85.
94. Azam TA, Ishihama A. Twelve species of the nucleoid-associated protein from *Escherichia coli*. Sequence recognition specificity and DNA binding affinity. J Biol Chem. 1999;274:33105–13.
95. Dorman CJ. Nucleoid-associated proteins and bacterial physiology. Adv Appl Microbiol. 2009;67:47–64. https://doi.org/10.1016/S0065-2164(08)01002-2.
96. Baltrus DA. Exploring the costs of horizontal gene transfer. Trends Ecol Evol. 2013;28:489–95. https://doi.org/10.1016/j.tree.2013.04.002.
97. Cathelyn JS, Ellison DW, Hinchliffe SJ, Wren BW, Miller VL. The RovA regulons of Yersinia enterocolitica and Yersinia pestis are distinct: evidence that many RovA-regulated genes were acquired more recently than the core genome. Mol Microbiol. 2007;66(1):189–205.
98. Heroven AK, Dersch P. RovM, a novel LysR-type regulator of the virulence activator gene rovA, controls cell invasion, virulence and motility of Yersinia pseudotuberculosis. Mol Microbiol. 2006;62(5):1469–83.
99. Ares MA, et al. H-NS nucleoid protein controls virulence features of Klebsiella pneumoniae by regulating the expression of type 3 pili and the capsule polysaccharide. Front Cell Infect Microbiol. 2016;6:13. https://doi.org/10.3389/fcimb.2016.00013.
100. Castang S, Dove SL. Basis for the essentiality of H-NS family members in Pseudomonas aeruginosa. J Bacteriol. 2012;194(18):5101–9. https://doi.org/10.1128/JB.00932-12.
101. Wade JT, Grainger DC. Spurious transcription and its impact on cell function. Transcription. 2018;9:182–9. https://doi.org/10.1080/21541264.2017.1381794.
102. Landick R, Wade JT, Grainger DC. H-NS and RNA polymerase: a love-hate relationship? Curr Opin Microbiol. 2015;24:53–9. https://doi.org/10.1016/j.mib.2015.01.009.
103. Rangarajan AA, Schnetz K. Interference of transcription across H-NS binding sites and repression by H-NS. Mol Microbiol. 2018;108:226–39. https://doi.org/10.1111/mmi.13926.
104. Singh SS, Singh N, Bonocora RP, Fitzgerald DM, Wade JT, Grainger DC. Widespread suppression of intragenic transcription initiation by H-NS. Genes Dev. 2014;28:214–9. https://doi.org/10.1101/gad.234336.113.
105. Lucht JM, Dersch P, Kempf B, Bremer E. Interactions of the nucleoid-associated DNA-binding protein H-NS with the regulatory region of the osmotically controlled proU operon of *Escherichia coli*. J Biol Chem. 1994;269:6578–86.
106. Bouffartigues E, Buckle M, Badaut C, Travers A, Rimsky S. H-NS cooperative binding to high-affinity sites in a regulatory element results in transcriptional silencing. Nat Struct Mol Biol. 2007;14:441–8.
107. Dame RT, Noom MC, Wuite GJ. Bacterial chromatin organization by H-NS protein unravelled using dual DNA manipulation. Nature. 2006;444:387–90.
108. Lang B, et al. High-affinity DNA binding sites for H-NS provide a molecular basis for selective silencing within proteobacterial genomes. Nucleic Acids Res. 2007;35:6330–7.

109. Lim CJ, Lee SY, Kenney LJ, Yan J. Nucleoprotein filament formation is the structural basis for bacterial protein H-NS gene silencing. Sci Rep. 2012;2:509. https://doi.org/10.1038/srep00509.
110. Chen JM, et al. Lsr2 of *Mycobacterium tuberculosis* is a DNA-bridging protein. Nucleic Acids Res. 2008;36:2123–35 gkm1162 [pii]. https://doi.org/10.1093/nar/gkm1162.
111. Gordon BR, Imperial R, Wang L, Navarre WW, Liu J. Lsr2 of Mycobacterium represents a novel class of H-NS-like proteins. J Bacteriol. 2008;190(21):7052–9.
112. Tendeng C, Soutourina OA, Danchin A, Bertin PN. MvaT proteins in Pseudomonas spp.: a novel class of H-NS-like proteins. Microbiology. 2003;149:3047–50.
113. Winardhi RS, Fu W, Castang S, Li Y, Dove SL, Yan J. Higher order oligomerization is required for H-NS family member MvaT to form gene-silencing nucleoprotein filament. Nucleic Acids Res. 2012;40(18):8942–52. https://doi.org/10.1093/nar/gks669.
114. Navarre WW, McClelland M, Libby SJ, Fang FC. Silencing of xenogeneic DNA by H-NS-facilitation of lateral gene transfer in bacteria by a defense system that recognizes foreign DNA. Genes Dev. 2007;21:1456–71.
115. Kotlajich MV, Hron DR, Boudreau BA, Sun Z, Lyubchenko YL, Landick R. Bridged filaments of histone-like nucleoid structuring protein pause RNA polymerase and aid termination in bacteria. Elife. 2015;4:e04970. https://doi.org/10.7554/eLife.04970.
116. Peters JM, Mooney RA, Grass JA, Jessen ED, Tran F, Landick R. Rho and NusG suppress pervasive antisense transcription in *Escherichia coli*. Genes Dev. 2012;26:2621–33. https://doi.org/10.1101/gad.196741.112.
117. Peters JM, Mooney RA, Kuan PF, Rowland JL, Keles S, Landick R. Rho directs widespread termination of intragenic and stable RNA transcription. Proc Natl Acad Sci U S A. 2009;106:15406–11. https://doi.org/10.1073/pnas.0903846106.
118. Ueguchi C, Seto C, Suzuki T, Mizuno T. Clarification of the dimerization domain and its functional significance for the *Escherichia coli* nucleoid protein H-NS. J Mol Biol. 1997;274:145–51.
119. Ueguchi C, Suzuki T, Yoshida T, Tanaka K, Mizuno T. Systematic mutational analysis revealing the functional domain organization of *Escherichia coli* nucleoid protein H-NS. J Mol Biol. 1996;263:149–62.
120. Badaut C, Williams R, Arluison V, Bouffartigues E, Robert B, Buc H, Rimsky S. The degree of oligomerization of the H-NS nucleoid structuring protein is related to specific binding to DNA. J Biol Chem. 2002;277:41657–66. https://doi.org/10.1074/jbc.M206037200.
121. Arold ST, Leonard PG, Parkinson GN, Ladbury JE. H-NS forms a superhelical protein scaffold for DNA condensation. Proc Natl Acad Sci U S A. 2010;107:15728–32. https://doi.org/10.1073/pnas.1006966107.
122. Amit R, Oppenheim AB, Stavans J. Increased bending rigidity of single DNA molecules by H-NS, a temperature and osmolarity sensor. Biophys J. 2003;84:2467–73.
123. Dame RT, Luijsterburg MS, Krin E, Bertin PN, Wagner R, Wuite GJ. DNA bridging: a property shared among H-NS-like proteins. J Bacteriol. 2005;187:1845–8. https://doi.org/10.1128/JB.187.5.1845-1848.2005.
124. Liu Y, Chen H, Kenney LJ, Yan J. A divalent switch drives H-NS/DNA-binding conformations between stiffening and bridging modes. Genes Dev. 2010;24:339–44. https://doi.org/10.1101/gad.1883510.
125. Baumann CG, Smith SB, Bloomfield VA, Bustamante C. Ionic effects on the elasticity of single DNA molecules. Proc Natl Acad Sci U S A. 1997;94:6185–90.
126. Williams LD, Maher LJ. Electrostatic mechanisms of DNA deformation. Annu Rev Biophys Biomol Struct. 2000;29:497–521. https://doi.org/10.1146/annurev.biophys.29.1.497.
127. Walthers D, Li Y, Liu Y, Anand G, Yan J, Kenney LJ. Salmonella enterica response regulator SsrB relieves H-NS silencing by displacing H-NS bound in polymerization mode and directly activates transcription. J Biol Chem. 2011;286:1895–902. https://doi.org/10.1074/jbc.M110.164962.
128. Will WR, Whitham PJ, Reid PJ, Fang FC. Modulation of H-NS transcriptional silencing by magnesium. Nucleic Acids Res. 2018;46:5717–25. https://doi.org/10.1093/nar/gky387.

129. Ma J, Bai L, Wang MD. Transcription under torsion. Science. 2013;340:1580–3.
130. Boudreau BA, Hron DR, Qin L, van der Valk RA, Kotlajich MV, Dame RT, Landick R. StpA and Hha stimulate pausing by RNA polymerase by promoting DNA-DNA bridging of H-NS filaments. Nucleic Acids Res. 2018;46:5525–46. https://doi.org/10.1093/nar/gky265.
131. Stoebel DM, Free A, Dorman CJ. Anti-silencing: overcoming H-NS-mediated repression of transcription in gram-negative enteric bacteria. Microbiology. 2008;154:2533–45. https://doi.org/10.1099/mic.0.2008/020693-0.
132. Zwir I, Latifi T, Perez JC, Huang H, Groisman EA. The promoter architectural landscape of the Salmonella PhoP regulon. Mol Microbiol. 2012;84:463–85. https://doi.org/10.1111/j.1365-2958.2012.08036.x.
133. Will WR, Navarre WW, Fang FC. Integrated circuits: how transcriptional silencing and counter-silencing facilitate bacterial evolution. Curr Opin Microbiol. 2015;23:8–13. https://doi.org/10.1016/j.mib.2014.10.005.
134. Caramel A, Schnetz K. Lac and lambda repressors relieve silencing of the *Escherichia coli* bgl promoter. Activation by alteration of a repressing nucleoprotein complex. J Mol Biol. 1998;284:875–83.
135. Madhusudan S, Paukner A, Klingen Y, Schnetz K. Independent regulation of H-NS-mediated silencing of the bgl operon at two levels: upstream by BglJ and LeuO and downstream by DnaKJ. Microbiology. 2005;151:3349–59.
136. Navarre WW, et al. Co-regulation of Salmonella enterica genes required for virulence and resistance to antimicrobial peptides by SlyA and PhoP/PhoQ. Mol Microbiol. 2005;56:492–508. https://doi.org/10.1111/j.1365-2958.2005.04553.x.
137. Atlung T, Ingmer H. H-NS: a modulator of environmentally regulated gene expression. Mol Microbiol. 1997;24:7–17.
138. Rosenthal RS, Rodwell VW. Purification and characterization of the heteromeric transcriptional activator MvaT of the Pseudomonas mevalonii mvaAB operon. Protein Sci. 1998;7:178–84. https://doi.org/10.1002/pro.5560070118.
139. Perez-Rueda E, Ibarra JA. Distribution of putative xenogeneic silencers in prokaryote genomes. Comput Biol Chem. 2015;58:167–72. https://doi.org/10.1016/j.compbiolchem.2015.06.007.
140. Diggle SP, Winzer K, Lazdunski A, Williams P, Cámara M. Advancing the quorum in Pseudomonas aeruginosa: MvaT and the regulation of N-acylhomoserine lactone production and virulence gene expression. J Bacteriol. 2002;184:2576–86.
141. Vallet I, et al. Biofilm formation in Pseudomonas aeruginosa: fimbrial cup gene clusters are controlled by the transcriptional regulator MvaT. J Bacteriol. 2004;186:2880–90.
142. Vallet-Gely I, Sharp JS, Dove SL. Local and global regulators linking anaerobiosis to cupA fimbrial gene expression in Pseudomonas aeruginosa. J Bacteriol. 2007;189:8667–76. https://doi.org/10.1128/JB.01344-07.
143. Westfall LW, Carty NL, Layland N, Kuan P, Colmer-Hamood JA, Hamood AN. mvaT mutation modifies the expression of the Pseudomonas aeruginosa multidrug efflux operon mexEF-oprN. FEMS Microbiol Lett. 2006;255:247–54. https://doi.org/10.1111/j.1574-6968.2005.00075.x.
144. Westfall LW, et al. The Pseudomonas aeruginosa global regulator MvaT specifically binds to the ptxS upstream region and enhances ptxS expression. Microbiology. 2004;150:3797–806. https://doi.org/10.1099/mic.0.27270-0.
145. Vallet-Gely I, Donovan KE, Fang R, Joung JK, Dove SL. Repression of phase-variable cup gene expression by H-NS-like proteins in Pseudomonas aeruginosa. Proc Natl Acad Sci U S A. 2005;102:11082–7. https://doi.org/10.1073/pnas.0502663102.
146. Winardhi RS, Castang S, Dove SL, Yan J. Single-molecule study on histone-like nucleoid-structuring protein (H-NS) paralogue in Pseudomonas aeruginosa: MvaU bears DNA organization mode similarities to MvaT. PLoS One. 2014;9:e112246. https://doi.org/10.1371/journal.pone.0112246.
147. Castang S, Dove SL. High-order oligomerization is required for the function of the H-NS family member MvaT in Pseudomonas aeruginosa. Mol Microbiol. 2010;78:916–31. https://doi.org/10.1111/j.1365-2958.2010.07378.x.

148. Renzi F, Rescalli E, Galli E, Bertoni G. Identification of genes regulated by the MvaT-like paralogues TurA and TurB of Pseudomonas putida KT2440. Environ Microbiol. 2010;12:254–63. https://doi.org/10.1111/j.1462-2920.2009.02064.x.
149. Rescalli E, Saini S, Bartocci C, Rychlewski L, De Lorenzo V, Bertoni G. Novel physiological modulation of the Pu promoter of TOL plasmid: negative regulatory role of the TurA protein of Pseudomonas putida in the response to suboptimal growth temperatures. J Biol Chem. 2004;279:7777–84. https://doi.org/10.1074/jbc.M310580200.
150. Suzuki-Minakuchi C, et al. Structural similarities and differences in H-NS family proteins revealed by the N-terminal structure of TurB in Pseudomonas putida KT2440. FEBS Lett. 2016;590:3583–94. https://doi.org/10.1002/1873-3468.12425.
151. Suzuki-Minakuchi C, et al. Effects of three different nucleoid-associated proteins encoded on IncP-7 plasmid pCAR1 on host Pseudomonas putida KT2440. Appl Environ Microbiol. 2015;81:2869–80. https://doi.org/10.1128/AEM.00023-15.
152. Summers EL, et al. The structure of the oligomerization domain of Lsr2 from Mycobacterium tuberculosis reveals a mechanism for chromosome organization and protection. PLoS One. 2012;7:e38542. https://doi.org/10.1371/journal.pone.0038542.
153. Ding P, et al. A novel AT-rich DNA recognition mechanism for bacterial xenogeneic silencer MvaT. PLoS Pathog. 2015;11:e1004967. https://doi.org/10.1371/journal.ppat.1004967.
154. Gordon BR, et al. Structural basis for recognition of AT-rich DNA by unrelated xenogeneic silencing proteins. Proc Natl Acad Sci U S A. 2011;108:10690–5. https://doi.org/10.1073/pnas.1102544108.
155. Shindo H, et al. Solution structure of the DNA binding domain of a nucleoid-associated protein, H-NS, from *Escherichia coli*. FEBS Lett. 1995;360:125–31.
156. Laal S, et al. Recombinant fusion protein identified by lepromatous sera mimics native Mycobacterium leprae in T-cell responses across the leprosy spectrum. Proc Natl Acad Sci U S A. 1991;88:1054–8.
157. Sela S, Thole JE, Ottenhoff TH, Clark-Curtiss JE. Identification of Mycobacterium leprae antigens from a cosmid library: characterization of a 15-kilodalton antigen that is recognized by both the humoral and cellular immune systems in leprosy patients. Infect Immun. 1991;59:4117–24.
158. Colangeli R, et al. The multifunctional histone-like protein Lsr2 protects mycobacteria against reactive oxygen intermediates. Proc Natl Acad Sci U S A. 2009;106:4414–8. https://doi.org/10.1073/pnas.0810126106.
159. Chen JM, German GJ, Alexander DC, Ren H, Tan T, Liu J. Roles of Lsr2 in colony morphology and biofilm formation of Mycobacterium smegmatis. J Bacteriol. 2006;188:633–41. https://doi.org/10.1128/JB.188.2.633-641.2006.
160. Colangeli R, et al. Transcriptional regulation of multi-drug tolerance and antibiotic-induced responses by the histone-like protein Lsr2 in M. tuberculosis. PLoS Pathog. 2007;3:e87. https://doi.org/10.1371/journal.ppat.0030087.
161. Dorman CJ. H-NS-like nucleoid-associated proteins, mobile genetic elements and horizontal gene transfer in bacteria. Plasmid. 2014;75:1–11. https://doi.org/10.1016/j.plasmid.2014.06.004.
162. Takeda T, Yun CS, Shintani M, Yamane H, Nojiri H. Distribution of genes encoding nucleoid-associated protein homologs in plasmids. Int J Evol Biol. 2011;2011:685015. https://doi.org/10.4061/2011/685015.
163. Doyle M, Fookes M, Ivens A, Mangan MW, Wain J, Dorman CJ. An H-NS-like stealth protein aids horizontal DNA transmission in bacteria. Science. 2007;315:251–2.
164. Beloin C, Deighan P, Doyle M, Dorman CJ. Shigella flexneri 2a strain 2457T expresses three members of the H-NS-like protein family: characterization of the Sfh protein. Mol Gen Genomics. 2003;270:66–77. https://doi.org/10.1007/s00438-003-0897-0.
165. Deighan P, Beloin C, Dorman CJ. Three-way interactions among the Sfh, StpA and H-NS nucleoid-structuring proteins of Shigella flexneri 2a strain 2457T. Mol Microbiol. 2003;48:1401–16.

166. Sherburne CK, Lawley TD, Gilmour MW, Blattner FR, Burland V, Grotbeck E, Rose DJ, Taylor DE. The complete DNA sequence and analysis of R27, a large IncHI plasmid from Salmonella typhi that is temperature sensitive for transfer. Nucleic Acids Res. 2000;28(10):2177–86. https://doi.org/10.1093/nar/28.10.2177.
167. Forns N, Baños RC, Balsalobre C, Juárez A, Madrid C. Temperature-dependent conjugative transfer of R27: role of chromosome- and plasmid-encoded Hha and H-NS proteins. J Bacteriol. 2005;187:3950–9. https://doi.org/10.1128/JB.187.12.3950-3959.2005.
168. Baños RC, Vivero A, Aznar S, García J, Pons M, Madrid C, Juárez A. Differential regulation of horizontally acquired and core genome genes by the bacterial modulator H-NS. PLoS Genet. 2009;5:e1000513. https://doi.org/10.1371/journal.pgen.1000513.
169. Nojiri H. Structural and molecular genetic analyses of the bacterial carbazole degradation system. Biosci Biotechnol Biochem. 2012;76:1–18. https://doi.org/10.1271/bbb.110620.
170. Lee S, et al. Effects of carbazole-degradative plasmid pCAR1 on biofilm morphology in Pseudomonas putida KT2440. Environ Microbiol Rep. 2016;8:261–71. https://doi.org/10.1111/1758-2229.12376.
171. Suzuki C, et al. Oligomerization mechanisms of an H-NS family protein, Pmr, encoded on the plasmid pCAR1 provide a molecular basis for functions of H-NS family members. PLoS One. 2014;9:e105656. https://doi.org/10.1371/journal.pone.0105656.
172. Suzuki C, Yun CS, Umeda T, Terabayashi T, Watanabe K, Yamane H, Nojiri H. Oligomerization and DNA-binding capacity of Pmr, a histone-like protein H1 (H-NS) family protein encoded on IncP-7 carbazole-degradative plasmid pCAR1. Biosci Biotechnol Biochem. 2011;75:711–7. https://doi.org/10.1271/bbb.100841.
173. Yun CS, et al. Pmr, a histone-like protein H1 (H-NS) family protein encoded by the IncP-7 plasmid pCAR1, is a key global regulator that alters host function. J Bacteriol. 2010;192:4720–31. https://doi.org/10.1128/JB.00591-10.
174. Yun CS, et al. MvaT family proteins encoded on IncP-7 plasmid pCAR1 and the host chromosome regulate the host transcriptome cooperatively but differently. Appl Environ Microbiol. 2016;82:832–42. https://doi.org/10.1128/AEM.03071-15.

Functions of Archaeal Nucleoid Proteins: Archaeal Silencers are Still Missing

2

Hugo Maruyama, Nicholas A. Kent, Hiromi Nishida, and Taku Oshima

Abstract

It is well known that horizontally transferred genes (HTGs) contribute to the adaptation of archaea to their living environment. Archaea have acquired HTGs not only from other archaea but also from bacteria. HTGs should be integrated into the host archaeal transcriptional networks to become functional. In bacteria, the nucleoid proteins, such as H-NS and Lsr2, are well known as the major factors to enhance horizontal transfer of foreign DNAs into the bacterial cells by silencing the expression of HTGs which permit bacteria to safely acquire the HTGs and give time to modify the regulatory elements and promoters of HTGs and to integrate the HTGs into host transcriptional networks. In archaea, although those factors have not been identified, the importance of silencers would be the same. Interestingly, there is a protein which has highly similar characteristics with H-NS in the hyperthermophilic archaeon *Thermococcus kodakarensis*. In this chapter, we summarize the known biological importance of the nucleoid proteins and histones in archaea and discuss the possible role of the archaeal nucleoid proteins in horizontal gene transfer in archaeal cells.

Keywords

Archaea · Bacteria · Horizontal gene transfer · Horizontally transferred gene · H-NS · Silencer · Histone

H. Maruyama
Department of Bacteriology, Osaka Dental University, Hirakata, Osaka, Japan
e-mail: maruyama@cc.osaka-dent.ac.jp

N. A. Kent
School of Biosciences, Cardiff University, Cardiff, UK
e-mail: KentN@cardiff.ac.uk

H. Nishida · T. Oshima (✉)
Biotechnology Research Center & Department of Biotechnology, Toyama Prefectural University, Toyama, Japan
e-mail: hnishida@pu-toyama.ac.jp; taku@pu-toyama.ac.jp

H. Nishida, T. Oshima (eds.), *DNA Traffic in the Environment*,
https://doi.org/10.1007/978-981-13-3411-5_2

2.1 Archaea Have Acquired Genes from Bacteria to Adapt to Their Living Environment

Archaea are single-cell microorganisms living in nearly all habitats on earth. They live not only in extreme circumstances, such as high temperature, low pH, and high salinity, but also more moderate mesophilic circumstances [1, 2]. Archaea live in various ecological niches and play a major role to drive the biological fluxes as a part of the biogeochemical cycles on earth with bacteria [3]. Major building blocks of life, such as hydrogen, carbon, nitrogen, sulfur, oxygen, and phosphorous are supplied from earth as inorganic compounds which are converted to the available forms for many organisms, and the resultant organic materials are degraded and recycled by archaea and bacteria [3, 4]. Thus, it is critical for many organisms to live that archaea and bacteria acquire the genes from other organisms to adapt to the highly variable habitats.

The horizontal gene transfer (HGT) has had major roles for archaea to adapt to different habitats [2]. Archaea transfer DNAs by various mechanisms similar to those in bacteria, natural competence, vesicle transport, virus transduction, conjugation, and cell fusions [2]. HGT of archaeal domain not only in intraspecies but also in interspecies has driven the archaeal evolution [5–8]. Furthermore, the metagenomics of the fosmid archaeal clones constructed from deep-Mediterranean samples followed by pangenome analysis of Thaumoarchaeota and Group II/III Euryarchaeota revealed that 23.9% of Thaumoarchaeota genes and 29.7% of Group II/III Euryarchaeota genes are horizontally transferred genes from bacteria [9]. The functions of these bacterial genes are related to metabolism and cell envelope biogenesis, suggesting that archaea which had lived in the higher thermophilic environment have acquired bacterial genes to adapt to cold and oligotrophic ocean environment [9]. These findings strongly suggest that horizontally transferred genes are currently active in archaeal cells. With much larger scale, comprehensive gene comparison analysis revealed that 2264 protein families were bacterial origin in which most frequent functions were metabolism [10].

The HGT from archaea is also important for bacteria living in extreme circumstances. *Aquifex aeolicus* and *Thermotoga maritima* had acquired genes from hyperthermophilic bacteria to adapt to high-temperature circumstances [1, 11–13]. Thus, these observations clearly suggested that archaea and bacteria use those interspecies horizontally transferred genes (HTGs) to survive in new living environments.

2.2 The Difference of General Transcriptional Regulation Between Archaea and Bacteria

Archaea have the hybrid transcriptional regulatory system; the eukaryote-type basal transcription system works with the prokaryote-type transcriptional regulator [14, 15]. While archaeal RNA polymerase is more similar to eukaryotic RNA-polymerase II than bacterial RNA polymerase, several subunits are homologous to the subunits composed of bacterial RNA polymerase; for instance, bacterial α subunit is homologous to Rpo3 and 11 subunits of archaeal RNA polymerase [16].

Archaeal promoters are composed of AT-rich TATA-like elements (TATA box) and purine-rich factor B recognition elements (BRE). TATA-binding protein (TBP) and TFIIB-related transcriptional factor B (TFB) bind to the TATA-box and BRE [14, 15, 17–20]. Genome-wide identification of transcriptional start sites (TSSs) and the search of the conserved sequences upstream of TSSs indicated that TATA box and BRE are highly conserved (especially TATA box) in the genome of various archaeal species including *Thermococcus kodakarensis* and *Sulfolobus solfataricus*, and *Haloferax volcanii* [21–23]. Thus, the RNA polymerase and promoter architecture in archaea are completely different from those in bacteria.

In contrast to the RNA polymerase and promoter architecture, a majority of transcriptional factors (TFs) in archaea are of the bacteria-like TFs containing helix-turn-helix (HTH), winged HTH, and ribbon-helix-helix (RHH) motifs and zinc ribbon [14, 15]. The eukaryotic homolog of transcriptional regulator was also found in the genomes of archaea, but rare [14, 15]. It was estimated that the origin of bacterial-like TFs in bacteria and archaea would be possessed by a common ancestor of them and vertically transmitted during the evolution of bacteria and archaea [24].

The regulatory mechanisms of archaeal TFs in transcription should be different from those of bacterial TFs, because in archaea pre-initiation complex is initially formed on TATA box and BRE to recruit RNA polymerase to promoters, while RNA polymerase directly binds with promoters in bacteria [16, 25]. In bacteria, the subunit of RNA polymerase can specifically recognize promoter sequence (−10 and −35 sequences) [25]. In contrast, archaeal RNA polymerase cannot directly recognize the promoter, instead TBP initially binds to TATA box, and TFB then interacts with BRE and stabilizes TATA–TBP complex. This TFB–TBP–TATA–BRE complex is the pre-initiation complex and recruits RNA polymerase to the promoter [14, 15].

The archaeal repressor-binding sites are overlapped with TATA box and/or BRE, or downstream of them. The binding of the repressor inhibits pre-initiation complex formation or recruitment of RNA polymerase. The archaeal activator binds upstream of or partially overlapped with BRE and stimulates to pre-initiation complex formation through the protein–protein interaction [14, 15]. Thus, the activator-binding sites have the optimal spacer length with the TATA box and BRE (15–17 bp) to bind to the same helical phase of the binding sites of TBP and BRE. In contrast to the archaeal TFs, many bacterial repressors bind to promoters and directly inhibit RNA polymerase–promoter interactions, while bacterial activators bind upstream of promoters and directly interact with α and/or subunits and recruit RNA polymerase to promoters [25]. Thus, it is clear that the factors that interact with TFs and regulatory sequences of transcription are completely different between archaea and bacteria, although the principles of transcriptional regulation of prokaryotic TFs have been conserved in archaea and bacteria. These facts clearly suggest that the HTGs transferred from bacteria need to acquire the new transcriptional regulation to be expressed in archaeal cells, such as archaeal promoters and transcriptional regulatory elements.

2.3 The Importance of Silencers for the Integration of Horizontally Transferred Genes in the Host Transcriptional Network in Bacteria

Inappropriate expression (basically a higher expression) of newly acquired genes is problematic for host cells [26]. Therefore, it is critical to regulate HTGs under the host transcriptional network and safely express HTGs whenever HTGs' functions are required (see Chap. 1). To integrate HTGs into the host transcriptional network, there have been several steps: (1) HTGs were integrated into the host genome with lower expression than native host genes, (2) a host activator would be acquired to express HGTs (if needed), and (3) the expression of HGTs has been fine-tuned for the host cells to adapt to their living environments [27].

Even for bacteria, to acquire and use HTGs from other bacteria, acquisitions and adaptations of transcriptional regulatory elements for HTGs have been required, when HTGs are integrated in the host transcriptional regulatory network [27]. This process would have been needed for millions of years. Actually, the phylogenetic analysis revealed that recently transferred genes are not regulated by known TFs, while HTGs present in host bacteria for a long time after their transfers tend to be regulated by known TFs [27]. When transferred from bacteria to archaea, there are three possibilities to express HTGs under the host transcriptional regulatory network: type 1, integrate at downstream of the archaeal active promoters; type 2, initially integrate at downstream of inactive archaeal promoter or the region which don't have any relations to transcriptional regulatory regions and acquire promoter and TFs; type 3, integrate with bacterial promoters and regulatory elements which will be modified to be recognized with archaeal RNAP and TFs. Actually, HGT on type I mechanism was reported; L29 gene encoding a ribosomal protein of *S. solfataricus* P2 is displaced with a HTG in the ribosomal protein operon. This type of HGT has been observed in operons encoding ribosomal proteins among bacteria [28]. In this case, because the functions of the ribosomal protein may be highly conserved and able to replace the same subunit in other organisms, the direct control of the expression of the HTG, the ribosomal protein gene, with existing promoter and significant expression of the HTGs may not be problematic for host archaea. However, if HTGs are transferred from bacteria which are lesser conserved in archaea and expressed with the existing archaeal promoter when those HTGs are integrated at downstream of active promoters, host archaeal cells would have high risk getting the fatal effects by the expression of the HTGs. In contrast, when HTGs from bacteria are integrated at downstream of lower expression promoters or promoterless regions into archaeal genome; for instance, downstream of terminator, HTGs should have no or lower expression and have low risk for host archaea to have fatal expressions of HTGs. However, for the host archaeal strains to use those HTGs, the host strains should acquire and/or generate promoters and regulatory elements to express those HTGs. Then, the host archaea again have risks of inappropriate expression of HTGs during acquisition and modification of the promoters and regulatory elements to get the optimal expression of HTGs. It was indicated that the laterally transferred genes are regulated by the host TF, TrmB, in *Halobacterium salinarum* NRC-1, which senses nutritional availability and maintains a balance of redox and energy [29]. Therefore, in archaea, the HTGs might really acquire the regulatory elements and promoters to be regulated by the host TF. Thus, it is obvious that the archaea requires the mechanisms to reduce risks during the integration of HTGs into their transcriptional networks.

In bacteria, to guard the host cells from the fatal expression of HTGs during the evolution of HTGs' promoters or transcriptional regulators (in host cells or HTGs), several nucleoid proteins function as the transcriptional silencers [30–32]. In *Escherichia coli* and *Salmonella*, the H-NS protein preferentially binds to HTGs and represses expression of HTGs during the evolution, which would permit higher frequency of mutations in the regulatory regions [30, 31]. For archaea, the silencing system is also useful and may be more important than bacteria, because it requires a long time to acquire completely new promoters and/or regulatory elements. However, it was still unknown whether those types of nucleoid proteins exist in archaea. We will discuss the possible candidates for silencers for archaea to safely integrate HTGs, in the next section.

2.4 The Archaeal Nucleoid Proteins May Be Silencers Supporting the Integration of Horizontal Transfer Genes in Host Genomes

As described earlier in this chapter, archaea possess a mixture of bacterial and eukaryotic features. The DNA replication, transcription, and translation machineries of archaea are more similar to those of eukaryotes, while the genes involved in metabolic processes show more similarities to their bacterial counterparts [33]. This feature is also adaptable to the nucleoid proteins in archaea. Depending on species, archaea encode various nucleoid proteins, including histone, Alba, TrmBL2, HTa (archaeal homolog of bacterial HU), Cren7, CC1, and so on [34]. Among the two major phyla in Archaea, Euryarchaeota and Crenarchaeota, histone is encoded by almost all euryarchaea, while Alba is generally encoded in crenarchaea and a part of euryarchaea. Each archaeal species usually encode more than one nucleoid protein and these proteins are considered to organize nucleoid architecture and gene regulation [34].

Apart from relatively well-studied nucleoid proteins such as histone and Alba, a novel class of archaeal nucleoid protein, TrmBL2 (TrmB-like 2) was discovered in *Thermococcales* (including genera *Thermococcus*, *Pyrococcus*, and *Palaeococcus*) [35]. TrmBL2 is a member of TrmB (transcriptional regulator of mal operon) family proteins that have helix-turn-helix (HTH) DNA-binding motif. TrmB is a transcription factor that regulates the gene cluster containing the ATP-binding cassette (ABC) transport system for maltose and trehalose (TM operon) in *Pyrococcus furiosus* and *Thermococcus litoralis* [36, 37]. Other groups of proteins homologous to TrmB were found in *Thermococcales* and among these proteins, TrmBL1 (TrmB-like 1; also called Tgr for *Thermococcales* glycolytic regulator) is a transcription regulator that controls genes involved in sugar metabolism [38–40]. TrmB and TrmBL1/Tgr both have a C-terminal sugar-binding domain and have a consensus binding sequence of DNA [38, 40]. These proteins dissociate from DNA when sugar is bound to this domain, resulting in gene activation or repression depending on the position of its binding site relative to the BRE-TATA box of the target gene [38].

Among the TrmB family proteins, TrmBL2 in *Thermococcales* has a striking difference compared to other TrmB or TrmB-like proteins and is considered a nucleoid protein rather than a simple transcription factor. The amount of this protein in the cell is extraordinarily abundant compared to other TrmB-like transcription factors [35]. It lacks the C-terminal sugar-binding domain, does not have a consensus binding sequence of DNA, and it binds to both intergenic and protein-coding sequences [35]. Furthermore, unlike TrmB or other TrmB-like proteins that are encoded only by a part of *Thermococcales* species, TrmBL2 is perfectly conserved among all *Thermococcales* species sequenced so far [41], suggesting its role other than a simple transcription factor. Structural studies suggest that the apparent nonspecific binding of TrmBL2 to DNA results from its unusual nonoptimal protein–DNA contact made by its HTH domains [42].

TrmBL2 has a great similarity to bacterial silencer H-NS and its functional homologs present in various bacteria. TrmBL2 occupies both the promoter and coding sequences along the genome, and the distribution is not biased to either of them, compared to the composition of the genomic sequence [35], while H-NS and its functional homologs bind both to promoter and coding sequences with the preference of the adenine and thymine (AT)-rich regions in the genome [30, 32, 43–47] and lower sequence specificity except nucleation sites [48, 49]. According to sequence analysis of TrmBL2-associated DNA in vivo, TrmBL2 does not have a specific binding sequence, although certain sites seem to exist where TrmBL2 prefers to bind first with relatively high affinity [50]. Deletion of the gene encoding TrmBL2 in *T. kodakarensis* (TK0471) results in upregulation of nearly a hundred of genes and these upregulated genes have positive correlation with TrmBL2 binding to its promoter (Fig. 2.1a, c). Unlike TrmB or TrmBL1/Tgr that directly controls genes involved in sugar metabolism, TrmBL2 does not seem to regulate a specific pathway [35]. In *E. coli*, the double-deletion strain of H-NS and StpA, a homolog of H-NS, hundreds of genes are upregulated too [51], similar to the case of TrmBL2 deletion in *T. kodakarensis* (Fig. 2.1b). The H-NS homolog, MvaT, in *Pseudomonas aeruginosa* makes higher-order oligomer, which forms a gene-silencing nucleoprotein filament [52, 53]. The mutations to inhibit the formation of higher-order oligomer reduce the activity to repress the expression of at least *cupA* gene and to form chromatin nucleoprotein filaments. The similarity of TrmBL2 with H-NS homolog is also found in this feature. When TrmBL2 bound to DNA, TrmBL2 forms a thick nucleoprotein filament that increases the stiffness of the DNA molecule [50]. Those facts strongly support that TrmBL2 is a silencer.

How does TrmBL2 determine its binding sites without a specific binding sequence? Possibly, it binds to the region where there is less histone. Similar to eukaryotic histone, archaeal histone preferentially binds to nucleosome positioning sequence with alternating G (guanine)/C (cytosine)—and A/T-di- and trinucleotide tracts capable of periodic major and minor groove compaction [54]. ChIP-seq

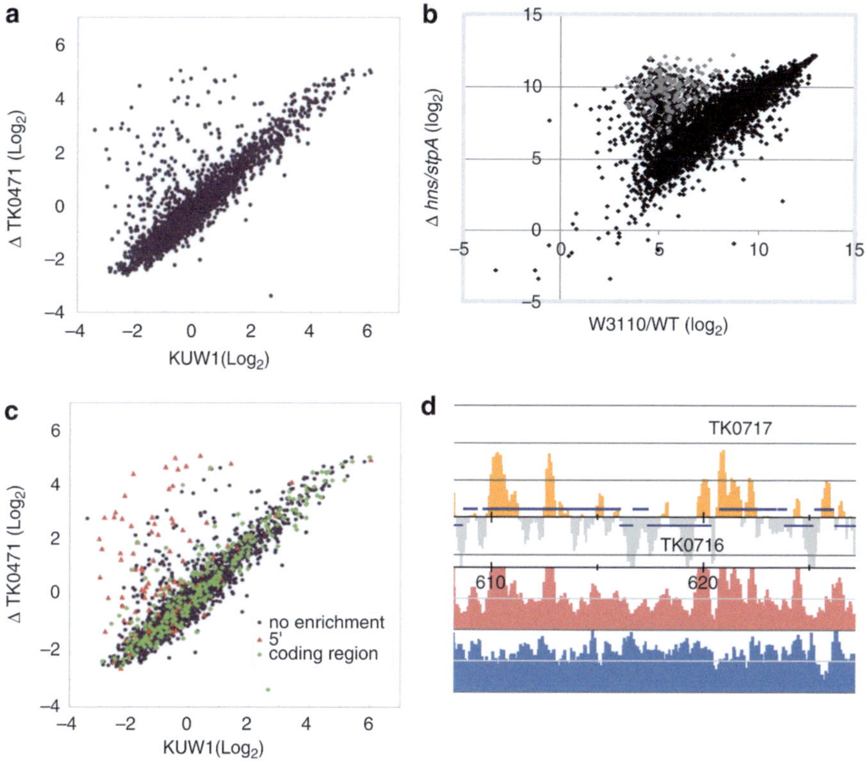

Fig. 2.1 (**a**, **c**) Scatterplot of the RNA abundance of all ORFs in *Thermococcus kodakarensis* measured by microarray. The fluorescence intensity of each ORF is plotted in log scale. X-axis represents the fluorescence intensity of all ORFs in the parental strain of delta-TK0471 strain, KUW1. Y-axis represents the fluorescence intensity of all ORFs in delta-TK0471 strain with the deletion of the whole coding sequence of TrmBL2 (TK0471). In (**c**), the color code classifies the ORFs depending on whether TrmBL2 bind is bound near its start codon (red; −150 to +50 relative to the start codon), to its coding sequence (green), or no TrmBL2 binding detected (blue). TrmBL2-binding sites were judged by high-throughput sequencing of DNA co-enriched with TrmBL2 in sucrose gradient sedimentation of MNase-digested *T. kodakarensis* chromatin. (**b**) Scatterplot of the RNA abundance of all ORFs in *Escherichia coli* K-12 and *hns/stpA* double-deletion strains. The fluorescence intensity of each ORF is plotted in log scale. X-axis represents the fluorescence intensity of all ORFs in the parental strain, W3110. Y-axis represents the fluorescence intensity of all ORFs in *hns* and *stpA* double-deletion mutant (*hns/stpA*). In *E. coli*, H-NS cooperatively represses several hundreds of genes with an H-NS homolog, StpA. To completely repress about 20% of those genes, H-NS and StpA require other H-NS homologs, Hha and YdgT. Gray dots represent the genes which require Hha and YdgT for their repression (reproduced from [51]) (**d**), TrmBL2 enrichment in the region containing TK0716 (Iron(II) transport protein A) and TK0717 (ABC-type molybdate transport system, periplasmic component), genes possibly transferred from bacteria [98]. Positive TrmBL2 binding is detected near the 5′ end of these ORFs. The positive value in the top profile shows the enrichment of the genomic region in TrmBL2-rich fraction in sucrose density gradient sedimentation of MNase-digested *T. kodakarensis* chromatin in the log phase. Blue horizontal lines represent predicted ORFs. The histogram in the middle shows the DNA sequence enriched in TrmBL2-rich fraction, and the histogram at the bottom indicates the DNA sequences taken from the whole cell extract

analysis shows that histone-binding sites in vivo perfectly match with the histone-binding sites when bare genomic DNA and recombinant histone is mixed [55]. Thus, histone has much more strict sequence requirement for its binding site, and TrmBL2 probably can locate to where there is less histone. Indeed, histone and TrmBL2 competes each other for DNA binding in vitro [50]. Supposedly, in histone encoding euryarchaeal species, TrmBL2 would bind to DNA sequences that do not have a histone-binding signal, most likely including the genes transferred from bacteria. Once such DNA is integrated into the genome, TrmBL2 would bind in a relatively random manner to the bacterial sequence, resulting in suppression of the gene when bound to the promoter. This could be an effective way to suppress genes transferred from other species, at least in histone-coding euryarchaea. Figure 2.1d shows a genomic region in *T. kodakarensis* where such suppression possibly occurs.

Is it possible for a histone-coding euryarchaea to suppress genes transferred from other histone-coding archaea? As described later in this chapter, GC content of archaeal genomic sequences correlates with the pI of its histone protein [56]. Thus, genes transferred from distant euryarchaea are expected to have lower histone binding in a host species with different GC content and pI of its histone. Besides, it has been shown that archaeal histone binds to intergenic or promoter region with relatively lower frequency [55, 57]. Thus, genes horizontally transferred from other archaea could also be suppressed by binding of protein such as TrmBL2 to the intergenic region.

Since there is not a single nucleoid protein that is encoded by all species in archaea, they seem to utilize different combinations of nucleoid proteins to mark its own genome and to suppress HTGs. Although the above hypothesis only explains suppression of HTGs in the combination of histone and TrmBL2, other nucleoid proteins may serve the same function. Outside of *Thermococcales*, TrmB family protein is found in other archaea such as *Halobacterium* and methanogenic archaea, and in some bacteria [29, 41, 58]. Although it needs to be clarified which of these proteins are simple transcription factors and which could be a nucleoid protein, suppression of HTGs by such nucleoid-associated TrmB family protein can be a general mechanism. In the case of *T. kodakarensis*, TrmBL2 was first found by its extraordinary abundance in the cell [35]. Abundance in the cell and nonspecific DNA binding could be a key to search for proteins that are involved in regulation of HTGs.

Most species in the phylum *Crenarchaeota* encode a protein called Alba (acetylation lowers binding affinity). Alba can constitute up to ~4% of the total cellular proteins in *Sulfolobus* [59]. The affinity of Alba to DNA increases upon deacetylation by the silencing protein Sir2 [60]. It has been proposed that its dynamic association to and dissociation from DNA is involved in transcriptional regulation and modulation of chromatin structure [60]. Although Alba was initially considered as a DNA-binding protein, recent studies have revealed the association of Alba with RNA species, ribonucleo-protein complexes, and RNA-binding proteins [61–63]. Alba is also encoded in the genome of some euryarchaeal species, although the protein is less abundant in these cells. Although it needs to be studied, it is possible that Alba plays a role in the regulation of HTGs in a way similar either to histone to mark its own genome or similar to TrmBL2 to suppress transcription of transferred genes.

In 1990, a protein homologous to eukaryotic histones was reported in the methanogen, *Methanothermus fervidus* [64]. This was the first example of a histone homolog in prokaryotes. Searches in genomic sequences as well as biochemical studies revealed that almost all euryarchaeal species encode histone proteins. Rare exceptions include the species in the order *Thermoplasmatales*, which encode the bacterial HU homolog (HTa) but not histones. This could be a result of horizontal gene transfer of HU from bacteria and subsequent loss of histone gene [65]. Although the structure of the histone fold motif is quite similar between eukaryotic and archaeal histones, there are several differences between the two (see below, Sect. 2.5). The archaeal hailstones consist solely of the histone-fold motif [66] and lack the N- and C-terminal tails. Eukaryotic histone forms an octamer, where two H2A/H2B dimers are attached on both sides of H3/H4 tetramer. Unlike eukaryotic histones, archaeal histones were shown to form a tetramer [67]. Approximately, 120 bp of DNA can wrap around an archaeal histone tetramer in vitro [68], which is shorter than that in the eukaryotic nucleosome.

Although many studies have supported the conclusion that archaeal histones and genomic DNA form nucleosome structures in vivo [67, 69], it was recently proposed that archaeal histone is not a static structure fixed to a tetramer but is rather flexible and can form multimers determined by the DNA sequence, based on sequencing of histone-bound DNA and mapping on *T. kodakarensis* genome [34, 57]. In this model, histone dimer binds to 30 bp DNA and this unit can pile up to form tetramer, hexamer, etc. that binds multiples of 30 bp DNA [57]. This could be explained by the existence of unrestricted 4-helix bundle (4HD) domain that connects each histone dimer. In contrast to the eukaryotic histone that is strongly restricted to the fixed octameric structure of H3/H4 tetramer and two H2A/H2B dimers determined by the combination of "handshake" of 4HD domains, unrestricted 4HD domain in archaeal histone allows the dimers to pile up to form a multimer [57].

The existence of such histone multimer was experimentally supported by the result of crystal structure. Substitution of a conserved glycine that lies on the histone dimer–dimer interface results in destabilization of histone multimer formation, reduced growth rate, and impairs transcription regulation in vivo [70]. This multimeric histone structure could also be a way to mark its own genomic DNA and selectively suppress HTGs from bacteria. If there is a difference in how histone forms multimer on the genome between archaeal species, it could also selectively suppress other archaeal sequences acquired horizontally. In next section, we will discuss this possibility.

2.5 Archaeal Histones Adapt to the GC Content of the Genome

Genomic DNA is packaged with histones to form chromatin in eukaryotes [71, 72]. The most fundamental repeating unit of chromatin is the nucleosome consisting of an octamer of histones (2 copies of each histone: H2A, H2B, H3, and H4) and the genomic DNA wrapped around the octamer [73]. DNA sequence preference exists for nucleosome occupancy and positioning [74–84]. Posttranslational modification of the histones have an important role in the interaction between histones and DNA [85, 86]. Thus, histone has a structural variation, which depends on not amino acid sequence but the posttranscriptional modification. It may induce a high conservation of amino acid sequence of histones. The distribution of DNA base composition (GC content) of eukaryotic genomes is similar to a normal (Gaussian) distribution with a peak of approximately 40% of GC, which is completely different from those of bacterial and archaeal genomes [78].

As described earlier in this section, H-NS and its homologs recognize AT-rich DNA regions and inhibit the expression from the regions [45–47, 87–90]. Considering that the percentage of AT in the horizontally transferred elements is greater than that in the host DNA [91, 92], gene expression from the horizontally transferred elements is selectively repressed by H-NS and its functional homologs in a base composition-dependent manner. The distribution of GC content of bacterial (and archaeal) genomes is rather distinct from a Gaussian distribution [93].

As mentioned in the previous Sect. 2.4, a subpopulation of archaea (in particular, phylum euryarchaeota) possesses histones, which are similar to and smaller than the eukaryotic histones H3 and H4 [34]. Archaeal histones lack amino- and carboxy-terminal tails that are modified in eukaryotic H3 or H4 [34]. Among the histone-possessing archaeal species, most (62%, 33 of 53 species) of them have a single histone gene, and archaeal histones are not posttranslationally modified [56, 94]. Phylogenetic analysis revealed that the distribution of archaeal histones is associated not with their species-evolutionary relationship, but with their genomic DNA base composition [56].

Eukaryotic histones have high isoelectric points (pI) due to high contents of basic amino acid residues arginine and lysine, indicating the involvement of ionic bonds in nucleosome formation [95]. The distribution of theoretical pIs of archaeal histones is completely different from that of fungal (eukaryotic) histones (Fig. 2.2). The arginine and lysine contents of archaeal histones differ from those of eukaryotic histones [56]. If archaeal histones have also the same functional constraint to bind DNA as well as eukaryotic histones, they should have high pIs. If archaea do not have such a functional constraint, arginine and lysine contents tend to be higher and lower, respectively, as the GC content of genomic DNA increases [96]. These biases are influenced by the genetic code (codon table). The correlation coefficient between the arginine content of archaeal histones and GC content of archaeal histone genes is 0.51 and that between the lysine content and GC content is −0.73 (Fig. 2.3), showing that archaeal histones do not have functional constraint to increase arginine and lysine contents of histones.

These results showed that archaeal histones have an amino acid sequence variation, which is associated with the genomic DNA base composition. Based on the phylogenetic analysis of the archaeal histones, some histone genes had been horizontally transferred [56]. Thus, archaea may have obtained histones as the genomic DNA changed its base composition. It is important that all eukaryotes have histones, but a part of archaea have histones. For the subpopulation of archaea,

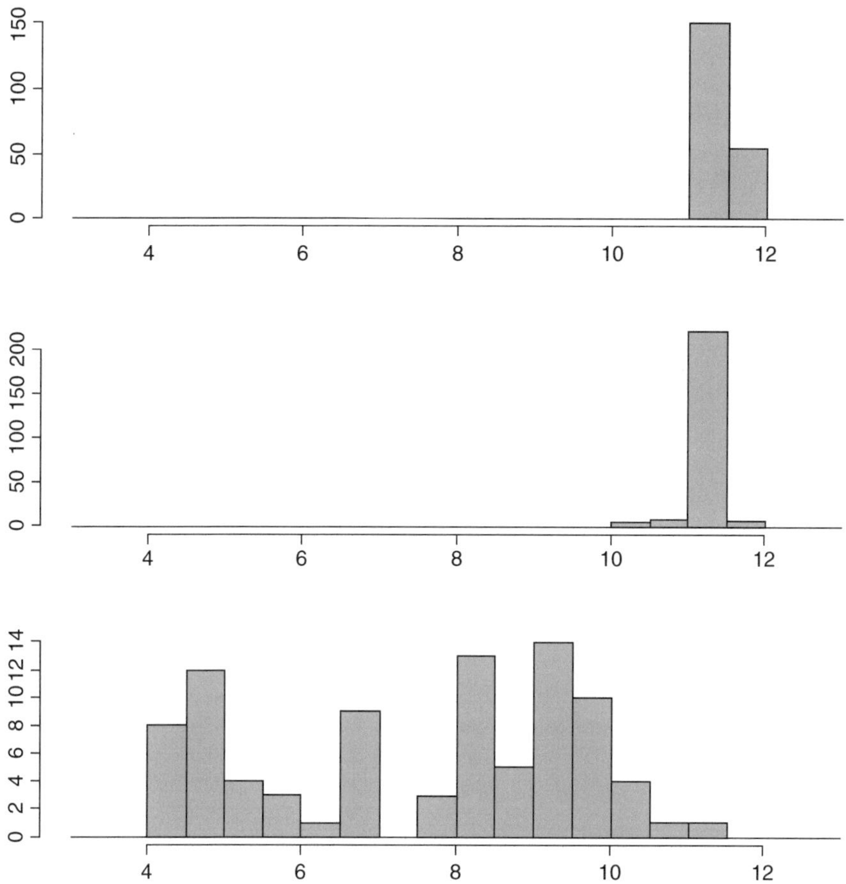

Fig. 2.2 Histograms of theoretical isoelectronic points of archaeal histones (bottom), and fungal histones H3 (top) and H4 (middle). *X*-axis, theoretical isoelectric point (pI). *Y*-axis, number of histones

histones may be an important protein, gene of which was sometimes obtained by horizontal transfer in the course of archaeal evolution. On the other hand, eukaryotes have maintained histones by vertical gene transfer from the parents. In the past, an ancestor of eukaryotes (which is thought to be close to archaea) might have a histone modification system, which is a diverging point between archaea and eukaryotes. Although the present archaeal histones are not posttranslationally modified [94], archaea have a posttranslational protein modification system [97]. Interestingly, the archaeal chromatin protein, Alba, is regulated by acetylation in *Sulfolobus* [60].

The GC adaptation of archaeal histones raises the interesting hypothesis that the core host genome distinguished by the histone binding and histone-nucleoid protein competition leads to the preferential binding of archaeal silencer of nucleoid proteins. In addition, the DNA base composition preferences of nucleoid-associated proteins (including histones) play a crucial role in global regulations of genes

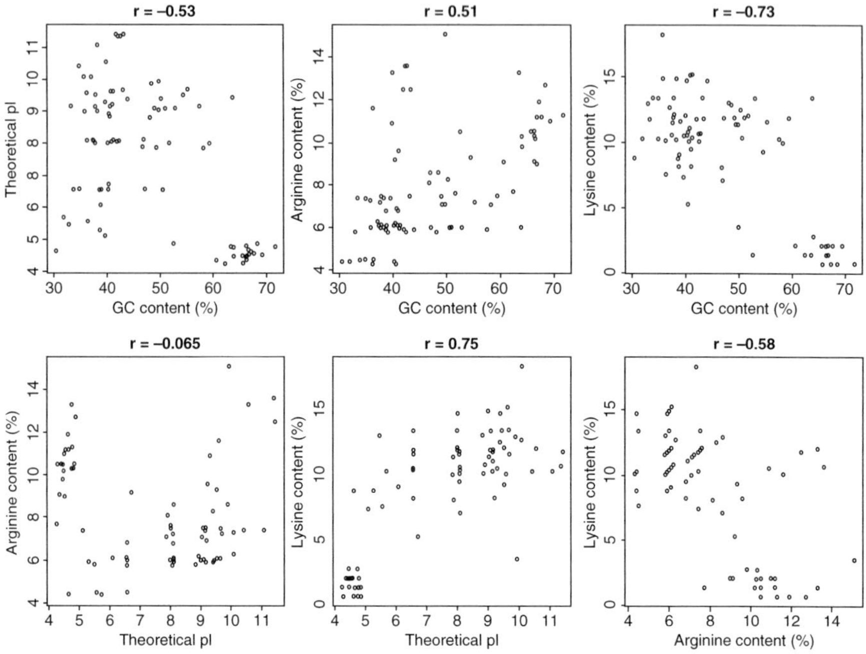

Fig. 2.3 Scatterplots between theoretical isoelectric points (pI) of archaeal histones and GC contents of archaeal histone genes, between arginine contents of archaeal histones and GC contents, between lysine contents of archaeal histones and GC contents, between arginine contents and theoretical pI, between lysine contents and theoretical pI, and between arginine contents and lysine contents. *r* indicates correlation coefficient. The correlation coefficient between the GC contents of archaeal histone genes and genomes is 0.94

among archaea as well as bacteria and eukaryotes. Further analyses of molecular mechanisms and genomic analyses of the binding mode of nucleoid proteins and histones are required for the understanding of how HTGs are silenced in archaea.

Acknowledgment This work was supported by Grant-in-Aid for Scientific Research on Innovative Areas (26106001 to TO).

References

1. Lopez-Garcia P, Zivanovic Y, Deschamps P, Moreira D. Bacterial gene import and mesophilic adaptation in archaea. Nat Rev Microbiol. 2015;13:447–56.
2. Wagner A, Whitaker RJ, Krause DJ, Heilers JH, van Wolferen M, van der Does C, Albers SV. Mechanisms of gene flow in archaea. Nat Rev Microbiol. 2017;15:492–501.
3. Offre P, Spang A, Schleper C. Archaea in biogeochemical cycles. Annu Rev Microbiol. 2013;67:437–57.
4. Falkowski PG, Fenchel T, Delong EF. The microbial engines that drive Earth's biogeochemical cycles. Science. 2008;320:1034–9.

5. Cadillo-Quiroz H, Didelot X, Held NL, Herrera A, Darling A, Reno ML, Krause DJ, Whitaker RJ. Patterns of gene flow define species of thermophilic Archaea. PLoS Biol. 2012;10:e1001265.
6. Naor A, Lapierre P, Mevarech M, Papke RT, Gophna U. Low species barriers in halophilic archaea and the formation of recombinant hybrids. Curr Biol. 2012;22:1444–8.
7. Papke RT, Koenig JE, Rodriguez-Valera F, Doolittle WF. Frequent recombination in a saltern population of Halorubrum. Science. 2004;306:1928–9.
8. Whitaker RJ, Grogan DW, Taylor JW. Recombination shapes the natural population structure of the hyperthermophilic archaeon Sulfolobus islandicus. Mol Biol Evol. 2005;22:2354–61.
9. Deschamps P, Zivanovic Y, Moreira D, Rodriguez-Valera F, Lopez-Garcia P. Pangenome evidence for extensive interdomain horizontal transfer affecting lineage core and shell genes in uncultured planktonic thaumarchaeota and euryarchaeota. Genome Biol Evol. 2014;6:1549–63.
10. Nelson-Sathi S, Sousa FL, Roettger M, Lozada-Chavez N, Thiergart T, Janssen A, Bryant D, Landan G, Schonheit P, Siebers B, McInerney JO, Martin WF. Origins of major archaeal clades correspond to gene acquisitions from bacteria. Nature. 2015;517:77–80.
11. Aravind L, Tatusov RL, Wolf YI, Walker DR, Koonin EV. Evidence for massive gene exchange between archaeal and bacterial hyperthermophiles. Trends Genet. 1998;14:442–4.
12. Nelson KE, Clayton RA, Gill SR, Gwinn ML, Dodson RJ, Haft DH, Hickey EK, Peterson JD, Nelson WC, Ketchum KA, McDonald L, Utterback TR, Malek JA, Linher KD, Garrett MM, Stewart AM, Cotton MD, Pratt MS, Phillips CA, Richardson D, Heidelberg J, Sutton GG, Fleischmann RD, Eisen JA, White O, Salzberg SL, Smith HO, Venter JC, Fraser CM. Evidence for lateral gene transfer between Archaea and bacteria from genome sequence of Thermotoga maritima. Nature. 1999;399:323–9.
13. Zhaxybayeva O, Swithers KS, Lapierre P, Fournier GP, Bickhart DM, DeBoy RT, Nelson KE, Nesbo CL, Doolittle WF, Gogarten JP, Noll KM. On the chimeric nature, thermophilic origin, and phylogenetic placement of the Thermotogales. Proc Natl Acad Sci U S A. 2009;106:5865–70.
14. Martinez-Pastor M, Tonner PD, Darnell CL, Schmid AK. Transcriptional regulation in Archaea: from individual genes to global regulatory networks. Annu Rev Genet. 2017;51:143–70.
15. Peeters E, Peixeiro N, Sezonov G. Cis-regulatory logic in archaeal transcription. Biochem Soc Trans. 2013;41:326–31.
16. Werner F. Molecular mechanisms of transcription elongation in archaea. Chem Rev. 2013;113:8331–49.
17. Hausner W, Frey G, Thomm M. Control regions of an archaeal gene. A TATA box and an initiator element promote cell-free transcription of the tRNA(Val) gene of Methanococcus vannielii. J Mol Biol. 1991;222:495–508.
18. Hausner W, Wettach J, Hethke C, Thomm M. Two transcription factors related with the eucaryal transcription factors TATA-binding protein and transcription factor IIB direct promoter recognition by an archaeal RNA polymerase. J Biol Chem. 1996;271:30144–8.
19. Palmer JR, Daniels CJ. In vivo definition of an archaeal promoter. J Bacteriol. 1995;177:1844–9.
20. Qureshi SA, Jackson SP. Sequence-specific DNA binding by the S. shibatae TFIIB homolog, TFB, and its effect on promoter strength. Mol Cell. 1998;1:389–400.
21. Babski J, Haas KA, Nather-Schindler D, Pfeiffer F, Forstner KU, Hammelmann M, Hilker R, Becker A, Sharma CM, Marchfelder A, Soppa J. Genome-wide identification of transcriptional start sites in the haloarchaeon Haloferax volcanii based on differential RNA-Seq (dRNA-Seq). BMC Genomics. 2016;17:629.
22. Jager D, Forstner KU, Sharma CM, Santangelo TJ, Reeve JN. Primary transcriptome map of the hyperthermophilic archaeon Thermococcus kodakarensis. BMC Genomics. 2014;15:684.
23. Wurtzel O, Sapra R, Chen F, Zhu Y, Simmons BA, Sorek R. A single-base resolution map of an archaeal transcriptome. Genome Res. 2010;20:133–41.
24. Aravind L, Anantharaman V, Balaji S, Babu MM, Iyer LM. The many faces of the helix-turn-helix domain: transcription regulation and beyond. FEMS Microbiol Rev. 2005;29:231–62.
25. Browning DF, Busby SJ. The regulation of bacterial transcription initiation. Nat Rev Microbiol. 2004;2:57–65.

26. Sorek R, Zhu Y, Creevey CJ, Francino MP, Bork P, Rubin EM. Genome-wide experimental determination of barriers to horizontal gene transfer. Science. 2007;318:1449–52.
27. Lercher MJ, Pal C. Integration of horizontally transferred genes into regulatory interaction networks takes many million years. Mol Biol Evol. 2008;25:559–67.
28. Yoon SH, Reiss DJ, Bare JC, Tenenbaum D, Pan M, Slagel J, Moritz RL, Lim S, Hackett M, Menon AL, Adams MW, Barnebey A, Yannone SM, Leigh JA, Baliga NS. Parallel evolution of transcriptome architecture during genome reorganization. Genome Res. 2011;21:1892–904.
29. Schmid AK, Reiss DJ, Pan M, Koide T, Baliga NS. A single transcription factor regulates evolutionarily diverse but functionally linked metabolic pathways in response to nutrient availability. Mol Syst Biol. 2009;5:282.
30. Dorman CJ. H-NS, the genome sentinel. Nat Rev Microbiol. 2007;5:157–61.
31. Higashi K, Tobe T, Kanai A, Uyar E, Ishikawa S, Suzuki Y, Ogasawara N, Kurokawa K, Oshima T. H-NS facilitates sequence diversification of horizontally transferred DNAs during their integration in host chromosomes. PLoS Genet. 2016;12:e1005796.
32. Navarre WW, McClelland M, Libby SJ, Fang FC. Silencing of xenogeneic DNA by H-NS-facilitation of lateral gene transfer in bacteria by a defense system that recognizes foreign DNA. Genes Dev. 2007;21:1456–71.
33. Koonin EV, Mushegian AR, Galperin MY, Walker DR. Comparison of archaeal and bacterial genomes: computer analysis of protein sequences predicts novel functions and suggests a chimeric origin for the archaea. Mol Microbiol. 1997;25:619–37.
34. Peeters E, Driessen RP, Werner F, Dame RT. The interplay between nucleoid organization and transcription in archaeal genomes. Nat Rev Microbiol. 2015;13:333–41.
35. Maruyama H, Shin M, Oda T, Matsumi R, Ohniwa R, Itoh T, Shirahige K, Imanaka T, Atomi H, Yoshimura SH, Takeyasu K. Histone and TK0471/TrmBL2 form a novel heterogeneous genome architecture in the hyperthermophilic archaeon Thermococcus kodakarensis. Mol Biol Cell. 2011;22:386–98.
36. Lee SJ, Engelmann A, Horlacher R, Qu Q, Vierke G, Hebbeln C, Thomm M, Boos W. TrmB, a sugar-specific transcriptional regulator of the trehalose/maltose ABC transporter from the hyperthermophilic archaeon Thermococcus litoralis. J Biol Chem. 2003;278:983–90.
37. Lee SJ, Moulakakis C, Koning SM, Hausner W, Thomm M, Boos W. TrmB, a sugar sensing regulator of ABC transporter genes in Pyrococcus furiosus exhibits dual promoter specificity and is controlled by different inducers. Mol Microbiol. 2005;57:1797–807.
38. Kanai T, Akerboom J, Takedomi S, van de Werken HJ, Blombach F, van der Oost J, Murakami T, Atomi H, Imanaka T. A global transcriptional regulator in Thermococcus kodakaraensis controls the expression levels of both glycolytic and gluconeogenic enzyme-encoding genes. J Biol Chem. 2007;282:33659–70.
39. Lee SJ, Surma M, Seitz S, Hausner W, Thomm M, Boos W. Characterization of the TrmB-like protein, PF0124, a TGM-recognizing global transcriptional regulator of the hyperthermophilic archaeon Pyrococcus furiosus. Mol Microbiol. 2007;65:305–18.
40. Reichelt R, Gindner A, Thomm M, Hausner W. Genome-wide binding analysis of the transcriptional regulator TrmBL1 in Pyrococcus furiosus. BMC Genomics. 2016;17:40.
41. Gindner A, Hausner W, Thomm M. The TrmB family: a versatile group of transcriptional regulators in archaea. Extremophiles. 2014;18:925–36.
42. Ahmad MU, Waege I, Hausner W, Thomm M, Boos W, Diederichs K, Welte W. Structural insights into nonspecific binding of DNA by TrmBL2, an archaeal chromatin protein. J Mol Biol. 2015;427:3216–29.
43. Castang S, McManus HR, Turner KH, Dove SL. H-NS family members function coordinately in an opportunistic pathogen. Proc Natl Acad Sci U S A. 2008;105:18947–52.
44. Gordon BR, Li Y, Wang L, Sintsova A, van Bakel H, Tian S, Navarre WW, Xia B, Liu J. Lsr2 is a nucleoid-associated protein that targets AT-rich sequences and virulence genes in Mycobacterium tuberculosis. Proc Natl Acad Sci U S A. 2010;107:5154–9.
45. Grainger DC, Hurd D, Goldberg MD, Busby SJ. Association of nucleoid proteins with coding and non-coding segments of the Escherichia coli genome. Nucleic Acids Res. 2006;34:4642–52.

46. Lucchini S, Rowley G, Goldberg MD, Hurd D, Harrison M, Hinton JC. H-NS mediates the silencing of laterally acquired genes in bacteria. PLoS Pathog. 2006;2:e81.
47. Oshima T, Ishikawa S, Kurokawa K, Aiba H, Ogasawara N. Escherichia coli histone-like protein H-NS preferentially binds to horizontally acquired DNA in association with RNA polymerase. DNA Res. 2006;13:141–53.
48. Bouffartigues E, Buckle M, Badaut C, Travers A, Rimsky S. H-NS cooperative binding to high-affinity sites in a regulatory element results in transcriptional silencing. Nat Struct Mol Biol. 2007;14:441–8.
49. Lang B, Blot N, Bouffartigues E, Buckle M, Geertz M, Gualerzi CO, Mavathur R, Muskhelishvili G, Pon CL, Rimsky S. High-affinity DNA binding sites for H-NS provide a molecular basis for selective silencing within proteobacterial genomes. Nucleic Acids Res. 2007;35:6330–7.
50. Efremov AK, Qu Y, Maruyama H, Lim CJ, Takeyasu K, Yan J. Transcriptional repressor TrmBL2 from Thermococcus kodakarensis forms filamentous nucleoprotein structures and competes with histones for DNA binding in a salt- and DNA supercoiling-dependent manner. J Biol Chem. 2015;290:15770–84.
51. Ueda T, Takahashi H, Uyar E, Ishikawa S, Ogasawara N, Oshima T. Functions of the Hha and Ydg T proteins in transcriptional silencing by the nucleoid proteins, H-NS and StpA, in Escherichia coli. DNA Res. 2013;20:263–71.
52. Castang S, Dove SL. High-order oligomerization is required for the function of the H-NS family member MvaT in Pseudomonas aeruginosa. Mol Microbiol. 2010;78:916–31.
53. Winardhi RS, Fu W, Castang S, Li Y, Dove SL, Yan J. Higher order oligomerization is required for H-NS family member MvaT to form gene-silencing nucleoprotein filament. Nucleic Acids Res. 2012;40:8942–52.
54. Bailey KA, Pereira SL, Widom J, Reeve JN. Archaeal histone selection of nucleosome positioning sequences and the procaryotic origin of histone-dependent genome evolution. J Mol Biol. 2000;303:25–34.
55. Nalabothula N, Xi L, Bhattacharyya S, Widom J, Wang JP, Reeve JN, Santangelo TJ, Fondufe-Mittendorf YN. Archaeal nucleosome positioning in vivo and in vitro is directed by primary sequence motifs. BMC Genomics. 2013;14:391.
56. Nishida H, Oshima T. Archaeal histone distribution is associated with archaeal genome base composition. J Gen Appl Microbiol. 2017;63:28–35.
57. Maruyama H, Harwood JC, Moore KM, Paszkiewicz K, Durley SC, Fukushima H, Atomi H, Takeyasu K, Kent NA. An alternative beads-on-a-string chromatin architecture in Thermococcus kodakarensis. EMBO Rep. 2013;14:711–7.
58. Reichlen MJ, Vepachedu VR, Murakami KS, Ferry JG. MreA functions in the global regulation of methanogenic pathways in Methanosarcina acetivorans. MBio. 2012;3:e00189–12.
59. Xue H, Guo R, Wen Y, Liu D, Huang L. An abundant DNA binding protein from the hyperthermophilic archaeon Sulfolobus shibatae affects DNA supercoiling in a temperature-dependent fashion. J Bacteriol. 2000;182:3929–33.
60. Bell SD, Botting CH, Wardleworth BN, Jackson SP, White MF. The interaction of Alba, a conserved archaeal chromatin protein, with Sir2 and its regulation by acetylation. Science. 2002;296:148–51.
61. Aravind L, Iyer LM, Anantharaman V. The two faces of Alba: the evolutionary connection between proteins participating in chromatin structure and RNA metabolism. Genome Biol. 2003;4:R64.
62. Goyal M, Banerjee C, Nag S, Bandyopadhyay U. The Alba protein family: structure and function. Biochim Biophys Acta. 2016;1864:570–83.
63. Guo R, Xue H, Huang L. Ssh10b, a conserved thermophilic archaeal protein, binds RNA in vivo. Mol Microbiol. 2003;50:1605–15.
64. Sandman K, Krzycki JA, Dobrinski B, Lurz R, Reeve JN. HMf, a DNA-binding protein isolated from the hyperthermophilic archaeon Methanothermus fervidus, is most closely related to histones. Proc Natl Acad Sci U S A. 1990;87:5788–91.

65. DeLange RJ, Williams LC, Searcy DG. A histone-like protein (HTa) from Thermoplasma acidophilum. II. Complete amino acid sequence. J Biol Chem. 1981;256:905–11.
66. Sandman K, Reeve JN. Archaeal chromatin proteins: different structures but common function? Curr Opin Microbiol. 2005;8:656–61.
67. Pereira SL, Grayling RA, Lurz R, Reeve JN. Archaeal nucleosomes. Proc Natl Acad Sci U S A. 1997;94:12633–7.
68. Tomschik M, Karymov MA, Zlatanova J, Leuba SH. The archaeal histone-fold protein HMf organizes DNA into bona fide chromatin fibers. Structure. 2001;9:1201–11.
69. Shioda M, Sugimori K, Shiroya T, Takayanagi S. Nucleosomelike structures associated with chromosomes of the archaebacterium Halobacterium salinarium. J Bacteriol. 1989;171:4514–7.
70. Mattiroli F, Bhattacharyya S, Dyer PN, White AE, Sandman K, Burkhart BW, Byrne KR, Lee T, Ahn NG, Santangelo TJ, Reeve JN, Luger K. Structure of histone-based chromatin in archaea. Science. 2017;357:609–12.
71. Igo-Kemenes T, Horz W, Zachau HG. Chromatin. Annu Rev Biochem. 1982;51:89–121.
72. Kornberg RD. Structure of chromatin. Annu Rev Biochem. 1977;46:931–54.
73. Luger K, Mader AW, Richmond RK, Sargent DF, Richmond TJ. Crystal structure of the nucleosome core particle at 2.8 A resolution. Nature. 1997;389:251–60.
74. Drew HR, Travers AA. DNA bending and its relation to nucleosome positioning. J Mol Biol. 1985;186:773–90.
75. Ioshikhes I, Hosid S, Pugh BF. Variety of genomic DNA patterns for nucleosome positioning. Genome Res. 2011;21:1863–71.
76. Kaplan N, Hughes TR, Lieb JD, Widom J, Segal E. Contribution of histone sequence preferences to nucleosome organization: proposed definitions and methodology. Genome Biol. 2010;11:140.
77. Lowary PT, Widom J. New DNA sequence rules for high affinity binding to histone octamer and sequence-directed nucleosome positioning. J Mol Biol. 1998;276:19–42.
78. Nishida H. Nucleosome positioning. ISRN Mol Biol. 2012;2012:245706.
79. Peckham HE, Thurman RE, Fu Y, Stamatoyannopoulos JA, Noble WS, Struhl K, Weng Z. Nucleosome positioning signals in genomic DNA. Genome Res. 2007;17:1170–7.
80. Satchwell SC, Drew HR, Travers AA. Sequence periodicities in chicken nucleosome core DNA. J Mol Biol. 1986;191:659–75.
81. Segal E, Fondufe-Mittendorf Y, Chen L, Thastrom A, Field Y, Moore IK, Wang JP, Widom J. A genomic code for nucleosome positioning. Nature. 2006;442:772–8.
82. Stein A, Takasuka TE, Collings CK. Are nucleosome positions in vivo primarily determined by histone-DNA sequence preferences? Nucleic Acids Res. 2010;38:709–19.
83. Tsankov A, Yanagisawa Y, Rhind N, Regev A, Rando OJ. Evolutionary divergence of intrinsic and trans-regulated nucleosome positioning sequences reveals plastic rules for chromatin organization. Genome Res. 2011;21:1851–62.
84. Valouev A, Johnson SM, Boyd SD, Smith CL, Fire AZ, Sidow A. Determinants of nucleosome organization in primary human cells. Nature. 2011;474:516–20.
85. Kouzarides T. Chromatin modifications and their function. Cell. 2007;128:693–705.
86. Strahl BD, Allis CD. The language of covalent histone modifications. Nature. 2000;403:41–5.
87. Dillon SC, Dorman CJ. Bacterial nucleoid-associated proteins, nucleoid structure and gene expression. Nat Rev Microbiol. 2010;8:185–95.
88. Dorman CJ. H-NS: a universal regulator for a dynamic genome. Nat Rev Microbiol. 2004;2:391–400.
89. Gordon BR, Li Y, Cote A, Weirauch MT, Ding P, Hughes TR, Navarre WW, Xia B, Liu J. Structural basis for recognition of AT-rich DNA by unrelated xenogeneic silencing proteins. Proc Natl Acad Sci U S A. 2011;108:10690–5.
90. Navarre WW, Porwollik S, Wang Y, McClelland M, Rosen H, Libby SJ, Fang FC. Selective silencing of foreign DNA with low GC content by the H-NS protein in Salmonella. Science. 2006;313:236–8.

91. Nishida H. Comparative analyses of base compositions, DNA sizes, and dinucleotide frequency profiles in archaeal and bacterial chromosomes and plasmids. Int J Evol Biol. 2012;2012:342482.
92. Rocha EP, Danchin A. Base composition bias might result from competition for metabolic resources. Trends Genet. 2002;18:291–4.
93. Nishida H. Genome DNA sequence variation, evolution, and function in bacteria and archaea. Curr Issues Mol Biol. 2013;15:19–24.
94. Forbes AJ, Patrie SM, Taylor GK, Kim YB, Jiang L, Kelleher NL. Targeted analysis and discovery of posttranslational modifications in proteins from methanogenic archaea by top-down MS. Proc Natl Acad Sci U S A. 2004;101:2678–83.
95. O'Farrell PZ, Goodman HM, O'Farrell PH. High resolution two-dimensional electrophoresis of basic as well as acidic proteins. Cell. 1977;12:1133–41.
96. Sueoka N. Correlation between base composition of deoxyribonucleic acid and amino acid composition of protein. Proc Natl Acad Sci U S A. 1961;47:1141–9.
97. Eichler J, Adams MW. Posttranslational protein modification in archaea. Microbiol Mol Biol Rev. 2005;69:393–425.
98. Fukui T, Atomi H, Kanai T, Matsumi R, Fujiwara S, Imanaka T. Complete genome sequence of the hyperthermophilic archaeon Thermococcus kodakaraensis KOD1 and comparison with Pyrococcus genomes. Genome Res. 2005;15:352–63.

Acquired and Innate Immunity in Prokaryotes Define Their Evolutionary Story

3

Toshihiro Ito, Masatoshi Okura, and Fumito Maruyama

Abstract

Prokaryotes have various defense systems, such as clustered regularly interspaced short palindromic repeats (CRISPR)-CRISPR associated (Cas) adaptive immune systems, to protect themselves from invading foreign DNA, in particular mobile genetic elements (MGEs). In prokaryotic genomes, various classes of the genes encoding prokaryotic defense systems often cluster in specific genomic regions, referred to as defense islands, which are involved in the evolution and diversification of prokaryotic defense systems. In this chapter, we review the functions of prokaryotic defense systems, their evolutionary dynamics, and their co-evolutionary arms race with invading foreign DNA. We also introduce our previous works related to the comparative genomic analyses of *Streptococcus* species and oral bacterial species, in particular focusing on restriction-modification (R-M) systems and CRISPR-Cas adaptive immune systems.

Keywords

Prokaryotic defense systems · Clustered regularly interspaced short palindromic repeats (CRISPR)-CRISPR associated (Cas) adaptive immune systems · Invading foreign DNA · Mobile genetic elements (MGEs) · Defense islands · Prokaryotic genome evolution · Co-evolutionary arms race

T. Ito · F. Maruyama (✉)
Department of Microbiology, Graduate School of Medicine, Kyoto University, Sakyo-ku, Kyoto, Japan
e-mail: ito.toshihiro.7c@kyoto-u.ac.jp; maruyama.fumito.5e@kyoto-u.ac.jp

M. Okura
Division of Bacterial and Parasitic Diseases, National Institute of Animal Health, National Agriculture and Food Research Organization, Tsukuba, Ibaraki, Japan
e-mail: mokura@affrc.go.jp

H. Nishida, T. Oshima (eds.), *DNA Traffic in the Environment*,
https://doi.org/10.1007/978-981-13-3411-5_3

3.1 The Roles of Mobile Genetic Elements in Prokaryotic Genome Evolution

Prokaryotes (bacteria and archaea) are ubiquitous in the natural environment, and encounter invading foreign DNA, in particular mobile genetic elements (MGEs), such as plasmids, DNA transposons, and bacteriophages (or simply phages) (Figs. 3.1a–c) [1, 2]. Plasmids are self-replicating extrachromosomal DNA molecules, which can be transmitted horizontally (referred to as horizontal gene transfer) between prokaryotic cells by direct cell-to-cell contact, referred to as conjugative transfer [3, 4]. The conjugative transfer of transmissible plasmids, such as F plasmids (Fig. 3.1a; the F plasmids enable prokaryotes to transfer DNA from the donor cells harboring the F plasmids (referred to as F^+ strains) to the recipient cells (referred to as F^- strains)) [5, 6], is thought to be one of the most important pathways for the transmission of virulence genes and antibiotic resistance genes in prokaryotes [7, 8]. Several studies have reported that virulence genes and antibiotic resistance genes of *Enterobacteriaceae*, including *Escherichia coli*, *Enterococcus faecalis*, and *Enterococcus faecium*, could be transferred through plasmids [8–10].

DNA transposons, which are self-transmissible DNA elements that excise themselves and transfer into another genomic location, also mediate the horizontal gene transfer of antibiotic resistance genes in prokaryotes through the mating of prokaryotic cells [11, 12] (see also Chap. 8). The mobilization of DNA transposons is typically facilitated by transposable elements composed of a transposase-encoding gene and two terminal inverted repeats (TIRs) flanked by target site duplications (TSDs) (Fig. 3.1b) [13]. A transposase-encoding gene in transposable elements is first transcribed and translated to a transposase, which recognizes and binds to TIRs of the transposable elements and then catalyzes the excision of the transposable elements from the donor site flanked by TSDs [13]. The transposable elements bound to the transposase integrate themselves into another genomic location, and finally are flanked by other TSDs [13]. Transposons can be divided into two major classes (Class I and Class II) [14] and other distinct classes, including self-synthesizing DNA transposons [15], based on the nature of the transposition intermediate. Class I transposons, known as retrotransposons, move through an RNA intermediate and are reverse transcribed by a transposable element-encoded reverse transcriptase (RT) before their integration at another genomic location by a "copy and paste" mechanism [14]. Class II DNA transposons encode a transposase as described above, and simply move as DNA segments into another genomic location by a "cut and paste" mechanism [14]. Self-synthesizing DNA transposons (alternatively, called Polintons [15, 16] or Mavericks [17]), first discovered in eukaryotic genome sequences [15], encode a protein-primed family B DNA polymerase and a retroviral-like integrase, and are more likely involved in transposon replication [18, 19]. DNA transposons in bacteria, particularly in Gram-negative bacteria, often contain integrons, which were first characterized in 1989 [20] and are currently classified into five classes, class 1, class 2, class 3, class 4, and class 5, according to their integrase (*intI*) gene sequences [21, 22]. Among the five classes of integrons, the class 1 integrons are common in Gram-negative bacteria, and are captured by a Tn*402*

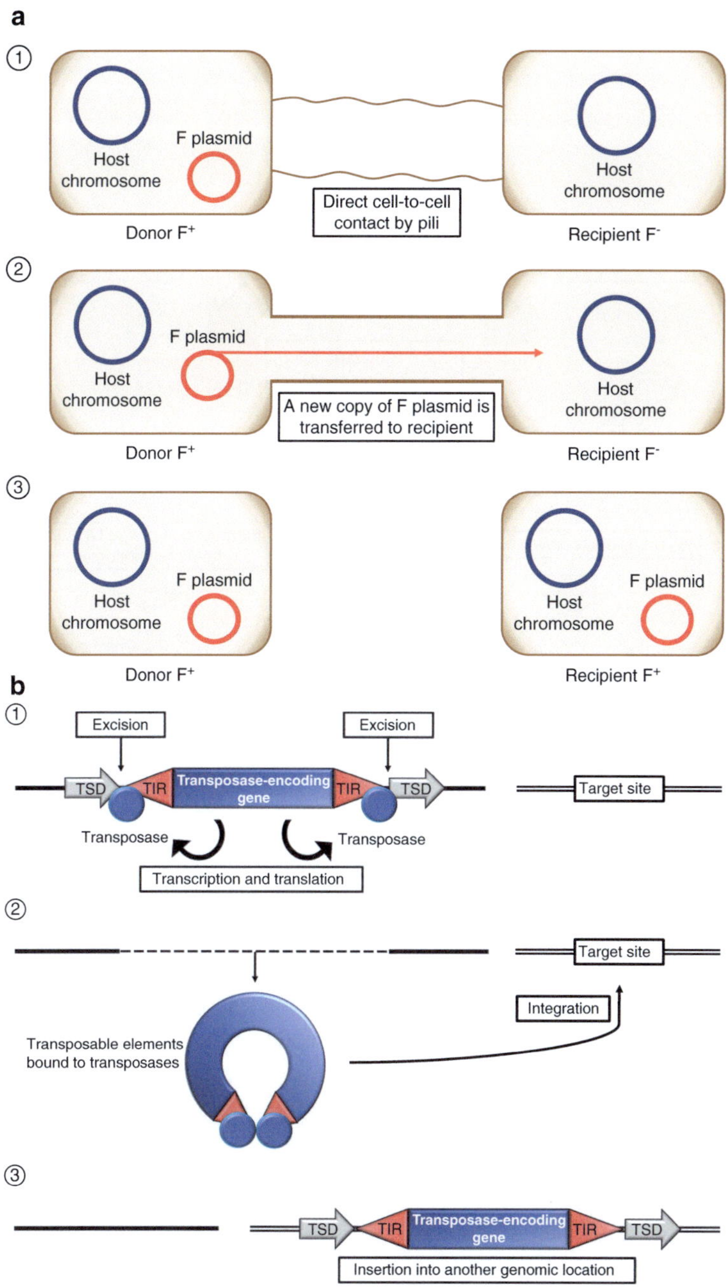

Fig. 3.1 Schematic description of the horizontal gene transfer by mobile gene elements (MGEs); (**a**) the F plasmid-mediated conjugation between F^+ and F^- prokaryotic cells; (**b**) typical Class II DNA transposons by a "cut and paste" mechanism (site-specific transposition is shown); (**c**) lytic and lysogenic phage life cycles (solid bold arrows indicate general lytic and lysogenic phage life cycles and dotted arrows indicate prophage induction under specific conditions, such as environmental stimulation or stress)

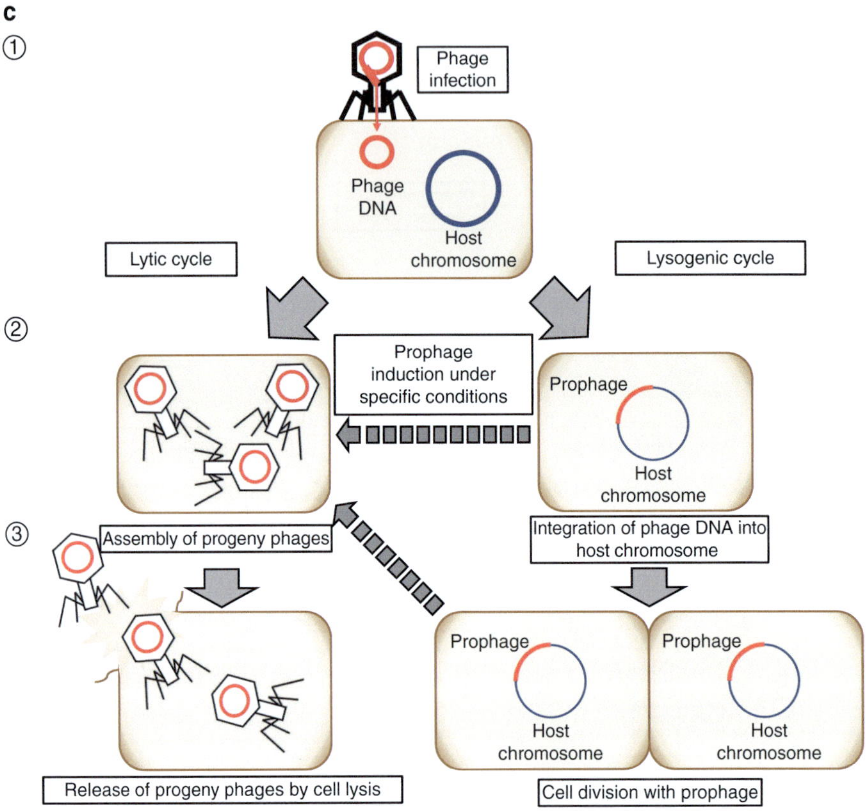

Fig. 3.1 (continued)

transposon to generate a transposon/integron hybrid carrying an antibiotic and antiseptic resistance-encoding cassette [22]; hence, the integrons are thought to be one of the key players in the dissemination of antibiotic resistance genes within a bacterial community [23]. Domingues et al. [24] have experimentally demonstrated that exposure of *Acinetobacter baylyi* to integron-containing *Salmonella* DNA led to the horizontal gene transfer of the integron by natural genetic transformation facilitated by the transposition of a Tn*21*-like transposon that contains the integron [24]. In Domingues et al. [24], exposure of *A. baylyi* to integron-containing *Acinetobacter baumannii* DNA also led to the horizontal gene transfer of the integron by natural genetic transformation, indicating that the interspecies transfer of transposons and integrons is not limited by the genetic relatedness of donor and recipient cells [24]. These findings strongly support that natural genetic transformation provides the widespread of MGEs, including transposons, in divergent species of bacteria [24].

Phages, which are obligate intracellular parasites of prokaryotes, present a serious threat to the life cycles of prokaryotes, and invade their host cells and propagate themselves in several ways, through lytic, lysogenic, pseudolysogenic, and chronic infection (Fig. 3.1c; only lytic and lysogenic life cycles are shown) [25, 26]. Lytic phages (alternatively, called virulent phages) use their host cell's translational machinery to synthesize their own proteins, and can kill the host cells to release their progenies by cell lysis [27]. Lysogenic phages (alternatively, called temperate phages) integrate their DNA into their host chromosomes and replicate passively along with the hosts [28]. In general, the infection of lysogenic phages does not initially cause cell lysis and remains in a dormant state in their host genomes (referred to as prophages) [26, 28]. However, in response to specific environmental stimulation or stress, the prophages are excised from the host genomes and can start reproducing progeny phages (referred to as prophage induction) [28, 29]. Very recently, Erez et al. [30] have experimentally demonstrated that *Bacillus subtilis* phage phi3T encodes various communication peptides, denoted as the *arbitrium* system, for a lysis-lysogeny switch [30]. In the *arbitrium* system, the probability that *B. subtilis* phage phi3T enters into a lysogenic cycle is proportional to the increase in the concentration of *arbitrium* peptides produced during its lytic cycle as host cell disruption, and is then inversely proportional to the number of living host cells [30]. The findings indicate that the *arbitrium* system enables the progenies of *B. subtilis* phage phi3T to prevent the overkill of their host cells. Pseudolysogeny (alternatively, called a phage carrier state) may be another survival strategy of phages in starved prokaryotic cells, in which, unlike the lysogenic infection, lytic phages postpone their lytic life cycles, until the infected host cells are exposed to nutrient-rich environment [26, 31]. Most of the pseudolysogeny remains unclear, yet may be important to better understand the complexity of phage life cycles and their roles in the natural environment. In chronic infection, rod-shaped single-stranded DNA phages, such as filamentous bacteriophage M13, can replicate and release their progenies for a long period without killing their host cells [25, 32]. Their ability to chronically replicate the progenies without host cell disruption has been used for laboratory experiments, such as M13 cloning [33] and phage display technology [34].

Among the phage life cycles described above, the horizontal gene transfer between prokaryotic cells typically occurs via their lysogenic cycles [27, 35] (see also Chap. 5). During the lysogenic cycles, the prophages occasionally package various pieces of the donor genomes, and allow the recipients to acquire new functions in a process, referred to as lysogenic conversion [28, 35, 36]. In the early 1950s, Freeman [37] has experimentally demonstrated that non-virulent strains of *Corynebacterium diphtheriae* infected with specific phages could be converted to virulent strains [37]. Other researchers discovered phage-inducible chromosomal islands (PICIs) in Gram-positive bacteria, in particular *Staphylococcus aureus* [35, 38–40]. *S. aureus* pathogenicity islands (SaPIs), which are known as the prototypical members of PICI family, are widely spread among staphylococcal species [41] and their mobility has been experimentally demonstrated [38, 42]. Recently, the PICIs are defined as a family of MGEs that might be responsible for the horizontal gene transfer of antibiotic resistance, and phage resistance, and in particular virulence genes [35, 40, 41]. Hence, phage-mediated transduction is considered as one of the important contributors to the widespread of virulence genes, antibiotic resistance genes, and phage resistance genes in the environment [43, 44].

Since the integration of foreign DNA into prokaryotic genomes makes an impact on prokaryotic life cycles, prokaryotes have several defense systems to protect themselves from invasion by foreign DNA. For instance, restriction-modification (R-M) systems, DNA phosphorothioate (PT) modification, toxin-antitoxin (TA) systems, phage abortive infection (Abi) systems, Argonaute proteins, bacteriophage exclusion (BREX) systems, and clustered regularly interspaced short palindromic repeats (CRISPR)-CRISPR associated (Cas) adaptive immune systems (henceforth called CRISPR-Cas systems) have been identified in the genomes of many prokaryotes (Figs. 3.2a–f and 3.3) [46]. These defense systems are roughly divided into relatively non-specific innate immune systems (R-M systems, DNA PT modification, TA systems, Abi systems, pAgo proteins, and BREX systems) and highly specific adaptive immune systems (CRISPR-Cas systems) [46]. This chapter introduces a brief overview of the principle of the innate immune systems in prokaryotes, followed by CRISPR-Cas systems, their evolutionary dynamics, and finally our investigation of CRISPR-Cas systems in *Streptococcus* species and oral bacterial species.

3.2 Prokaryotic Innate Immune Systems Against Foreign DNA Invasion

R-M systems (Fig. 3.2a), first discovered in the early 1950s [47, 48], are one of the best-characterized prokaryotic defense systems. The defense mechanism of R-M systems allows prokaryotes to distinguish between their own DNA (methylated DNA) and invading foreign DNA (non-methylated DNA) by a pair of enzymes, a restriction endonuclease (REase) and a DNA methyltransferase (MTase) [49, 50]. An REase recognizes methylated bases (5-methylcytosine (m5C), *N*4-methylcytosine (m4C), and *N*6-methyladenine (m6A)) in host DNA, and cleaves non-methylated DNA as non-self DNA at specific sites [51]. Meanwhile, a DNA MTase acts as a host DNA methylation enzyme, which transfers a methyl group (CH_3) from methyl donating compounds, such as S-adenosylmethionine (SAM), into the potential REase target sites in the host DNA for the prevention of self DNA cleavage [52–54]. R-M systems are classified into four types, Type I, Type II, Type III, and Type IV, based on their enzyme composition, sequence recognition, cleavage position, co-factor requirements, and substrate specificity [53]. Among the four types of R-M systems, Type II R-M systems are the most typical and most studied R-M systems [46]. The Type II R-M enzymes, such as EcoRI isolated from *E. coli* species [55], have been commonly used for laboratory experiments, such as restriction fragment length polymorphisms (RFLPs) and DNA cloning [56], because of its highly specific DNA cleavage ability. Meanwhile, Type IV R-M systems are composed of only an REase (referred to as a Type IV restriction enzyme), which can recognize and cleave only methylated DNA, unlike other Type R-M systems [57, 58].

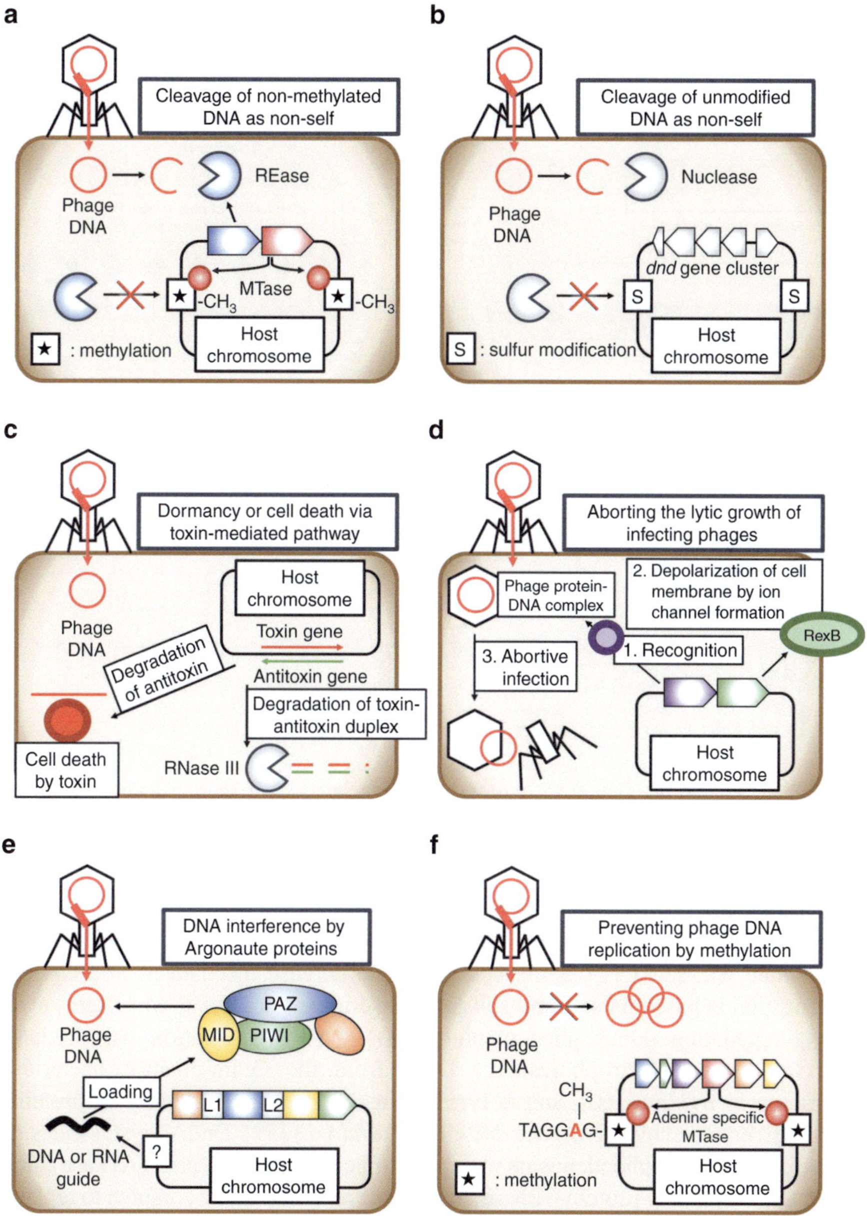

Fig. 3.2 Schematic description of prokaryotic innate immune systems; (**a**) restriction-modification (R-M) systems (typical Type II R-M systems are shown); (**b**) DNA phosphorothioate (PT) modification; (**c**) toxin-antitoxin (TA) systems (a *hok*/*sok* system is shown as Type I TA system); (**d**) phage abortive infection (Abi) systems (Rex system, which is carried out in ascending order of numbers); (**e**) prokaryotic Argonaute (pAgo) proteins (typical long pAgo proteins are shown); (**f**) bacteriophage exclusion (BREX) systems

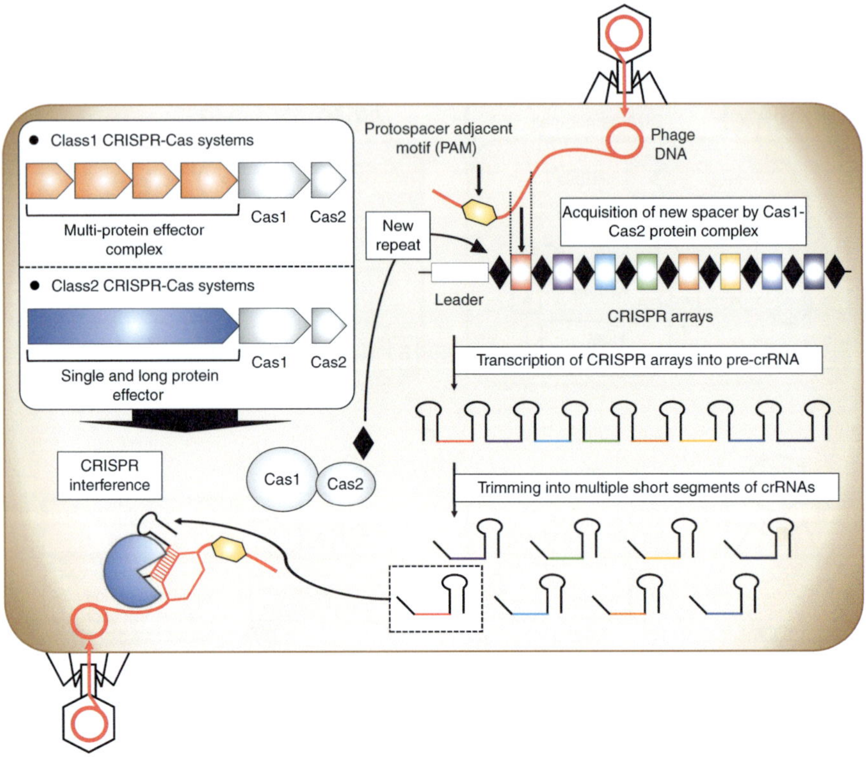

Fig. 3.3 Schematic description of clustered regularly interspaced short palindromic repeats (CRISPR)-CRISPR associated (Cas) adaptive immune systems (referring to Barrangou and Horvath [45] with slight modification)

DNA PT modification (Fig. 3.2b) is a sulfur modification of self DNA, in which a non-bridging oxygen atom in self DNA is substituted with a sulfur atom, and unmodified non-self DNA is cleaved [59]. The defense mechanism of DNA PT modification is based on self/non-self DNA discrimination, like R-M systems. The DNA degradation (Dnd) phenomenon of DNA PT modification was initially observed during electrophoresis of DNA from the Gram-positive bacterium *Streptomyces lividans* [60], and is typically associated with a family of proteins (DndA-E) encoded by a five-gene *dnd* cluster (*dndA-E*) [59]. DndA (cysteine desulfurase), DndC (phosphoadenosine phosphosulfate reductase), DndD (ATPase), and DndE (DNA-binding protein with a distinct fold) proteins form a complex to replace an oxygen atom with sulfur, whereas a DndB protein acts as a negative transcriptional regulator for PT-modifying genes, and is not essential for DNA PT modification [46, 59, 61, 62]. A DndD protein acts as an ATPase possibly associated with DNA structure alteration or DNA nicking during sulfur incorporation [63–65]. Xu et al. [66] have reported that *Salmonella enterica* serovar Cerro 87 possesses a four-gene cluster homologous to *dndB-E* genes, termed *dptB-E*, and its cognate three-gene *dnd* cluster (*dndF-H*) (alternatively, called *dptF-H* genes) [59, 66]. The DndF-H proteins encoded by *dndF-H* genes make double-stranded DNA damage as a trigger for SOS response, cell filamentation, and prophage induction [59]. The combination of the DNA PT modification-related genes found in the *S. enterica* serovar Cerro 87 has been proposed as DNA PT modification-dependent R-M systems [67], which are composed of modification-related genes (*dndB-E*) and restriction-related genes (*dndF-H*) [67], and have been experimentally demonstrated as temperature-dependent defense systems [59].

TA systems (Fig. 3.2c) are also one of the most studied prokaryotic defense systems, as well as R-M systems, and are closely linked to dormancy induction or programmed cell death (PCD) [68]. The genetic modules of TA systems are composed of stable toxin genes and unstable antitoxin genes in multiple copies, and activated when prokaryotic cells possessing the genetic modules are exposed to stress, such as antibiotic treatment and phage infection [68, 69]. TA system-related genes, known as post-segregational killing (PSK) systems, were first discovered in the 1980s [70–72], such as a hok/sok system composed of *hok* (host killing), *sok* (suppression of killing), and *mok* (modulation of killing) in plasmid R1 of *E. coli* [70]. In the hok/sok system (Fig. 3.2c), the activity of *hok* as stable toxin-encoding mRNA is suppressed by being bound to *sok* as non-coding unstable neutralize antisense RNA that is complementary to *hok*, resulting in the formation of an RNA duplex that is degraded by RNase III [73, 74]. However, when the plasmids encoding TA systems are not transmitted to daughter cells, the *sok* is rapidly degraded and highly toxic Hok proteins are synthesized from the *hok*, ultimately leading to cell death by depolarization of the cell membrane [73, 74]. TA systems are classified into six types, Type I, Type II, Type III, Type IV, Type V, and Type VI, based on the nature and the mechanism of antitoxins [75]. Among the six types, Type II TA systems are well-characterized TA systems, in which protein antitoxins neutralize the toxin activity of TA systems by directly binding to their cognate toxins [76]. In most Type II TA systems, the Type II toxins are a potent endoribonuclease and cleave cellular mRNA at specific sequences to inhibit translation in response to stress [75]. For instance, drug-tolerant *Mycobacterium tuberculosis*, which possesses a large number of TA systems [77], shows antibiotic tolerance by a remarkable downregulation of genes associated with growth, metabolism, and lipid synthesis, and an upregulation of stress-associated sigma factors, transcription factors, drug efflux pumps, and toxin-antitoxin genes during a prolonged antibiotic exposure [78]. Another example of toxin-antitoxin genes in Type II TA systems is higher eukaryotes and prokaryotes nucleotide-binding (HEPN)-minimal nucleotidyltransferase (MNT) genetic modules, which are the most abundant genes in hyperthermophilic archaea [79–81]. The HEPN-MNT genetic modules are composed of HEPN as RNA-cleaving toxins and MNT as predicted antitoxins, and are predicted as antibiotic resistance systems [80]. These TA systems are widely present in chromosomes or plasmids of prokaryotes as TA loci, and are stably maintained in the populations of viable cells [68, 74, 76].

Abi systems (Fig. 3.2d) lead to the death of infected cells (referred to as altruistic suicide), to terminate the production of progeny phages, and then protect clonal cells in the prokaryotic populations [82, 83]. Many of Abi phenotypes, denoted by

"Abi" and capital alphabet letters (e.g., AbiA), have been found in the plasmids of *Lactococcus lactis* [84], which is one of the model microorganisms for Abi system studies. Although Abi systems are thought to be distinct from TA systems, many of Abi systems share the genetic modules with TA systems, such as HEPN domains described above [46, 85]. A recent study has shown that Type III TA systems, $ToxIN_{pa}$ composed of ToxN as endoribonuclease toxins and $ToxI_{pa}$ as RNA antitoxins, are employed by the phytopathogenic bacterium *Pectobacterium atrosepticum* for anti-phage activity [82, 86]. On the other hand, the Rex system found in λ-lysogenic *E. coli* strains [83, 87] is a well-studied Abi system that has the different mechanism from $ToxIN_{pa}$ system. The Rex system is composed of two proteins, RexA protein as an intracellular sensor and RexB protein as an ion channel [83, 88]. When phage infection occurs, RexA protein is first activated by recognizing a phage protein-DNA complex, followed by RexB protein activation [88, 89]. The activated RexB protein induces a drop in the cellular ATP level, thereby stopping cell multiplication and aborting the lytic growth of the infecting phages [88, 89].

Prokaryotic Argonaute (pAgo) proteins (Fig. 3.2e) are nucleic acid-guided proteins, which use RNA or DNA guides and provide specific cleavage of complementary nucleic acid targets as a host defense mechanism [90]. Argonaute proteins, first discovered in eukaryotes [91], are essential components of an RNA-induced silencing complex (RISC) responsible for RNA interference (RNAi) (alternatively, called RNA silencing) [92–94]. Small non-coding RNA, such as microRNA (miRNA), small interfering RNA (siRNA), and P-element induced wimpy testis (PIWI)-interacting RNA (piRNA), guides Argonaute proteins to complementary RNA targets in RNAi [92, 95, 96], leading to translational inhibition, mRNA destabilization, or RNA target cleavage [97]. In contrast to eukaryotic Argonaute (eAgo) proteins, some pAgo proteins are key players in specific DNA target cleavage, referred to as DNA interference (DNAi) [98, 99]. Based on the domain structure, pAgo proteins are roughly divided into two groups, long and short pAgo proteins [100]. Among the two groups of pAgo proteins, the domain structure of long pAgo proteins is structurally similar to that of eAgo proteins [101]. Both eAgo proteins and long pAgo proteins are composed of a PIWI endonuclease domain, a PIWI-Argonaute-Zwille (PAZ) domain, a middle (MID) domain, and an amino-terminal (N-terminal) domain, along with two domain linkers (L1 and L2) [46, 100]. Meanwhile, short pAgo proteins are composed of only a PIWI domain and a MID domain [100]. Instead of a PAZ domain and a N-terminal domain, short pAgo proteins are associated with the genes encoding an analogue of PAZ (APAZ) domain, which is fused to a putative nuclease domain, such as Sir2, Mrr, or TIR proteins [90, 101]. It is assumed that short pAgo proteins are responsible for nucleic acid-guided target recognition, and their associated nuclease-APAZ domain is responsible for guide generation and/or target degradation, respectively [101]. Recent bioinformatic analysis has revealed that pAgo proteins are encoded in ~32% and ~9% of the sequenced archaeal and bacterial genomes, respectively [100], and the phylogenetic tree of pAgo-encoding genes does not follow the prokaryotic phylogeny based on ribosomal RNA and other universal genes [100]. These indicated that horizontal gene transfer results in the widespread of the pAgo protein-encoding genes in prokaryotes. Other recent studies have demonstrated that most of the characterized long pAgo proteins, including *Thermus thermophilus* (*Tt*Ago) [99] and *Pyrococcus furiosus* (*Pf*Ago) [102], can target DNA using either DNA or RNA guides [101], and can act as prokaryotic defense systems against invading foreign DNA as well as RNAi in eukaryotes. Moreover, it has been proposed that the features of long pAgo proteins described above can be used as genome-editing tools, because they can selectively cleave double-stranded DNA targets [101].

BREX systems (Fig. 3.2f) are a relatively novel defense mechanism against phage infection, and are composed of a combination of six genes among 15 BREX-related genes [103]. The BREX-related genes include five *pgl* genes (*pglW, pglX, pglXI, pglY*, and *pglZ*) and ten *brx* genes (*brxA, brxB, brxC, brxD, brxE, brxF, brxHI, brxHII, brxL*, and *brxP*) [103]. Each of these genes encodes a serine/threonine kinase (*pglW*), an adenine-specific DNA methyltransferase (*pglX, pglXI*), an ATP-binding P-loop protein posessing ATPase activity (*pglY*), an alkaline phosphatase (*pglZ*), an NusB-like RNA-binding anti-termination protein (*brxA*), a protein of unknown function (*brxB, brxE, brxF*), an ATP-binding protein (*brxC, brxD*), an Lhr-like helicase (*brxHI*), a DNA/RNA helicase (*brxHII*), a Lon-like protease (*brxL*), and a phosphoadenosine phosphosulfate (PAPS) reductase (*brxP*), respectively [103–105]. In the early 1980s, BREX-like systems have been already discovered in *Streptomyces coelicolor* A3(2), denoted as "phage growth limitation (Pgl) systems" [106], which are composed of *pglWXYZ* genes [107]. Moreover, two of the five *pgl* genes in BREX systems, *pglXZ* genes, show sequence homology to genes in Pgl systems [103]. The phage resistance mechanism of BREX systems is based on self/non-self DNA discrimination, like R-M systems [103]. Prokaryotes possessing BREX systems methylate their own genomes, typically at the fifth position of a non-palindromic 5′-TAGG$\underline{A}$G-3′ hexameric sequence, and can prevent phage DNA replication, but they do not cleave phage DNA, unlike R-M systems [103]. BREX systems are currently classified into six types, Type 1, Type 2, Type 3, Type 4, Type 5, and Type 6, based on the six-gene combination and the order of the six genes, all of which contain a *pglZ* gene [103]. Goldfarb et al. [103] analyzed the genomes of approximately 1500 bacteria and archaea and revealed that *pglZ* genes as a putative member of an alkaline phosphatase superfamily are present in approximately 10% of these microbial genomes [103], suggesting that BREX systems are widely distributed as phage resistance systems.

3.3 CRISPR-Cas Adaptive Immune Systems

CRISPR-Cas is protein-encoding genes located in hypervariable genetic loci (referred to as CRISPR-*cas* loci), which are composed of multiple repeat-spacer arrays and the CRISPR-associated (*cas*) genes encoding CRISPR-associated (Cas) proteins located in close-proximity to the repeat-spacer arrays [45]. CRISPR-*cas* loci in prokaryotes memorize past encounters with invading foreign DNA as CRISPR spacers and act as prokaryotic adaptive immune systems [45]. The repeat-spacer arrays, later referred to as CRISPR arrays [108, 109], were first discovered in the 3′-end flanking region of alkaline phosphatase isozyme (*iap*) genes in *E. coli* [110]. Soon after, similar repeats were discovered in the genome sequences of many prokaryotes [111]. In 2000, the prokaryotic repeat-spacer arrays were named as short regularly spaced repeats (SRSR) [112]; in 2002, the arrays were renamed as spacers interspersed direct repeats (SPIDR) [108, 109], and finally as clustered regularly interspaced short palindromic repeats (CRISPR) on the basis of the structural feature [108, 109]. At the same time, four *cas* genes (*cas1*, *cas2*, *cas3*, and *cas4* genes encoding Cas1, Cas2, Cas3, and Cas4 proteins, respectively) were also identified nearby CRISPR arrays, and the structural coherence of the CRISPR arrays and the *cas* genes was characterized as CRISPR-*cas* loci [108, 109]. In 2005, the involvement of CRISPR-*cas* loci in adaptive immune systems against invading foreign DNA began to be speculated on the basis of some scientific evidence; (1) CRISPR spacers derive from the genomic fragments of invading foreign DNA including MGEs (later referred to as protospacers) [113–116], (2) phages and plasmids harboring the protospacers fail to infect their hosts yielding antisense RNA (later referred to as CRISPR RNA (crRNA)) [114, 115, 117]. The experimental demonstration of CRISPR-Cas systems against phage infection was carried out by Barrangou et al. [118], in which phage-sensitive *Streptococcus thermophilus* strains became resistant to Streptococcal phages after acquiring new CRISPR spacers identical to the protospacer sequences in the phage DNA [118]. Soon after, the adaptive immune activity of CRISPR-Cas systems against plasmid transfer was experimentally demonstrated by Marraffini and Sontheimer [119], in which the CRISPR-*cas* loci in *Staphylococcus epidermidis* strains acted as DNAi machinery, and prevented the transfer of staphylococcal conjugative plasmids, as well as RNAi in eukaryotes [119].

CRISPR-Cas systems act as DNA-encoded, RNA-guided, and nucleic acid-targeting interference (CRISPR interference) in prokaryotes (Fig. 3.3) [45]. The first step of CRISPR-Cas systems is to detect the motifs associated with protospacers of invading foreign DNA, referred to as protospacer adjacent motifs (PAMs) [120], and cut the protospacers at the site a few base pairs upstream or downstream from the PAMs by nuclease and integrase activity of Cas proteins, such as a Cas1–Cas2 protein complex or Cas9 proteins, respectively [121–123]. The shortly cut protospacers as new CRISPR spacers are integrated between multiple repeats of CRISPR arrays adjacent to a leader sequence in CRISPR-*cas* loci, and the CRISPR arrays are transcribed into a long precursor, referred to as precursor CRISPR RNA (pre-crRNA) [121]. The pre-crRNA is further trimmed into multiple short segments of crRNA by the endonuclease activity of Cas proteins, such as Cas6 proteins [122, 124]. The crRNA binds to the complementary nucleic acid sequences, and the crRNA-guided CRISPR interference is then carried out by the crRNA-guided nuclease activity of Cas proteins (referred to as an effector), such as Cas9 proteins [125], which cleave the crRNA-guided nucleic acid targets.

To date, CRISPR-*cas* loci are present in approximately 50% of bacterial genomes and approximately 90% of archaeal genomes [122]. CRISPR-Cas systems are currently classified into two classes (class 1 and class 2), including six major types (Type I, Type II, Type III, Type IV, Type V, and Type VI) and 28 subtypes (e.g., Type I-A), based on the CRISPR-Cas machinery, in particular the combination of Cas proteins, and its mode of action (Table 3.1) [126]. A recent report has described that approximately 90% of all identified CRISPR-*cas* loci belong to class 1 CRISPR-Cas systems, and the remaining 10% belong to class 2 CRISPR-Cas systems [128]. Class 1 CRISPR-Cas systems with multi-protein effector complexes include Type I, Type III, and Type IV CRISPR-Cas systems [126]. Among the three types in class 1 CRISPR-Cas systems, Type I and Type III CRISPR-Cas systems are

Table 3.1 Current classification of CRISPR-Cas systems (referring to [126, 127])

Major type	Subtype	Combination of Cas protein-encoding genes and other genes	Target
Type I	I-A	*cas1*, *cas2*, *cas3′*, *cas3″*, *cas4* (×2), *cas5*, *cas6*, *cas7*, *cas8a1*, *cas11*	DNA
	I-B	*cas1*, *cas2*, *cas3*, *cas4*, *cas5*, *cas6*, *cas7*, *cas8b1*	DNA
	I-C	*cas1*, *cas2*, *cas3*, *cas4*, *cas5*, *cas7*, *cas8c*	DNA
	I-D	*cas1*, *cas2*, *cas3′*, *cas3″*, *cas4*, *cas5*, *cas6*, *cas7*, *cas10d*	DNA
	I-E	*cas1*, *cas2*, *cas3*, *cas5*, *cas6*, *cas7*, *cas8e*, *cas11*	DNA
	I-F	*cas1*, *cas2*, *cas3*, *cas5*, *cas6f*, *cas7*, *cas8f*	DNA
	I-U	*cas1*, *cas2*, *cas3*, *cas4*, *cas5*, *cas6*, *cas7*, *cas8u2*	DNA
Type II	II-A	*cas1*, *cas2*, *cas9*, *csn2*, tracrRNA	DNA
	II-B	*cas1*, *cas2*, *cas4*, *cas9*, tracrRNA	DNA
	II-C	*cas1*, *cas2*, *cas9*, tracrRNA	DNA
	II-C variant	*cas1*, *cas2*, *cas4*, *cas9*, tracrRNA	DNA
Type III	III-A	*cas1*, *cas2*, *cas5*, *cas6*, *cas7* family (*csm3*, *csm5*), *cas10*, *cas11*, *csm6*	DNA, RNA
	III-B	*cas5*, *cas6*, *cas7* family (*cmr1*, *cmr4*, *cmr6*), *cas10*, *cas11*	DNA?, RNA?
	III-B variant	*cas1*, *cas2*, *cas5*, *cas6*, *cas7* family (*cmr1*, *cmr4*, *cmr6*), *cas10*, *cas11*, reverse transcriptase, PD-DExK nuclease	DNA, RNA
	III-C	*cas5*, *cas7* family (*cmr1*, *cmr4*, *cmr6*), *cas10*, *cas11*	–
	III-D	*cas5*, *cas7* family (*csm3*, *csm5*), *cas10*, *cas11*, all1473	DNA?, RNA?
Type IV	IV	*cas5*, *cas6-like*, *cas7*, *cas8-like*, *dinG*	DNA?
	IV variant	*cas5*, *cas7*, *cas8-like*, *cas11*	DNA?
Type V	V-A	*cas1*, *cas2*, *cas4*, *cas12a*	DNA
	V-B	*cas1*, *cas2*, *cas4*, *cas12b*, tracrRNA	DNA
	V-C	*cas1*, *cas12c*	DNA
	V-D	*cas1*, *cas12d* (*casY*)	DNA
	V-E	*cas1*, *cas2*, *cas4*, *cas12e* (*casX*), tracrRNA	DNA
	V-U (tentative)	–	–
Type VI	VI-A	*cas1*, *cas2*, *cas13a*	RNA
	VI-B1	*cas13b*, *csx28*	RNA?
	VI-B2	*cas13b*, *csx27*	RNA?
	VI-C	*cas13c*	RNA?

Class 1 CRISPR-Cas systems include Type I, Type III, and Type IV CRISPR-Cas systems, and Class 2 CRISPR-Cas systems include Type II, Type V, and Type VI CRISPR-Cas systems

the most common CRISPR-Cas systems and are distributed in many archaea and a few bacteria, whereas Type IV CRISPR-Cas systems are a relatively rare [127, 129]. The multi-protein effector complexes of class 1 CRISPR-Cas systems are composed of multiple Cas proteins, such as CRISPR-associated complex for antiviral defense (Cascade) in Type I CRISPR-Cas systems [117, 130] and a CRISPR-Cas subtype Mtube (Csm)-CRISPR RAMP module (Cmr) complex in Type III CRISPR-Cas systems [131, 132]. Both of the multiple Cas proteins, such as Cas5, Cas6, Cas7, and Cmr proteins, include paralogous repeat-associated mysterious proteins (RAMPs) responsible for RNA binding and/or ribonuclease activity [133]. Several structural analyses have revealed that RAMPs contain single- or multi-domains of RNA recognition motif (RRM), known as a ferredoxin-like fold, which is a common protein fold often found in the structure of nucleic acid-binding proteins [133–135]. Recently, RT-associated CRISPR-*cas* loci encoding an RT-Cas1 fusion protein have been found by several bioinformatic analyses [136–138], and were classified as Type III CRISPR-Cas systems [139]. Soon after, it was experimentally demonstrated that RT-associated Type III CRISPR-Cas systems can acquire CRISPR spacers directly from RNA [140], implying that the Type III CRISPR-Cas multi-protein effector modules with an RT-Cas1 fusion protein may have the ability to adapt to invaders with both DNA and RNA [141].

Class 2 CRISPR-Cas systems with single and long protein effector modules include Type II, Type V, and Type VI CRISPR-Cas systems [126, 128]. Among the three types in class 2 CRISPR-Cas systems, Type II CRISPR-Cas systems with Cas9 proteins as a multi-domain effector are one of the best-characterized CRISPR-Cas systems [142], and are widely used for genome editing [143–146]. In Type II CRISPR-Cas systems, pre-crRNA is bound to trans-acting CRISPR RNA (tracrRNA), which is complementary to the repeat sequence in the pre-crRNA, and the pre-crRNA-tracrRNA duplex is processed into a crRNA-tracrRNA duplex by a double-stranded RNA-specific ribonuclease RNase III [147–149]. The crRNA-tracrRNA duplex forms a complex with Cas9 proteins, and the Cas9-crRNA-tracrRNA complex finally introduces site-specific double-stranded DNA cleavage at the specific target site [148, 150]. Cas9 proteins contain two nuclease domains, RuvC and HNH, to generate paired double nicks [151], which can be re-programmed by single-guide RNA (sgRNA) to cleave its specific DNA target [145, 146, 148]. In Type V and Type VI CRISPR-Cas systems, Cas12 and Cas13 proteins are predicted as multi-domain protein effector, respectively [126, 128]. All Cas12 proteins contain RuvC-like endonuclease domains, and in some cases a putative novel nuclease (Nuc) domain [152]. Very recently, new CRISPR-Cas systems, referred to as CRISPR-CasX and CRISPR-CasY, were discovered from uncultivated bacteria [153], and were putatively classified as new subtypes of Type V CRISPR-Cas systems, because both CasX and CasY proteins contain RuvC-like domains (renamed as Cas12e (CasX) and Cas12d (CasY) proteins, respectively) [127]. All Cas13 proteins contain two HEPN domains, which may possess RNase activity, as well as HEPN toxins in Type II TA systems (see Sect. 3.2) [80, 81], suggesting that Cas13 proteins would have evolved from HEPN toxins [127].

The CRISPR-Cas systems described above show a remarkable diversity in terms of the combination of Cas proteins and its mode of action [126]. Meanwhile, recent comparative genomic and structural analyses have revealed that Cas1 and Cas2 proteins, in particular Cas1 proteins, are universally conserved, and are thought to play key roles in almost all CRISPR-Cas systems [19, 127]. Krupovic et al. [19] discovered "genomic islands" in the genome sequences of many archaea and some bacteria, in which some *cas1* genes encoding homologous Cas1 proteins are not associated with CRISPR-*cas* loci, referred to as *cas*1-solo genes [19]. The *cas*1-solo genes are classified into two groups, group 1 and group 2, based on *cas1* gene phylogeny [19]. Among the two groups, group 2 *cas*1-solo genes are co-localized with protein-primed family B DNA polymerase-encoding genes (henceforth called *polB* genes) in genomic islands [19], as well as self-synthesizing DNA transposons in eukaryotes [15]. The MGEs composed of *cas1*-solo genes, *polB* genes, TIRs, and TSDs (see Sect. 3.1) were denoted as "casposons," [19] which employ Cas1-solo proteins as an integrase (alternatively, called a casposase) [154]. Casposons are currently classified into four families, family 1, family 2, family 3, and family 4, based on the gene composition and phylogeny of Cas1 proteins [155]. The experimental demonstration of casposon mobilization has not been performed yet, but comparative genomic analysis of various strains of *Methanosarcina mazei* has shown the recent mobility of casposons as potentially active MGEs [155]. Béguin et al. [156] have revealed the close similarities between the insertion mechanisms of casposons and CRISPR spacers facilitated by a casposase and a Cas1–Cas2 protein complex, respectively [156]. These findings strongly support the evolutionary relationship between the adaptive modules of CRISPR-Cas systems and casposons [156].

The evolution and diversification of CRISPR-Cas systems appear to be driven by the co-evolutionary arms race between prokaryotes and phages. Phages possess the genes encoding anti-CRISPR proteins, referred to as *Acr* genes, which encode anti-CRISPR proteins that show the ability to inhibit the functions of diverse effector complexes in CRISPR-Cas systems [157]. As described above, lysogenic phages integrate their own genomes into their host chromosomes, and then form prophage regions [27], and might express the anti-CRISPR genes to maintain the prophage regions during lysogeny [158]. To date, more than 20 families of *Acr* genes have been identified in previously published reports, and can inhibit the anti-phage activity in some subtypes of Type I and Type II CRISPR-Cas systems [157]. Bondy-Denomy et al. [159] have experimentally demonstrated the two mechanisms of anti-CRISPR activity by AcrF1, AcrF2, and AcrF3 proteins, which specifically inhibit the activity of Type I-F CRISPR-Cas system [159]. AcrF1 and AcrF2 proteins could inhibit the DNA-binding activity of the multi-protein effector complex in Type I-F CRISPR-Cas system, whereas AcrF3 proteins directly bind to Cas3 proteins and prevent their recruitment to the DNA-bound multi-protein effector complex [159]. It has also been known that phages adapt and evolve against some prokaryotic innate immune systems, in particular R-M systems [160]. Several phages possess MTase-encoding genes, and modify their own genomes by the MTase activity to protect themselves from the REase activity of R-M systems [161]. Another mechanism is that the phages possessing overcome classical restriction (*ocr*) genes produce Ocr proteins, which directly bind to both MTase and REase of Type I R-M systems, and then inhibit the endonuclease activity of the R-M systems [160, 161]. The co-evolution by the prokaryotic host-phage arms race remains unclear, but it could be the result of a diversification of prokaryotic defense systems [162].

3.4 Effect of the Evolutionary Dynamics of Defense Systems on Prokaryotic Genome Evolution

Makarova et al. [163] have performed the comparative genomic analysis of 1055 completely sequenced prokaryotic genomes focused on defense systems, and have indicated the involvement of transposable elements in the gain, loss, and exchange of the defense system-encoding genes (henceforth called defense genes) [163]. The results suggested that many species have intraspecific variation in defense systems through the horizontal gene transfer of defense genes. The authors also discovered specific genomic locations in prokaryotic genomes, referred to as defense islands [163], which are enriched in not only the genes encoding deferent classes of defense systems, in particular R-M systems, T-A systems, and BREX systems, but also uncharacterized genes as candidates for new types of defense systems [103, 163]. After that, the comparative analysis of 35 groups of closely related bacterial genomes and one group of archaeal genomes has demonstrated that the defense genes in more than half of the analyzed genomes tend to be co-localized in defense islands [164]. In some species, MGEs in defense islands are frequently involved in the evolution and diversification of defense systems, suggesting that defense islands are responsible for the enhanced dynamics of the evolution in prokaryotic defense systems [164].

Very recently, Ofir et al. [165] have reported new defense systems, denoted as defense island system associated with restriction modification (DISARM), as multi-gene R-M systems [165]. DISARM systems are composed of five genes: four genes encode a DNA MTase, a helicase domain, a phospholipase D (PLD) domain, a domain of unknown function (DUF) 1998 (a helicase-associated domain), and one remaining gene has an unknown function [165]. Among the five genes, at least the four genes encoding a DNA MTase, a helicase domain, a DUF1998, and a gene of unknown function are essential for the anti-phage activity of DISARM systems. Meanwhile, the genes encoding a PLD domain are not always essential, but are required for the anti-phage activity against myophage SPO1 and podophages Nf. Among the five genes, a DUF1998 is enriched in defense islands, and is then thought to be a part of anti-phage defense systems [163], also indicating that DISARM systems are widely spread prokaryotic defense systems in defense islands.

Alternative DNA methylation levels have been shown to cause changes in gene expression in bacteria [166], because bacteria use a DNA MTase as a switch to systematically change their transcriptome [167, 168]. It is conceivable that the variation in R-M systems may affect the DNA methylation patterns and also may ultimately split clonal populations into epigenetic lineages [169]. In fact, clustering based on the presence/absence of R-M systems in genomes accurately reproduced the core genome phylogenetic structure in *Neisseria meningitidis*, and each phylogenetic clade harbored a unique repertoire of R-M systems [170]. A genomic clade-distinctive combination of the R-M system enzymes was also observed in *Burkholderia pseudomallei*, which is the causative agent for melioidosis [171]. Furthermore, clade-specific DNA methylation patterns in *B. pseudomallei* were observed, suggesting that R-M systems contribute to the limiting intraspecific exchange of genetic material, and the genomic clades may represent functional units of genetic isolation in this species [171]. In *Streptococcus pneumoniae*, a genomic region, where different lineages code for variable R-M systems, was found [172]. This locus has been proposed to play a role in the fine-tuning of the extent of genomic plasticity [172]. Therefore, these findings suggested that the intraspecific variation in R-M systems plays a key role in the genome diversification. Prokaryotic defense systems other than R-M systems also appear to have additional functions besides the limitation/prevention against invasion by foreign DNA. A role in stabilizing genomic islands was proposed for TA systems [76, 166]. CRISPR-Cas systems are also able to control transcription endogenously, and regulate important lifestyle-based bacterial phenotypes, such as pathogenicity [173].

To analyze the intraspecific diversification of bacterial populations associated with prokaryotic defense systems, we have investigated 47 genome sequences of *Streptococcus suis*, an important swine pathogen and an emerging zoonotic agent, with a focus on the defense genes [174]. Our comparative genome analysis of *S. suis* indicated similar or identical profiles of the defense genes related to R-M systems, TA systems, Abi systems, and CRISPR-Cas systems in the same genomic clusters and several cassette-like defense system loci [174]. Among the defense system loci, one locus found in *S. suis* was a variable region. In the region, not only genetic elements in R-M systems but also those in Abi systems and CRISPR-Cas systems, prophages, and/or other genes were replaced with each other [174]. In addition, the shift of prokaryotic defense systems at the locus was coincident with the branching of the genomic clusters in many cases [174]. Taken together with our investigation and previously published researches, it is possible to hypothesize that some prokaryotic defense systems, in particular R-M systems, affect the extent of genomic plasticity and the intraspecific diversification of bacterial populations in certain species, and in some cases deferent classes of defense genes are located on the same locus. However, it remains unclear to what extent such prokaryotic defense systems are involved in the intraspecific speciation. Future investigation is needed to evaluate the authenticity of this hypothesis.

3.5 Functions of CRISPR-Cas Adaptive Immune Systems in *Streptococcus* Species and Oral Bacterial Species

We have investigated how CRISPR-Cas systems act in prokaryotic cells by the use of *Streptococcus* species and oral bacterial species, and are suspecting that CRISPR-Cas systems play key roles in the genomic evolution of the bacterial species investigated. *Streptococcus pyogenes*, also known as group A beta-hemolytic Streptococci (GAS), is one of the most virulent pathogens causing a broad spectrum of infectious diseases, such as pharyngitis as the most common bacterial disease in children and streptococcal toxic shock syndrome (STSS) as life-threatening illness [175]. Our investigation found numerous prophage regions in the genome sequences of *S. pyogenes*, although their genome size is a relatively small (approximately 1.9 mega base pairs) [176]. In addition, the streptococcal virulence genes encoding superantigens, a hyaluronidase, and a streptodornase were also found in the prophage regions of the *S. pyogenes* genome sequences investigated [176]. These results suggested that *S. pyogenes* takes advantage of prophages within its genome for survival. Subsequently, we have investigated the association between the number of prophages and CRISPR-*cas* loci in the genomic sequences of 13 *S. pyogenes* strains available in a public nucleotide database [177]. As a result, CRISPR-*cas* loci were found in 10 of the 13 strains investigated, and interestingly the number of CRISPR spacers was inversely proportional to the number of prophages in the ten strains [177], suggesting that *S. pyogenes* is able to control the acquisition of prophages based on the contents of CRISPR-*cas* loci, and also contribute to strain-specific pathogenesis of streptococcal phages (Fig. 3.4a).

As for other *Streptococcus* species, we have analyzed the complete genome sequences of several *Streptococcus mutans* serotype c strains, which are known as a major cause of dental caries (tooth decay) [178, 179]. As a result, CRISPR-*cas* loci were found in almost all of the analyzed *S. mutans* serotype c strains, where no prophage regions were found in all the *S. mutans* serotype c strains [178, 179]. In addition, the CRISPR spacer sequences in the analyzed *S. mutans* serotype *c* strains exhibited high similarity to the genome sequences from several streptococcal phages, including *S. mutans* phage M102 [179]. Our investigation suggested that *Streptococcus* species possess species-specific survival strategy; *S. mutans* may be attacked by streptococcal phages and then protected by CRISPR-Cas systems (Fig. 3.4b), whereas *S. pyogenes* acquires virulence-encoding genes by phage-mediated transduction.

We have also investigated the functions of CRISPR-Cas systems in three oral bacterial species referred to as "red complex," *Porphyromonas gingivalis*, *Tannerella forsythia*, and *Treponema denticola* [180–183]; they have been frequently detected in human periodontal pockets [184]. We have first sequenced the complete genome of *P. gingivalis* strain TDC60 isolated from a severe periodontal lesion in a Japanese patient for subsequent comparative genome analysis [181]. In the comparative genome analysis of the 3 *P. gingivalis* complete genome sequences including the complete genome of *P. gingivalis* strain TDC60, multiple CRISPR-Cas subtypes were identified in each of their genome sequences, although the potential targets were not determined for the CRISPR-Cas subtypes identified [181]. To identify the targets, our further investigation of CRISPR-Cas systems in *P. gingivalis* was carried out [182, 183]. In Watanabe et al. [182], 60 *P. gingivalis* isolates were used for genetic typing and intraspecific diversity analysis [182]. As a result, a total of 2150 CRISPR spacers were identified in the 60 *P. gingivalis* isolates, and only 29 of the 2150 CRISPR spacers exhibited high sequence similarity to the genome sequences available in public nucleotide databases [182]. Of the 29 CRISPR spacers analyzed, 19 CRISPR spacers exhibited high sequence

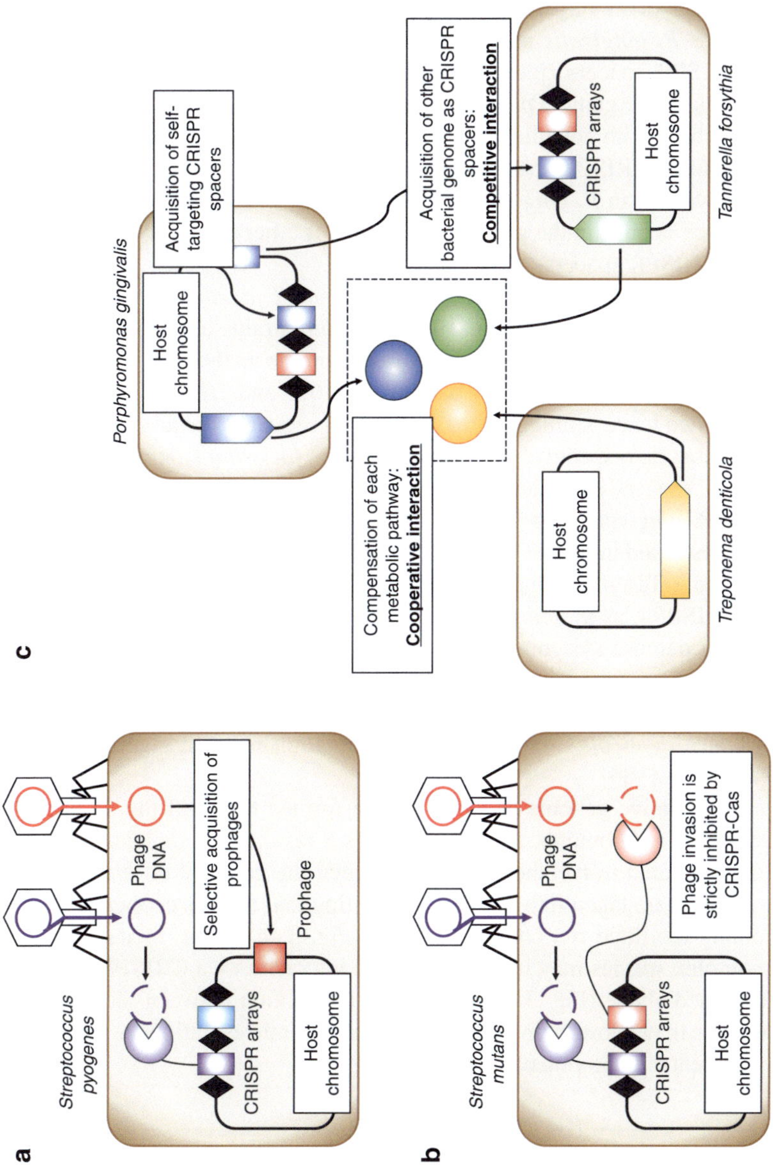

Fig. 3.4 Functions of CRISPR-Cas systems in *Streptococcus* species and oral bacterial species; (**a**) *Streptococcus pyogenes*; (**b**) *Streptococcus mutans*; (**c**) three red complex species (*Porphyromonas gingivalis*, *Tannerella forsythia*, and *Treponema denticola*), and their competitive and cooperative interaction

similarity to the three *P. gingivalis* genome sequences available in the databases; hence, we hypothesized that genetic recombination and rearrangement within *P. gingivalis* strains might be regulated by CRISPR-Cas systems [182]. The comparative genome analysis was then performed in Watanabe et al. [183] using the draft genome sequences of 51 *P. gingivalis* isolates, and the publicly available genome sequences of 13 *P. gingivalis* and 46 other *Porphyromonas* species [183]. As a result, the CRISPR spacers (identified from the 41 *P. gingivalis* isolates) with potential targets in the genus *Porphyromonas* were approximately 23 times more abundant than those with potential targets in other genus taxa (1720/6896 CRISPR spacers vs. 74/6896 CRISPR spacers) [183]. These results strongly suggested that CRISPR-Cas systems in *P. gingivalis* are able to limit genetic recombination and rearrangement by acquiring the genome fragments of other *P. gingivalis* strains as self-targeting CRISPR spacers.

As for other red complex species, *T. forsythia* and *T. denticola*, we have analyzed the draft genome sequences of 19 *T. forsythia* strains and 14 *T. denticola* strains for comparison [180]. In the comparative genome analysis of the 19 *T. forsythia* strains and the 14 *T. denticola* strains, 106/1631 and 7/78 CRISPR spacers exhibited high sequence similarity to the genome sequences available in public nucleotide databases, respectively [180]. In the 19 *T. forsythia* strains, 16/106 and 3/106 CRISPR spacers exhibited significant sequence similarity to the genome sequences of *P. gingivalis* and *T. denticola*, respectively [180]. In addition, the CRISPR spacers found in four of the 19 *T. forsythia* strains exhibited high sequence similarity to the MTase-encoding genes in *P. gingivalis* [180]. These results suggested that CRISPR-Cas systems in *T. forsythia* may attack MGEs including the genes encoding defense systems in other red complex species [180]. Meanwhile, 6/7 CRISPR spacers found in the 14 *T. denticola* strains exhibited significant sequence similarity to hypothetical genes in the genome sequences of *T. denticola*, suggesting that *T. denticola* limits genetic recombination and rearrangement within *T. denticola* strains [180].

Further comparative genome analysis was performed to reveal the association between the three red complex species [180]. As a result, gene deficiencies were mutually compensated in metabolic pathways when the genes of all the three red complex species were taken into account, suggesting that there is cooperative relationship among the three red complex species (Fig. 3.4c) [180]. Meanwhile, the three red complex species may have competitive interaction via CRISPR-Cas systems as described above (Fig. 3.4c) [180]. The association between the three red complex species may allow them to have different genomic evolutionary strategy to survive in the dental environment.

3.6 Future Perspectives

In this chapter, the functions of prokaryotic defense systems and their co-evolutionary arms race with invading foreign DNA were discussed. The prokaryote-phage co-evolution is thought to be one of the drivers of phenotypic and genotypic diversification [162]. A recent study has described that, on average, the origins of approximately 7% of the CRISPR spacers found in prokaryotic genome sequences were identifiable, whereas the remaining 93% have not yet been unidentifiable, termed "CRISPR dark matter" [185]. It is noteworthy that identifying the CRISPR dark matter would provide blueprints for better understanding of microbial complexity and robustness in the biosphere. Moreover, it has also been discussed that several phages may be able to control the community structure in the prokaryotic populations. For instance, CRISPR-*cas* loci have been found in two prophage regions in *Clostridium difficile* genome sequences [186], suggesting that the phages possessing CRISPR-*cas* loci enable their hosts to prevent the infection of other phage types for their interspecific competition.

In our previous investigation, the time-course metagenomic analysis of microbes in artificially polluted soils with four harmful aromatic compounds, 3-chlorobenzoate (3CB), phenanthrene, biphenyl, and carbazole, showed another potential capability of CRISPR-Cas systems (unpublished data). The time-course metagenomic analysis showed that the number of identified CRISPR repeats rapidly and proportionally changed according to those of bacteria in the polluted soil bacterial community. Meanwhile, the number of identified CRISPR spacers in the polluted soil bacterial community was stable throughout the analysis period. The time-course metagenomic analysis also showed that the species diversity and functions of the bacterial community were resilient against the chemical disturbance, whereas the contents of CRISPR arrays were altered and did not return to their original states. These results strongly suggested that the bacteria exposed to the chemicals newly inserted unidentifiable CRISPR spacers between the identifiable repeats for recording the environmental fluctuation; hence, we may call the phenomenon associated with the CRISPR arrays in the bacterial community as "memory of bacterial communities." From all the information presented in this chapter, we conclude that studying the prokaryotic defense systems allows us not only to understand the co-evolutionary arms race between prokaryotes and invading foreign DNA but also to predict microbial complexity and robustness.

Acknowledgements This work was supported by the Ministry of Education, Culture, Sports, Science and Technology of Japan and the Japan Society for the Promotion of Science under Grants-in-Aid for Scientific Research (KAKENHI) (grant numbers 16H05830/16H05501/16H01782/16H02767), a Kurita Water and Environment Foundation Grant, an Ichiro Kanehara Foundation Scholarship Grant for Research in Basic Medical Sciences and Medical Care, a Senri Life Science Foundation Kishimoto Grant, a grant from the Japan Agency for Medical Research and Development (project number 965144), and a grant by JST/JICA, SATREPS.

References

1. Fineran PC, Charpentier E. Memory of viral infections by CRISPR-Cas adaptive immune systems: acquisition of new information. Virology. 2012;434:202–9.
2. García-Aljaro C, Ballesté E, Muniesa M. Beyond the canonical strategies of horizontal gene transfer in prokaryotes. Curr Opin Microbio. 2017;38:95–105.

3. Actis LA, Tolmasky ME, Crosa JH. Bacterial plasmids: replication of extrachromosomal genetic elements encoding resistance to antimicrobial compounds. Front Biosci. 1999;4:D43–62.
4. Burrus V. Mechanisms of stabilization of integrative and conjugative elements. Curr Opin Microbiol. 2017;38:44–50.
5. Gubbins MJ, Will WR, Frost LS. The F-plasmid, a paradigm for bacterial conjugation. In: Mullany P, editor. The dynamic bacterial genome. Cambridge: Cambridge University Press; 2005. p. 151–206.
6. Lederberg J, Cavalli LL, Lederberg EM. Sex compatibility in *Escherichia coli*. Genetics. 1952;37:720–30.
7. Bedhomme S, Perez Pantoja D, Bravo IG. Plasmid and clonal interference during post-horizontal gene transfer evolution. Mol Ecol. 2017;26:1832–47.
8. Schubert S, Darlu P, Clermont O, et al. Role of intraspecies recombination in the spread of pathogenicity islands within the *Escherichia coli* species. PLoS Pathog. 2009;5:e1000257.
9. García-Solache M, Lebreton F, McLaughlin RE, et al. Homologous recombination within large chromosomal regions facilitates acquisition of β-lactam and vancomycin resistance in *Enterococcus faecium*. Antimicrob Agents Chemother. 2016;60:5777–86.
10. Manson JM, Hancock LE, Gilmore MS. Mechanism of chromosomal transfer of *Enterococcus faecalis* pathogenicity island, capsule, antimicrobial resistance, and other traits. Proc Natl Acad Sci. 2010;107:12269–74.
11. Marosevic D, Kaevska M, Jaglic Z. Resistance to the tetracyclines and macrolide-lincosamide-streptogramin group of antibiotics and its genetic linkage—a review. Ann Agric Environ Med. 2017;24:338–44.
12. Martínez JL, Coque TM, Lanza VF, et al. Genomic and metagenomic technologies to explore the antibiotic resistance mobilome. Ann N Y Acad Sci. 2017;1388:26–41.
13. Jangam D, Feschotte C, Betrán E. Transposable element domestication as an adaptation to evolutionary conflicts. Trends Genet. 2017;33:817–31.
14. Wicker T, Sabo F, Hua-Van A, et al. Reply: a unified classification system for eukaryotic transposable elements should reflect their phylogeny. Nat Rev Genet. 2009;10:276.
15. Kapitonov VV, Jurka J. Self-synthesizing DNA transposons in eukaryotes. Proc Natl Acad Sci. 2006;103:4540–5.
16. Krupovic M, Koonin EV. Polintons: a hotbed of eukaryotic virus, transposon and plasmid evolution. Nat Rev Microbiol. 2015;13:105–15.
17. Pritham EJ, Putliwala T, Feschotte C. Mavericks, a novel class of giant transposable elements widespread in eukaryotes and related to DNA viruses. Gene. 2007;390:3–17.
18. Krupovic M, Koonin EV. Self-synthesizing transposons: unexpected key players in the evolution of viruses and defense systems. Curr Opin Microbiol. 2016;31:25–33.
19. Krupovic M, Makarova KS, Forterre P, et al. Casposons: a new superfamily of self-synthesizing DNA transposons at the origin of prokaryotic CRISPR-Cas immunity. BMC Biol. 2014;12:36.
20. Stokes HT, Hall RM. A novel family of potentially mobile DNA elements encoding site-specific gene-integration functions: integrons. Mol Microbiol. 1989;3:1669–83.
21. AlAmri AM. Integron in Gram negative bacteria: classes and fitness cost. Asian J Med Sci. 2017;8:1–4.
22. Gillings MR. Integrons: past, present, and future. Microbiol Mol Biol Rev. 2014;78:257–77.
23. Yu G, Li Y, Liu X, et al. Role of integrons in antimicrobial resistance: a review. Afr J Microbiol Res. 2013;7:1301–10.
24. Domingues S, Harms K, Fricke WF, et al. Natural transformation facilitates transfer of transposons, integrons and gene cassettes between bacterial species. PLoS Pathog. 2012;8:e1002837.
25. Clokie MRJ, Millard AD, Letarov A, et al. Phages in nature. Bacteriophage. 2011;1:31–45.
26. Weinbauer MG. Ecology of prokaryotic viruses. FEMS Microbiol Rev. 2004;28:127–81.
27. Salmond GPC, Fineran PC. A century of the phage: past, present and future. Nat Rev Microbiol. 2015;13:777–86.

28. Howard-Varona C, Hargreaves KR, Abedon ST, et al. Lysogeny in nature: mechanisms, impact and ecology of temperate phages. ISME J. 2017;11:1511–20.
29. Lemire S, Figueroa-Bossi N, Bossi L. Bacteriophage crosstalk: coordination of prophage induction by trans-acting antirepressors. PLoS Genet. 2011;7:e1002149.
30. Erez Z, Steinberger-Levy I, Shamir M, et al. Communication between viruses guides lysis–lysogeny decisions. Nature. 2017;541:488–93.
31. Ripp S, Miller RV. The role of pseudolysogeny in bacteriophage-host interactions in a natural freshwater environment. Microbiology. 1997;143:2065–70.
32. Rakonjac J, Bennett NJ, Spagnuolo J, et al. Filamentous bacteriophage: biology, phage display and nanotechnology applications. Curr Issues Mol Biol. 2011;13:51–76.
33. Messing J, Gronenborn B, Müller-Hill B, et al. Filamentous coliphage M13 as a cloning vehicle: insertion of a *Hin*dII fragment of the *lac* regulatory region in M13 replicative form *in vitro*. Proc Natl Acad Sci. 1977;74:3642–6.
34. Smith GP. Filamentous fusion phage: novel expression vectors that display cloned antigens on the virion surface. Science. 1985;228:1315–7.
35. Penadés JR, Chen J, Quiles-Puchalt N, et al. Bacteriophage-mediated spread of bacterial virulence genes. Curr Opin Microbiol. 2015;23:171–8.
36. Bondy-Denomy J, Davidson AR. When a virus is not a parasite: the beneficial effects of prophages on bacterial fitness. J Microbiol. 2014;52:235–42.
37. Freeman VJ. Studies on the virulence of bacteriophage-infected strains of *Corynebacterium diphtheriae*. J Bacteriol. 1951;61:675.
38. Lindsay JA, Ruzin A, Ross HF, et al. The gene for toxic shock toxin is carried by a family of mobile pathogenicity islands in *Staphylococcus aureus*. Mol Microbiol. 1998;29:527–43.
39. Martínez-Rubio R, Quiles-Puchalt N, Martí M, et al. Phage-inducible islands in the Gram-positive cocci. ISME J. 2017;11:1029–42.
40. Novick RP, Christie GE, Penadés JR. The phage-related chromosomal islands of Gram-positive bacteria. Nat Rev Microbiol. 2010;8:541–51.
41. Penadés JR, Christie GE. The phage-inducible chromosomal islands: a family of highly evolved molecular parasites. Annu Rev Virol. 2015;2:181–201.
42. Úbeda C, Maiques E, Knecht E, et al. A suggested nomenclature for bacterial host modification and restriction systems and their enzymes. Mol Microbiol. 2005;56:836–44.
43. Colavecchio A, Cadieux B, Lo A, et al. Bacteriophages contribute to the spread of antibiotic resistance genes among foodborne pathogens of the *Enterobacteriaceae* family—a review. Front Microbiol. 2017;8:1108.
44. Touchon M, de Sousa JAM, Rocha EPC. Embracing the enemy: the diversification of microbial gene repertoires by phage-mediated horizontal gene transfer. Curr Opin Microbiol. 2017;38:66–73.
45. Barrangou R, Horvath P. A decade of discovery: CRISPR functions and applications. Nat Microbiol. 2017;2:17092.
46. Koonin EV, Makarova KS, Wolf YI. Evolutionary genomics of defense systems in archaea and bacteria. Annu Rev Microbiol. 2017a;71:233–61.
47. Luria SE. Host-induced modifications of viruses. Cold Spring Harb Symp Quant Biol. 1953;18:237–44.
48. Luria SE, Human ML. A nonhereditary, host-induced variation of bacterial viruses. J Bacteriol. 1952;64:557.
49. Blow MJ, Clark TA, Daum CG, et al. The epigenomic landscape of prokaryotes. PLoS Genet. 2016;12:e1005854.
50. Smith HO, Nathans D. A suggested nomenclature for bacterial host modification and restriction systems and their enzymes. J Mol Biol. 1973;81:419–23.
51. Pingoud A, Jeltsch A. Recognition and cleavage of DNA by type-II restriction endonucleases. FEBS J. 1997;246:1–22.
52. Loenen WAM, Dryden DTF, Raleigh EA, et al. Type I restriction enzymes and their relatives. Nucleic Acids Res. 2014;42:20–44.

53. Roberts RJ, Belfort M, Bestor T, et al. A nomenclature for restriction enzymes, DNA methyltransferases, homing endonucleases and their genes. Nucleic Acids Res. 2003;31:1805–12.
54. Smith RM, Jacklin AJ, Marshall JJT, et al. Organization of the BcgI restriction–modification protein for the transfer of one methyl group to DNA. Nucleic Acids Res. 2013;41:405–17.
55. Hedgpeth J, Goodman HM, Boyer HW. DNA nucleotide sequence restricted by the RI endonuclease. Proc Natl Acad Sci. 1972;69:3448–52.
56. Mesapogu S, Jillepalli CM, Arora DK. Restriction enzymes and their role in microbiology. In: Dilip KA, Surajit D, Mesapogu S, editors. Analyzing microbes manual of molecular biology techniques. Berlin: Springer; 2013. p. 31–6.
57. Ishikawa K, Fukuda E, Kobayashi I. Conflicts targeting epigenetic systems and their resolution by cell death: novel concepts for methyl-specific and other restriction systems. DNA Res. 2010;17:325–42.
58. Loenen WA, Raleigh EA. The other face of restriction: modification-dependent enzymes. Nucleic Acids Res. 2013;42:56–69.
59. Chen C, Wang L, Chen S, et al. Convergence of DNA methylation and phosphorothioation epigenetics in bacterial genomes. Proc Natl Acad Sci. 2017b;114:4501–6.
60. Zhou X, Deng Z, Firmin JL, et al. Site-specific degradation of *Streptomyces lividans* DNA during electrophoresis in buffers contaminated with ferrous iron. Nucleic Acids Res. 1988;16:4341–52.
61. Lan W, Hu Z, Shen J, et al. Structural investigation into physiological DNA phosphorothioate modification. Sci Rep. 2016;6:25737.
62. Zheng T, Jiang P, Cao B, et al. DndEi exhibits helicase activity essential for DNA phosphorothioate modification and ATPase activity strongly stimulated by DNA substrate with a GAAC/GTTC motif. J Biol Chem. 2016;291:1492–500.
63. Hu W, Wang C, Liang J, et al. Structural insights into DndE from *Escherichia coli* B7A involved in DNA phosphorothioation modification. Cell Res. 2012;22:1203–6.
64. Wang L, Chen S, Vergin KL. DNA phosphorothioation is widespread and quantized in bacterial genomes. Proc Natl Acad Sci. 2011;108:2963–8.
65. Yao F, Xu T, Zhou X, et al. Functional analysis of *spfD* gene involved in DNA phosphorothioation in *Pseudomonas fluorescens* Pf0-1. FEBS Lett. 2009;583:729–33.
66. Xu T, Yao F, Zhou X, et al. A novel host-specific restriction system associated with DNA backbone S-modification in *Salmonella*. Nucleic Acids Res. 2010;38:7133–41.
67. Cao B, Cheng Q, Gu C, et al. Pathological phenotypes and *in vivo* DNA cleavage by unrestrained activity of a phosphorothioate-based restriction system in *Salmonella*. Mol Microbiol. 2014;93:776–85.
68. Yamaguchi Y, Park JH, Inouye M. Toxin-antitoxin systems in bacteria and archaea. Annu Rev Genet. 2011;45:61–79.
69. Pandey DP, Gerdes K. Toxin–antitoxin loci are highly abundant in free-living but lost from host-associated prokaryotes. Nucleic Acids Res. 2005;33:966–76.
70. Gerdes K, Larsen JE, Molint S. Stable inheritance of plasmid R1 requires two different loci. J Bacteriol. 1985;161:292–8.
71. Gerdes K, Rasmussen PB, Molin S. Unique type of plasmid maintenance function: postsegregational killing of plasmid-free cells. Proc Natl Acad Sci. 1986;83:3116–20.
72. Ogura T, Hiraga S. Mini-F plasmid genes that couple host cell division to plasmid proliferation. Proc Natl Acad Sci. 1983;80:4784–8.
73. Gerdes K, Helin K, Christensen OW, et al. Translational control and differential RNA decay are key elements regulating postsegregational expression of the killer protein encoded by the *parB* locus of plasmid R1. J Mol Biol. 1988;203:119–29.
74. Harms A, Brodersen DE, Mitarai N, et al. Toxins, targets, and triggers: an overview of toxin-antitoxin biology. Mol Cell. 2018;70:768.
75. Page R, Peti W. Toxin-antitoxin systems in bacterial growth arrest and persistence. Nat Chem Biol. 2016;12:208–14.
76. Mruk I, Kobayashi I. To be or not to be: regulation of restriction–modification systems and other toxin–antitoxin systems. Nucleic Acids Res. 2014;42:70–86.

77. Ramage HR, Connolly LE, Cox JS. Comprehensive functional analysis of *Mycobacterium tuberculosis* toxin-antitoxin systems: implications for pathogenesis, stress responses, and evolution. PLoS Genet. 2009;5:e1000767.
78. Walter ND, Dolganov GM, Garcia BJ, et al. Transcriptional adaptation of drug-tolerant *Mycobacterium tuberculosis* during treatment of human tuberculosis. J Infect Dis. 2015;212:990–8.
79. Grynberg M, Erlandsen H, Godzik A. HEPN: a common domain in bacterial drug resistance and human neurodegenerative proteins. Trends Biochem Sci. 2003;28:224–6.
80. Makarova KS, Wolf YI, Koonin EV. Comprehensive comparative-genomic analysis of type 2 toxin-antitoxin systems and related mobile stress response systems in prokaryotes. Biol Direct. 2009a;4:19.
81. Yao J, Guo Y, Zeng Z, et al. Identification and characterization of a HEPN-MNT family type II toxin–antitoxin in *Shewanella oneidensis*. Microb Biotechnol. 2015;8:961–73.
82. Chen B, Akusobi C, Fang X, et al. Environmental T4-family bacteriophages evolve to escape abortive infection via multiple routes in a bacterial host employing "altruistic suicide" through type III toxin-antitoxin systems. Front Microbiol. 2017a;8:1006.
83. Labrie SJ, Samson JE, Moineau S. Bacteriophage resistance mechanisms. Nat Rev Microbiol. 2010;8:317–27.
84. Chopin MC, Chopin A, Bidnenko E. Phage abortive infection in lactococci: variations on a theme. Curr Opin Microbiol. 2005;8:473–9.
85. Anantharaman V, Makarova KS, Burroughs AM, et al. Comprehensive analysis of the HEPN superfamily: identification of novel roles in intra-genomic conflicts, defense, pathogenesis and RNA processing. Biol Direct. 2013;8:15.
86. Blower TR, Pei XY, Short FL, et al. A processed noncoding RNA regulates an altruistic bacterial antiviral system. Nat Struct Mol Bio. 2011;18:185–90.
87. Benzer S. Fine structure of a genetic region in bacteriophage. Proc Natl Acad Sci. 1955;41:344–54.
88. Parma DH, Snyder M, Sobolevski S, et al. The Rex system of bacteriophage lambda: tolerance and altruistic cell death. Genes Dev. 1992;6:497–510.
89. Snyder L. Phage-exclusion enzymes: a bonanza of biochemical and cell biology reagents? Mol Microbiol. 1995;15:415–20.
90. Makarova KS, Wolf YI, van der Oost J, et al. Prokaryotic homologs of Argonaute proteins are predicted to function as key components of a novel system of defense against mobile genetic elements. Biol Direct. 2009b;4:29.
91. Bohmert K, Camus I, Bellini C, et al. *AGO1* defines a novel locus of *Arabidopsis* controlling leaf development. EMBO J. 1998;17:170–80.
92. Hutvagner G, Simard MJ. Argonaute proteins: key players in RNA silencing. Nat Rev Mol Cell Biol. 2008;9:22–32.
93. Wang Y, Juranek S, Li H, et al. Structure of an argonaute silencing complex with a seed-containing guide DNA and target RNA duplex. Nature. 2008a;456:921–6.
94. Wang Y, Sheng G, Juranek S, et al. Structure of the guide-strand-containing argonaute silencing complex. Nature. 2008b;456:209–13.
95. Peters L, Meister G. Argonaute proteins: mediators of RNA silencing. Mol Cell. 2007;26:611–23.
96. Shabalina SA, Koonin EV. Origins and evolution of eukaryotic RNA interference. Trends Ecol Evol. 2008;23:578–87.
97. Ketting RF. The many faces of RNAi. Dev Cell. 2011;20:148–61.
98. Kaya E, Doxzen KW, Knoll KR, et al. A bacterial Argonaute with noncanonical guide RNA specificity. Proc Natl Acad Sci. 2016;113:4057–62.
99. Swarts DC, Jore MM, Westra ER, et al. DNA-guided DNA interference by a prokaryotic Argonaute. Nature. 2014a;507:258–61.
100. Swarts DC, Makarova K, Wang Y, et al. The evolutionary journey of Argonaute proteins. Nat Struct Mol Biol. 2014b;21:743–53.
101. Hegge JW, Swarts DC, van der Oost J. Prokaryotic Argonaute proteins: novel genome-editing tools? Nat Rev Microbiol. 2018;16:5–11.

102. Swarts DC, Hegge JW, Hinojo I, et al. Argonaute of the archaeon *Pyrococcus furiosus* is a DNA-guided nuclease that targets cognate DNA. Nucleic Acids Res. 2015;43:5120–9.
103. Goldfarb T, Sberro H, Weinstock E, et al. BREX is a novel phage resistance system widespread in microbial genomes. EMBO J. 2015;34:169–83.
104. Barrangou R, van der Oost J. Bacteriophage exclusion, a new defense system. EMBO J. 2014;34:134–5.
105. Hoskisson PA, Sumby P, Smith MC. The phage growth limitation system in *Streptomyces coelicolor* A(3)2 is a toxin/antitoxin system, comprising enzymes with DNA methyltransferase, protein kinase and ATPase activity. Virology. 2015;477:100–9.
106. Chinenova TA, Mkrtumian NM, Lomovskaia ND. Genetic characteristics of a new phage resistance trait in *Streptomyces coelicolor* A3 (2). Genetika. 1982;18:1945–52.
107. Sumby P, Smith MCM. Genetics of the phage growth limitation (Pgl) system of *Streptomyces coelicolor* A3(2). Mol Microbiol. 2002;44:489–500.
108. Jansen R, van Embden JDA, Gaastra W, et al. Identification of a novel family of sequence repeats among prokaryotes. OMICS. 2002a;6:23–33.
109. Jansen R, Van Embden JDA, Gaastra W, et al. Identification of genes that are associated with DNA repeats in prokaryotes. Mol Microbiol. 2002b;43:1565–75.
110. Ishino Y, Shinagawa H, Makino K, et al. Nucleotide sequence of the *iap* gene, responsible for alkaline phosphatase isozyme conversion in *Escherichia coli*, and identification of the gene product. J Bacteriol. 1987;169:5429–33.
111. Mojica FJM, Rodriguez-Valera F. The discovery of CRISPR in archaea and bacteria. FEBS J. 2016;283:3162–9.
112. Mojica FJ, Díez-Villaseñor C, Soria E, et al. Biological significance of a family of regularly spaced repeats in the genomes of Archaea, Bacteria and mitochondria. Mol Microbiol. 2000;36:244–6.
113. Deveau H, Barrangou R, Garneau JE, et al. Phage response to CRISPR-encoded resistance in *Streptococcus thermophilus*. J Bacteriol. 2008;190:1390–400.
114. Bolotin A, Quinquis B, Sorokin A, et al. Clustered regularly interspaced short palindrome repeats (CRISPRs) have spacers of extrachromosomal origin. Microbiology. 2005;151:2551–61.
115. Mojica FJ, Díez-Villaseñor C, García-Martínez J, et al. Intervening sequences of regularly spaced prokaryotic repeats derive from foreign genetic elements. J Mol Evol. 2005;60:174–82.
116. Pourcel C, Salvignol G, Vergnaud G. CRISPR elements in *Yersinia pestis* acquire new repeats by preferential uptake of bacteriophage DNA, and provide additional tools for evolutionary studies. Microbiology. 2005;151:653–63.
117. Brouns SJJ, Jore MM, Lundgren M, et al. Small CRISPR RNAs guide antiviral defense in prokaryotes. Science. 2008;321:960–4.
118. Barrangou R, Fremaux C, Deveau H, et al. CRISPR provides acquired resistance against viruses in prokaryotes. Science. 2007;315:1709–12.
119. Marraffini LA, Sontheimer EJ. CRISPR interference limits horizontal gene transfer in staphylococci by targeting DNA. Science. 2008;322:1843–5.
120. Mojica FJM, Díez-Villaseñor C, García-Martínez J, et al. Short motif sequences determine the targets of the prokaryotic CRISPR defence system. Microbiology. 2009;155:733–40.
121. Amitai G, Sorek R. CRISPR-Cas adaptation: insights into the mechanism of action. Nat Rev Microbiol. 2016;14:67–76.
122. Marraffini LA. CRISPR-Cas immunity in prokaryotes. Nature. 2015;526:55–61.
123. Nuñez JK, Lee AS, Engelman A, et al. Integrase-mediated spacer acquisition during CRISPR-Cas adaptive immunity. Nature. 2015;519:193–8.
124. Carte J, Wang R, Li H, et al. Cas6 is an endoribonuclease that generates guide RNAs for invader defense in prokaryotes. Genes Dev. 2008;22:3489–96.
125. Heler R, Samai P, Model JW, et al. Cas9 specifies functional viral targets during CRISPR-Cas adaptation. Nature. 2015;519:199–202.
126. Koonin EV, Makarova KS. Mobile genetic elements and evolution of CRISPR-Cas systems: all the way there and back. Genome Biol Evol. 2017;9:2812–25.

127. Koonin EV, Makarova KS, Zhang F. Diversity, classification and evolution of CRISPR-Cas systems. Curr Opin Microbiol. 2017b;37:67–78.
128. Shmakov S, Smargon A, Scott D, et al. Diversity and evolution of class 2 CRISPR-Cas systems. Nat Rev Microbiol. 2017b;15:169–82.
129. Zhang Q, Ye Y. Not all predicted CRISPR-Cas systems are equal: isolated *cas* genes and classes of CRISPR like elements. BMC Bioinf. 2017;18:92.
130. Sinkunas T, Gasiunas G, Waghmare SP, et al. *In vitro* reconstitution of Cascade-mediated CRISPR immunity in *Streptococcus thermophilus*. EMBO J. 2013;32:385–94.
131. Haft DH, Selengut J, Mongodin EF, et al. A guild of 45 CRISPR-associated (Cas) protein families and multiple CRISPR/Cas subtypes exist in prokaryotic genomes. PLoS Comput Biol. 2005;1:e60.
132. Tamulaitis G, Venclovas Č, Siksnys V. Type III CRISPR-Cas immunity: major differences brushed aside. Trends Microbiol. 2017;25:49–61.
133. Reeks J, Naismith JH, White MF. CRISPR interference: a structural perspective. Biochem J. 2013;453:155–66.
134. Maris C, Dominguez C, Allain FHT. The RNA recognition motif, a plastic RNA-binding platform to regulate post-transcriptional gene expression. FEBS J. 2005;272:2118–31.
135. Wang R, Li H. The mysterious RAMP proteins and their roles in small RNA-based immunity. Protein Sci. 2012;21:463–70.
136. Kojima KK, Kanehisa M. Systematic survey for novel types of prokaryotic retroelements based on gene neighborhood and protein architecture. Mol Biol Evol. 2008;25:1395–404.
137. Makarova KS, Grishin NV, Shabalina SA, et al. A putative RNA-interference-based immune system in prokaryotes: computational analysis of the predicted enzymatic machinery, functional analogies with eukaryotic RNAi, and hypothetical mechanisms of action. Biol Direct. 2006;1:7.
138. Simon DM, Zimmerly S. A diversity of uncharacterized reverse transcriptases in bacteria. Nucleic Acids Res. 2008;36:7219–29.
139. Toro N, Nisa-Martínez R. Comprehensive phylogenetic analysis of bacterial reverse transcriptases. PLoS One. 2014;9:e114083.
140. Silas S, Mohr G, Sidote DJ, et al. Direct CRISPR spacer acquisition from RNA by a natural reverse transcriptase-Cas1 fusion protein. Science. 2016;351:aad4234.
141. Silas S, Makarova KS, Shmakov S, et al. On the origin of reverse transcriptase-using CRISPR-Cas systems and their hyperdiverse, enigmatic spacer repertoires. MBio. 2017;8:e00897–17.
142. Sontheimer EJ, Barrangou R. The bacterial origins of the CRISPR genome-editing revolution. Hum Gene Ther. 2015;26:413–24.
143. Doudna JA, Charpentier E. The new frontier of genome engineering with CRISPR-Cas9. Science. 2014;346:1258096.
144. Mali P, Yang L, Esvelt KM, et al. RNA-guided human genome engineering via Cas9. Science. 2013;339:823–6.
145. Ran FA, Hsu PD, Lin C-Y, et al. Double nicking by RNA-guided CRISPR Cas9 for enhanced genome editing specificity. Cell. 2013a;154:1380–9.
146. Ran FA, Hsu PD, Wright J, et al. Genome engineering using the CRISPR-Cas9 system. Nat Protoc. 2013b;8:2281–308.
147. Deltcheva E, Chylinski K, Sharma CM, et al. CRISPR RNA maturation by *trans*-encoded small RNA and host factor RNase III. Nature. 2011;471:602–7.
148. Jinek M, Chylinski K, Fonfara I, et al. A programmable dual-RNA–guided DNA endonuclease in adaptive bacterial immunity. Science. 2012;337:816–21.
149. Karvelis T, Gasiunas G, Miksys A, et al. crRNA and tracrRNA guide Cas9-mediated DNA interference in *Streptococcus thermophilus*. RNA Biol. 2013;10:841–51.
150. Chylinski K, Le Rhun A, Charpentier E. The tracrRNA and Cas9 families of type II CRISPR-Cas immunity systems. RNA Biol. 2013;10:726–37.
151. Gasiunas G, Barrangou R, Horvath P, et al. Cas9–crRNA ribonucleoprotein complex mediates specific DNA cleavage for adaptive immunity in bacteria. Proc Natl Acad Sci. 2012;109:E2579–86.

152. Yamano T, Nishimasu H, Zetsche B, et al. Crystal structure of Cpf1 in complex with guide RNA and target DNA. Cell. 2016;165:949–62.
153. Burstein D, Harrington LB, Strutt SC, et al. New CRISPR-Cas systems from uncultivated microbes. Nature. 2017;542:237–41.
154. Hickman AB, Dyda F. The casposon-encoded Cas1 protein from *Aciduliprofundum boonei* is a DNA integrase that generates target site duplications. Nucleic Acids Res. 2015;43:10576–87.
155. Krupovic M, Shmakov S, Makarova KS, et al. Recent mobility of casposons, self-synthesizing transposons at the origin of the CRISPR-Cas immunity. Genome Biol Evol. 2016;8:375–86.
156. Béguin P, Charpin N, Koonin EV, et al. Casposon integration shows strong target site preference and recapitulates protospacer integration by CRISPR-Cas systems. Nucleic Acids Res. 2016;44:10367–76.
157. Sontheimer EJ, Davidson AR. Inhibition of CRISPR-Cas systems by mobile genetic elements. Curr Opin Microbiol. 2017;37:120–7.
158. Chaudhary K, Chattopadhyay A, Pratap D. Anti-CRISPR proteins: counterattack of phages on bacterial defense (CRISPR/Cas) system. J Cell Physiol. 2017;233:57–9.
159. Bondy-Denomy J, Garcia B, Strum S, et al. Multiple mechanisms for CRISPR-Cas inhibition by anti-CRISPR proteins. Nature. 2015;526:136–9.
160. Samson JE, Magadán AH, Sabri M, et al. Revenge of the phages: defeating bacterial defences. Nat Rev Microbiol. 2013;11:675–87.
161. Krüger DH, Bickle TA. Bacteriophage survival: multiple mechanisms for avoiding the deoxyribonucleic acid restriction systems of their hosts. Microbiol Rev. 1983;47:345–60.
162. Koskella B, Brockhurst MA. Bacteria–phage coevolution as a driver of ecological and evolutionary processes in microbial communities. FEMS Microbiol Rev. 2014;38:916–31.
163. Makarova KS, Wolf YI, Snir S, et al. Defense islands in bacterial and archaeal genomes and prediction of novel defense systems. J Bacteriol. 2011;193:6039–56.
164. Puigbò P, Makarova KS, Kristensen DM, et al. Reconstruction of the evolution of microbial defense systems. BMC Evol Biol. 2017;17:94.
165. Ofir G, Melamed S, Sberro H, et al. DISARM is a widespread bacterial defence system with broad anti-phage activities. Nat Microbiol. 2018;3:90–8.
166. Vasu K, Nagaraja V. Diverse functions of restriction-modification systems in addition to cellular defense. Microbiol Mol Biol Rev. 2013;77:53–72.
167. Sánchez-Romero MA, Cota I, Casadesús J. DNA methylation in bacteria: from the methyl group to the methylome. Curr Opin Microbiol. 2015;25:9–16.
168. Srikhanta YN, Maguire TL, Stacey KJ, et al. The phasevarion: a genetic system controlling coordinated, random switching of expression of multiple genes. Proc Natl Acad Sci. 2005;102:5547–51.
169. Casadesus J, Low D. Epigenetic gene regulation in the bacterial world. Microbiol Mol Biol Rev. 2006;70:830–56.
170. Budroni S, Siena E, Hotopp JCD, et al. *Neisseria meningitidis* is structured in clades associated with restriction modification systems that modulate homologous recombination. Proc Natl Acad Sci. 2011;108:4494–9.
171. Nandi T, Holden MTG, Didelot X, et al. *Burkholderia pseudomallei* sequencing identifies genomic clades with distinct recombination, accessory, and epigenetic profiles. Genome Res. 2015;25:129–41.
172. Eutsey RA, Powell E, Dordel J, et al. Genetic stabilization of the drug-resistant PMEN1 Pneumococcus lineage by its distinctive DpnIII restriction-modification system. MBio. 2015;6:e00173–15.
173. Barrangou R. The roles of CRISPR-Cas systems in adaptive immunity and beyond. Curr Opin Immunol. 2015;32:36–41.
174. Okura M, Nozawa T, Watanabe T, et al. A locus encoding variable defense systems against invading DNA identified in *Streptococcus suis*. Genome Biol Evol. 2017;9:1000–12.

175. Ralph AP, Carapetis JR. Group a streptococcal diseases and their global burden. In: Chhatwal GS, editor. Host-pathogen interactions in streptococcal diseases. Berlin: Springer; 2012. p. 1–27.
176. Nakagawa I, Kurokawa K, Yamashita A, et al. Genome sequence of an M3 strain of *Streptococcus pyogenes* reveals a large-scale genomic rearrangement in invasive strains and new insights into phage evolution. Genome Res. 2003;13:1042–55.
177. Nozawa T, Furukawa N, Aikawa C, et al. CRISPR inhibition of prophage acquisition in *Streptococcus pyogenes*. PLoS One. 2011;6:e19543.
178. Aikawa C, Furukawa N, Watanabe T, et al. Complete genome sequence of the serotype *k* *Streptococcus mutans* strain LJ23. J Bacteriol. 2012;194:2754–5.
179. Maruyama F, Kobata M, Kurokawa K, et al. Comparative genomic analyses of *Streptococcus mutans* provide insights into chromosomal shuffling and species-specific content. BMC Genomics. 2009;10:358.
180. Endo A, Watanabe T, Ogata N, et al. Comparative genome analysis and identification of competitive and cooperative interactions in a polymicrobial disease. ISME J. 2015;9:629–42.
181. Watanabe T, Maruyama F, Nozawa T, et al. Complete genome sequence of the bacterium *Porphyromonas gingivalis* TDC60, which causes periodontal disease. J Bacteriol. 2011;193:4259–60.
182. Watanabe T, Nozawa T, Aikawa C, et al. CRISPR regulation of intraspecies diversification by limiting IS transposition and intercellular recombination. Genome Biol Evol. 2013;5:1099–114.
183. Watanabe T, Shibasaki M, Maruyama F, et al. Investigation of potential targets of *Porphyromonas* CRISPRs among the genomes of *Porphyromonas* species. PLoS One. 2017;12:e0183752.
184. Paster BJ, Olsen I, Aas JA, et al. The breadth of bacterial diversity in the human periodontal pocket and other oral sites. Periodontol 2000. 2006;42:80–7.
185. Shmakov SA, Sitnik V, Makarova KS, et al. The CRISPR spacer space is dominated by sequences from species-specific mobilomes. MBio. 2017a;8:e01397–17.
186. Sebaihia M, Wren BW, Mullany P, et al. The multidrug-resistant human pathogen *Clostridium difficile* has a highly mobile, mosaic genome. Nat Genet. 2006;38:779–86.

RNA-Mediated Crosstalk Between Bacterial Core Genome and Foreign Genetic Elements

4

Masatoshi Miyakoshi

Abstract

Gene expression is regulated at both transcriptional and posttranscriptional levels. Foreign genetic elements have evolved to integrate their expression systems into the host regulatory network. Transcription of AT-rich foreign genes is globally silenced by chromosomal nucleoid proteins such as H-NS. Similarly, a global posttranscriptional regulator Hfq tends to bind AU-rich transcripts and plays an important role to optimize expression of foreign genes as well as its core genome. The RNA chaperone Hfq also facilitates base-pairing between RNA molecules in *trans* and thus contributes to the crosstalk between chromosomal and foreign genes. Moreover, other classes of RNA chaperones are also involved in the crosstalk. Taking the model pathogen *Salmonella* Typhimurium as an example, this chapter focuses on the RNA-mediated crosstalk between chromosome and foreign genetic elements.

Keywords

Posttranscriptional regulation · Small RNA · RNA-binding protein · *Salmonella*

4.1 General Mechanisms of Posttranscriptional Regulation in Bacteria

Genetic information encoded on DNA is first transcribed into RNA, most of which functions in translation (rRNA and tRNA) or is in turn translated into proteins (mRNA). Therefore, gene expression seems primarily regulated at the

M. Miyakoshi (✉)
Faculty of Medicine, University of Tsukuba, Tsukuba, Ibaraki, Japan
e-mail: mmiyakoshi@md.tsukuba.ac.jp

H. Nishida, T. Oshima (eds.), *DNA Traffic in the Environment*,
https://doi.org/10.1007/978-981-13-3411-5_4

transcriptional level but can also be controlled at the posttranscriptional level. Transcription is regulated on the basis of interaction between DNA and DNA-binding proteins. Transcription initiation is triggered by RNA polymerase (RNAP) holoenzyme binding at a promoter sequence, and its regulation largely relies on transcriptional factors (TFs) with a range of affinity and location of their binding sequences relative to the promoter. Before identification of the repressor of *lac* operon as the LacI protein, RNA was first conceived as the regulatory molecule which binds to the operator DNA because of its complementarity [1]. RNA has been thought suitable for regulatory functions since it adopts a variety of secondary and tertiary structures intramolecularly and also has a capability of intermolecular interactions.

RNA is subject to posttranscriptional regulation in its stability and, in case of mRNA, translation efficiency. RNA is turned over through degradation by numerous ribonucleases (RNase) [2, 3]. RNA decay rate is determined by accessibility of RNases, which is modulated by both accessory proteins and antisense RNAs. Moreover, translation efficiency of mRNA affects its half-life since translating ribosomes prevent mRNA degradation [4]. In bacteria, translation is initiated by base-pairing between Shine–Dalgarno (SD) sequence of mRNA and 3′-terminal anti-SD sequence of 16S rRNA [5]. Importantly, translation is coupled with transcription in bacteria, i.e., translating ribosome directly interacts with transcribing RNAP [6]. Therefore, translation is often regulated at the 5′ untranslated region (UTR) and early coding sequence (CDS), but rarely at the 3′ part in contrast to eukaryotic regulatory mechanism. Similar to transcriptional regulation, posttranscriptional regulation turns out to be both negative and positive, depending on the secondary structures of RNA molecules and location of target sequences relative to the translation initiation region of mRNA [7–10].

Antisense RNAs can be divided into two types, *trans*-acting or *cis*-acting, according to their genetic loci [11, 12]. *cis*-Acting antisense RNA is encoded in the same locus as its target RNA and thus holds complete complementarity. *trans*-Acting antisense RNA is expressed from a physically unlinked locus from the loci encoding target RNAs. By virtue of its only partial and incomplete complementarity, *trans*-acting antisense RNA can regulate multiple targets with the help of RNA chaperones. One of the most extensively studied class of *trans*-acting antisense RNAs is Hfq-dependent small RNAs (sRNAs) with the length from 40 to hundreds of nucleotides (nt) [13–15]. Some of them are not actually noncoding; they encode small peptides and thus associate with ribosomes [16, 17]. Moreover, a number of Hfq-dependent sRNAs are processed from mRNAs or share genetic loci with relatively large protein-coding genes [13, 18–20].

4.2 Discovery of Antisense RNAs and Their Chaperones

Historically, since the first discovery of antisense RNA in 1981 [21], findings on RNA-mediated regulation were largely obtained from studies on plasmids. Not surprisingly, plasmids encode regulatory RNAs to control themselves mostly *in cis*

[11, 12, 22]. Although the combination of *cis*-acting antisense RNA and its binding protein is specific for each plasmid, its activity is regulated by several housekeeping RNases through processing and subsequent degradation [23]. Recently, homologs of plasmid-encoded RBPs have been discovered in chromosomes, and more novel bacterial RBPs will be discovered through a nonbiased method [24].

This section will not cover the type I toxin–antitoxin (TA) systems, which are involved in plasmid maintenance, abortive bacteriophage infection, and bacterial persistence [25–27] (see Chap. 3). Antitoxin is basically specific for its toxin mRNA and independent of RBPs, but exceptionally an antitoxin sRNA RalA acts on toxin-encoding *ralR* mRNA *in trans* in an Hfq-dependent fashion [28].

4.2.1 ColE1 Plasmid Replication

ColE1 plasmid of *Escherichia coli* encodes colicin E1, and its backbone has frequently been utilized as a cloning vector. Replication of ColE1 initiates from an RNA primer whose precursor (RNA II) is transcribed from 555-nt upstream of the origin and is processed at the origin by RNase H. The primer formation is inhibited by RNA I, a ~108-nt-long *cis*-encoded antisense RNA which hybridizes with 5′ region of the nascent primer transcript [21]. RNA I interacts with RNA II via a loop–loop kissing interaction which is specifically stabilized by Rom, the ColE1 plasmid-encoded 63-aa protein [29]. Therefore, deletion of *rom* results in a high-copy-number plasmid, which is very well known as pUC cloning vectors [30].

4.2.2 F Plasmid Conjugation

F plasmid of *E. coli* has been extensively studied as a model for bacterial conjugation [31]. Expression of *tra* operon encoding the transfer apparatus is controlled by a transcriptional cascade composed of the primary activator TraJ and autoregulators TraM and TraY [32, 33]. Translation of TraJ is repressed by ~80-nt *cis*-encoded antisense RNA FinP, whose sequence is specific for each F-like plasmids [34]. Though disrupted in the natural F plasmid [35, 36], FinO protein protects FinP from degradation by RNase E [37, 38] and facilitates interaction between FinP and the 5′UTR of *traJ* mRNA [39–41]. The FinP-*traJ* double-stranded RNA is rapidly degraded by RNase III [37], repressing the TraJ expression irreversibly.

4.2.3 Host Factor for Qβ RNA Phage

Hfq is originally identified as a host factor for Qβ RNA phage in *E. coli* [42]. As Qβ phage can overcome the requirement of Hfq by at least four point mutations in its 3′-terminal nucleotides [43], the role of Hfq is likely to facilitate the secondary structure formation of 3′ end where Qβ synthesis initiates.

Hfq is now known to be the RNA chaperone for vast majority of small RNAs which act *in trans*. Hfq is conserved in α-, β-, and γ-Proteobacteria and also in Firmicutes. Though not essential for viability, deletion of *hfq* causes pleiotropic effects on various bacterial species, such as attenuated virulence [44, 45].

Hfq comprises a conserved N-terminal Sm domain and a less conserved and unstructured C-terminal domain. The Sm domain adopts a homohexameric toroidal structure with three RNA-binding faces, i.e., proximal, distal, and rim (lateral) faces. The proximal face specifically recognizes U-rich stretches at the 3′ end, which is often generated by Rho-independent transcription termination [46, 47]. The distal face binds AAN motifs with a high affinity [48–50]. The conserved arginine residues on the Hfq rim have an affinity for internal UA-rich sequences [51].

Co-immunoprecipitation of Hfq followed by microarray analyses [52, 53] or RNA immunoprecipitation followed by RNA-seq analyses (RIP-seq) [54–56] has comprehensively identified the in vivo ligands of Hfq both in *E. coli* and in *Salmonella*. More precisely, in vivo UV crosslinking immunoprecipitation followed by RNA-seq (CLIP-seq) in *E. coli* [57] and in *Salmonella* [58] has determined Hfq-binding sites on the transcripts in vivo. Importantly, the latter study has revealed that among the 640 binding sites in *Salmonella*, Hfq preferentially contacts with RNA molecules transcribed from pathogenicity islands, SPI1 and SPI2 [58].

The major function of Hfq is thought to facilitate RNA–RNA interactions *in trans* through relatively short and incomplete base-pairing. An Hfq-dependent sRNA generally contains at least one "seed" sequence that is complementary to target mRNAs [59] and the Rho-independent terminator at 3′end which binds the proximal face of Hfq [60, 61]. In addition, Hfq binds to the sRNA elsewhere with either rim or distal face [52]. The Hfq-mediated base-pairing between mRNA and sRNA has been extensively studied [62], and RNA interaction by ligation and sequencing (RIL-seq) analysis in *E. coli* has identified ~2800 chimeric RNAs generated via Hfq-mediated base-pairing *in trans* [63]. The binding partners are dominated by canonical sRNA–mRNA pairs but also included pairs of sRNA–sRNA and sRNA–tRNA precursors [64, 65].

4.2.4 CsrA

CsrA/RsmA (carbon storage regulator/regulator of secondary metabolism) is another class of posttranscriptional regulators [66]. CsrA/RsmA binds single-stranded GGA motifs mostly in the loop of RNA hairpin structure, which are likely found in SD sequence. Therefore, CsrA primarily functions as a negative regulator by modulating translation initiation and/or RNA stability. There are a few exceptions; in *E. coli*, CsrA activates *flhDC* mRNA by preventing 5′-end-dependent degradation by RNase E [67] or binds 5′UTR of *pgaABCD* operon mRNA to remodel its secondary structure and induce Rho-dependent premature termination [68].

Positive regulation is also achieved by sequestration of CsrA/RsmA by sRNAs containing multiple RBP-binding sites such as CsrB and CsrC of *E. coli* [69, 70]. By forming a globular complex with 18 CsrA proteins, CsrB sRNA sequesters CsrA to antagonize CsrA-mediated regulation. CLIP-seq analyses identified multiple direct CsrA-binding sites, 467 sites in *Salmonella* [58] and 457 sites in *E. coli* [71]. These studies revealed that in addition to CsrB and CsrC, CsrA binds multiple sRNAs including those bound with Hfq.

The CsrA system is conserved in γ-Proteobacteria but is absent in α- and β-Proteobacteria. In Firmicutes, Spirochaete, and Thermotogae, the *csrA* gene is located in a flagellum gene cluster, and the activity of CsrA is modulated by the flagellar assembly protein FliW [72]. In an ε-Proteobacterium *Campylobacter jejuni* that lacks CsrA-titrating sRNAs, the *flaA* mRNA encoding the major flagellin primarily binds CsrA and inhibits CsrA-mediated regulation of flagellar genes [73].

4.2.5 ProQ/FinO-Domain RNA-Binding Proteins

Genes encoding FinO-like proteins are distributed in prokaryotic chromosomes of at least β- and γ-Proteobacteria, emerging as another class of RNA chaperone [74, 75]. *proQ* was found by *E. coli* genetic screening for increased resistance to 3,4-dehydroproline [76, 77], which turned out to be an indirect effect on the osmoregulatory glycine betaine transporter ProP [78]. Biochemical study revealed that ProQ is indeed an RBP which efficiently associates with FinP RNA fragment as well as FinO [79].

Separately, *proQ* was found in *Legionella pneumophila* to regulate natural transformation [80], an important system for HGT as well as conjugative transfer and transduction. RIP-seq analysis of ProQ (renamed as RocC in *L. pneumophila*) revealed that a *trans*-acting regulatory sRNA RocR is the primary ligand of RocC [81]. RocR is folded into three stem loops and base-pairs with ribosome-binding sequences of multiple competence genes such as *comEA* via its first loop sequence to repress translation.

In *Salmonella*, a gradient profiling of RNA–RBP complexes by sequencing (Grad-seq) revealed ProQ as a global RBP [24]. One of the ProQ-dependent sRNAs RaiZ is derived from 3′ region of *raiA* mRNA encoding a cold-shock-inducible ribosome-inactivating protein through processing by RNase E [82]. ProQ stabilizes RaiZ and facilitates base-pairing *in trans* with a single target *hupA* mRNA to repress translation of the histone-like protein HU. Interestingly, RaiZ is also associated with Hfq, which has no effect on the stability and function of RaiZ.

While FinO and RocC are highly specialized with only single sRNA ligands, ProQ binds with multiple sRNAs and mRNAs in vivo. CLIP-seq analysis revealed that ProQ binds 467 sites in *Salmonella*, of which 21 and 5 transcripts are expressed from pCol1B9 and pSLT plasmids [83]. The common feature of RNA ligands for this class of RBPs is relatively long double-stranded stem-loop structures [75] and the absence of common sequence motifs [83]. Although the structures of ProQ-/FinO-like RNA-binding domains have been determined [84–86], it is so far unclear what determines the specificity for their RNA ligands. ProQ is able to bind with FinP-derived dsRNA fragment in vitro and can rescue the *finO* deficiency in F plasmid conjugation in *E. coli* [79, 84], but in *Salmonella* ProQ does not bind FinP of pSLT plasmid in vivo [83].

4.3 Crosstalk Between Chromosome and Foreign Genetic Elements

Salmonella enterica serovar Typhimurium (*S*. Typhimurium) has been a good example of interplay between the core and foreign genetic elements in the regulation of gene expression. The genome exhibits a mosaic structure consisting of five large *Salmonella* pathogenicity islands in addition to several small pathogenicity islets (collectively SPI regions), Gifsy-1, Gifsy −2, Fels-1, and Fels −2 prophages and an extrachromosomal F-like plasmid pSLT [87–89]. *S*. Typhimurium causes gastroenteritis in humans and systemic disease in mice. The critical virulence factors of this facultative intracellular pathogen are effector proteins which are delivered by type 3 secretion systems (T3SS). *S*. Typhimurium harbors two major SPIs, which encode distinct T3SSs and thus are critical for its virulence. During the course of infection, SPI1 is expressed in invasion into host epithelial cells and subsequently SPI2 is expressed in intracellular replication and survival, and thus these two clusters of genes are regulated temporally and spatially.

In a representative *S*. Typhimurium strain 4/74, 86% of the whole genome has been transcribed under 22 growth conditions, and a total of 3838 promoters have been identified [90]. The number of sRNAs has been raised to 280 in the single strain, which is comparable to that of transcriptional regulators. Moreover, during infection in macrophage, the strain expresses 3583 promoters and 246 sRNAs [91].

The SPI1 master regulator HilD activates HilC and RtsA, and a complex feed-forward loop of these three AraC/XylS family transcriptional regulators activates the OmpR/ToxR family regulator HilA, which directly activates *prg*/*org*, *inv*/*spa*, and *sic/sip* operons encoding SPI1 T3SS and downstream transcriptional regulators such as InvF [92]. All of these components are encoded on SPI1 except RtsA.

SPI-2 is regulated by interplay of multiple two-component transcription regulators, SPI2-encoded SsrA/SsrB and core genome-encoded PhoP/PhoQ and OmpR/EnvZ systems [93]. In addition, the SPI-1 regulator HilD is shown to activate transcription of *ssrA–ssrB* operon, suggesting a temporal regulation from SPI1 to SPI2 induction at the transcriptional level [94].

Transcription of AT-rich foreign genes in SPI1 and SPI2 regions is pervasively silenced by chromosomal nucleoid-associated proteins (NAPs) such as H-NS [95, 96] (see Chap. 1). Transcriptional silencing by NAPs is relieved by specific transcriptional regulators that are specified by these islands as described above [97]. Like H-NS, global posttranscriptional regulators Hfq and CsrA extensively bind RNA molecules transcribed from SPI regions [58]. Since Hfq also functions as a translational repressor by binding directly on mRNAs in several cases in *E. coli* [98–102], it is tempting to speculate that Hfq globally silences the AU-rich foreign transcripts at the posttranscriptional level and is relieved by specific sRNAs or RBPs in response to relevant signals.

The following sections focus on crosstalk between chromosome and foreign genetic elements with relevance to RNA-mediated regulation in *Salmonella*.

4.3.1 sRNAs Specified by Salmonella Pathogenicity Islands

The Hfq-dependent sRNA InvR is encoded on SPI1 (Fig. 4.1a) and induced under the control of HilD (Fig. 4.1b) and therefore is expected to be an important sRNA particularly relevant to the posttranscriptional regulation of *Salmonella* pathogenicity. So far, the only known target of InvR is the outer membrane protein OmpD, whose expression is directly repressed through extensive base-pairing [103]. Expression of OmpD is repressed in macrophages, and downregulation of OmpD supports *Salmonella* proliferation inside macrophages [104]. Although InvR does not seem to alter the expression pattern of secreted SPI1 effector proteins [103], more targets of InvR other than *ompD* will be discovered with high-resolution RNA-seq technologies.

In addition, SPI1 encodes the other three sRNAs, STnc1410, STnc3020, and STnc 4240, and SPI2 specifies one sRNA STnc1220 (Fig. 4.1a) [90], whose functions are still unknown. The 17-kb SPI3 expresses a *trans*-acting sRNA derived from 5′ leader mRNA of *mgtCBR* operon, which is required for intramacrophage survival and growth in low-Mg^{2+} media [105]. The leader RNA base-pairs with *fljBA* phase 2 flagellin operon mRNA to degrade in an Hfq- and RNase E-dependent manner (Fig. 4.1b). Since the *mgtCBR* operon is induced inside macrophages under the control of PhoP/PhoQ two-component system, the RNA-mediated regulation results in repression of *Salmonella* flagellin synthesis to escape from the host innate immune system.

There are also several island-encoded sRNAs (designated as Isr) outside of the major SPIs (Fig. 4.1a). Gifsy-1 prophage encodes IsrK and IsrJ [106]. IsrK is transcribed as the 77-nt short form of IsrK and the 910-nt-long IsrK-*orf45*-*anrP*-IsrJ transcript [107]. IsrK activates coupled translation of Orf45 and AnrP on the longer transcript *in trans*, and AnrP activates transcription of bacteriophage Q-like antiterminator protein AntQ, which is encoded divergently upstream of IsrK. AntQ affects transcription termination globally. The role of the downstream sRNA IsrJ is unknown.

Salmonella contains two copies of RyhB, i.e., RyhB-1 and RyhB-2 (IsrE), both of which respond to iron limitation under the control of Fur transcriptional regulator but are differentially regulated depending on growth phases [106]. Both RyhB-1 and RyhB-2 are involved in oxidative and nitrosative stress responses [108, 109]. Microarray analysis suggested that RyhB-2 specifically regulates motility and chemotaxis genes *in trans* [110], but it is unknown whether these mRNAs are directly regulated by base-pairing mechanism. RyhB-2 also acts on its overlapping gene STM1273.1n *in cis*, which results in downregulation of downstream *yeaQ* gene [111].

IsrM, which is conserved in *S.* Typhimurium but absent in *S. enterica* serovar Typhi and *Salmonella bongori*, directly hybridizes with translation initiation regions of *hilE* and *sopA* mRNAs to repress expression of a global transcriptional regulator of SPI1 and one of the SPI1 effector proteins, respectively (Fig. 4.1b) [112]. Importantly, deletion of *isrM* resulted in deficiencies in invasion into epithelial cells and intracellular survival, which are restored by ectopic expression of IsrM *in trans*.

RaoN encoded on SPI11 was identified by random transposon mutagenesis in the intergenic region between *cspH* and *envE* genes in the same direction [113].

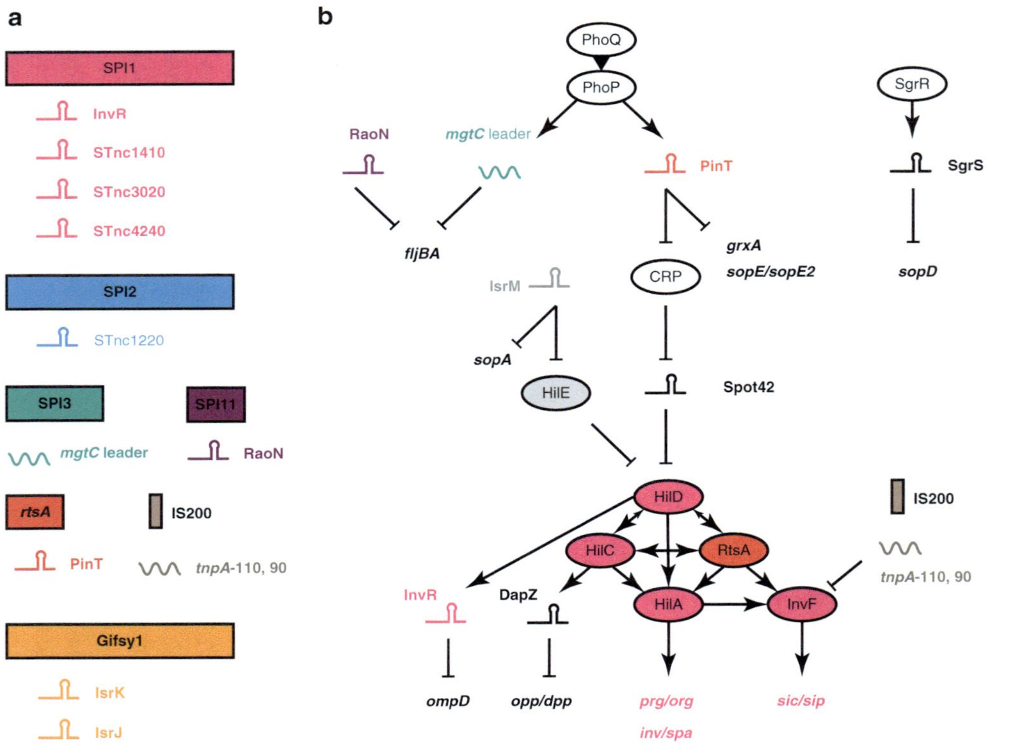

Fig. 4.1 (**a**) *trans*-Acting regulatory RNAs encoded on SPIs and other foreign genetic elements in *Salmonella*. (**b**) Crosstalk in SPI1 regulatory cascade by transcriptional and posttranscriptional regulators. Each SPI region or IS (rectangular) expresses transcriptional regulators (oval) and regulatory RNAs (curved line) in the corresponding color. The regulatory cascade is composed of mixture of factors encoded on non-SPI foreign regions (gray) and core genome (white oval or black curved line)

Expression of RaoN was impaired in an *hfq* mutant. In the *raoN* deletion mutant, lactate dehydrogenase gene *ldhA* and phase 2 flagellin gene *fljB* were upregulated (Fig. 4.1b), but direct regulation of these mRNAs by RaoN has not been clarified.

4.3.2 sRNAs Regulated by SPI-Encoded Transcriptional Regulators

From transcriptome profiling of *S.* Typhimurium grown under 22 conditions [90] and in macrophages [91], two sRNAs (InvR and STnc520) and eleven sRNAs (IsrH_1_2, STnc3730, STnc3090, STnc470, STnc3020, STnc3180, STnc440 (PinT), STnc1480, STnc3050, sRNA10, and STnc3170) are expressed under the conditions relevant for SPI1 and SPI2 transcriptional networks, respectively. STnc520 was originally identified in Hfq co-IP [56] and later shown to exhibit similar expression pattern with InvR through positive regulation by SPI-1-encoded transcriptional regulator SprB [114]. The targets of STnc520 have not been identified so far. Surprisingly, none of the latter 11 sRNAs are encoded on the SPI2 island, and only six of them are specific for *S. enterica* but are absent in *S. bongori* [115].

DapZ sRNA, which is embedded in the 3′UTR of *dapB* gene, is expressed through the direct regulation by RtsA and HilC in the HilD transcriptional network in *Salmonella* [55, 114]. DapZ regulon overlaps with GcvB regulon containing amino acid and peptide transporters, such as *oppA* and *dppA* (Fig. 4.1b) [116, 117]. Expression profile of DapZ is clearly different from that of GcvB; GcvB is induced by glycine under the control of GcvA/GcvR transcriptional regulators, and the decay rate of GcvB is controlled by SroC sRNA, which is derived from the 3′UTR of *gltI* mRNA [65]. The stability of DapZ is not affected by SroC (Miyakoshi et al. unpublished data). Transient downregulation of amino acid/peptide transporters by DapZ during the SPI-1 condition might be beneficial for efficient host invasion.

PinT (STnc440) is an 80-nt sRNA encoded in a 15-kb *Salmonella*-specific islet that encodes an AraC/XylS transcriptional activator RtsA and is strongly induced in SPI2-inducing medium and during infection under the control of PhoP/PhoQ transcriptional regulator [118]. Through direct base-pairing mechanism, PinT negatively regulates *sopE* and *sopE2* mRNAs encoding SPI1 effector, *grxA* mRNA encoding glutathione/glutaredoxin 1, and *crp* mRNA encoding the cAMP receptor protein (Fig. 4.1b). Time-course expression profile suggests that PinT temporally regulates the transition from SPI-1 to SPI-2 virulence stages [118].

4.3.3 Chromosomal sRNA Regulates Horizontally Acquired Genes

The SPI-1 master regulator of *Salmonella* invasion HilD is posttranscriptionally regulated by multiple factors. At SD sequence, the *hilD* mRNA is directly bound with CsrA [83], which represses translation and induces mRNA decay [119]. Hfq also binds the CDS of *hilD* mRNA [83], and overexpression of *hilD* rescued the reduced expression of SPI-1 effector proteins in *Salmonella* Δ*hfq* mutant [56]. The

310-nt-long 3′UTR of *hilD* mRNA is attacked by a degradosome complex containing RNase E and polynucleotide phosphorylase [120]. The Hfq-dependent sRNA Spot42 activates the expression of *hilD* probably by direct base-pairing with its 3′UTR to protect the mRNA from the degradosome [121]. Since the expression of Spot42 is repressed by CRP-cAMP, this regulation in part accounts for the CRP-mediated repression of SPI-1. Moreover, CRP is repressed by PinT under SPI-2 condition [118], suggesting CRP as a key factor for the temporal expression between the two pathogenicity islands of *Salmonella*.

In *Salmonella*, SgrS sRNA induced by glucose–phosphate stress specifically represses *sopD* mRNA encoding an effector protein which can be secreted by both SPI1 and SPI2 T3SSs (Fig. 4.1b), but not its duplicated gene *sopD2* [122]. It is concluded that the chromosomal sRNA is involved in the regulation of *Salmonella* virulence, but more studies are awaited to answer what is the physiological consequence of this differential regulation of SopD and SopD2.

IS200 is present in the *S*. Typhimurium genome with the copy number of 5–12, while *S*. Typhi genome harbors 26 copies. IS200 expresses two transcripts, the transposase *tnpA* mRNA and its antisense RNA art200 (STnc490). art200 regulates *tnpA* mRNA by binding at its overlapping 5′ region *in cis* [123]. Although art200 binds both Hfq [56] and ProQ [24], contribution of these RBPs remains obscure. The 5′ region of *tnpA* mRNA processes regulatory RNAs, tnpA-110 and tnpA-90, which represses *invF* mRNA [124]. The SPI1 transcriptional regulator InvF activates the transcription of *sicAsipBC* operon encoding translocon/effector proteins (Fig. 4.1b). This might account for the propagation of this dormant transposon in pathogenic *Salmonella* genomes.

pSLT is the ~95-kb virulence plasmid of *S*. Typhimurium belonging to IncFIB incompatibility group. In addition to FinP, pSLT encodes an sRNA IesR-1, which is strongly induced in a nongrowing dormant state in host cells and acts on PSLT047 gene *in cis* [125]. Transfer of pSLT is regulated at the transcriptional level; ArcAB directly activates the *traY* promoter in response to microaerobiosis [126] and Lrp activates the *traJ* promoter which is methylated by Dam [127, 128]. At the posttranscriptional level, FinO is the *cis*-acting posttranscriptional regulator of *traJ* and is also methylated by Dam [129]. The pSLT conjugation is indirectly regulated by a chromosomal Hfq-dependent sRNA RprA, which is one of the three activator sRNAs for *rpoS* regulation and is induced by host-produced bile salts [130]. By relieving translational inhibitory stem-loop structure of mRNA 5′UTR, RprA activates a conserved *Salmonella*-specific gene STM4242 (designated *ricI*), whose product interacts with the pSLT conjugation apparatus TraV to inhibit pSLT conjugation.

pSLT harbors a fimbriae operon *pefACDEF* which is only expressed when CsrA is available. The plasmid-specified fimbriae are normally repressed by the 5′UTR of chromosomal *fimAICDHF* mRNA encoding the major type I fimbriae, which contains multiple CsrA-binding sites and titrates CsrA [131]. Therefore, CsrA mediates crossregulation at the posttranscriptional level between chromosome- and plasmid-encoded fimbriae.

4.4 Perspectives

As represented by the regulation of *Salmonella* virulence, regulatory RNAs mediate crosstalk between core and foreign genetic elements as well as transcriptional regulators (Fig. 4.1b). While horizontally acquired sRNAs manipulate the core genome probably to optimize the expression level and timing, chromosomal sRNAs often utilize highly conserved seed regions to regulate horizontally acquired genes, which are likely to be integrated into the host regulatory network. Other pathogens such as *E. coli*, *Shigella*, *Yersinia*, *Vibrio*, and *Pseudomonas* are also armed with virulence islands and plasmids and have already been evolved to control their expression via regulatory RNAs [132, 133].

By analogy with the global transcriptional regulator H-NS, which silences expression of foreign genetic elements but is counteracted by plasmid-encoded H-NS homologs [134], there are a few reports of RBPs encoded on mobile genetic elements that are likely to have multiple ligands and modify chromosomal gene expression. The virulence plasmid pXO1 of *Bacillus anthracis* encodes an Hfq homolog [135], whose overexpression exerts toxicity specifically on the strain but not on *Bacillus subtilis* [136]. *L. pneumophila* harbors a CsrA homolog encoded on an integrative conjugative element (ICE), designated as CsrT, in addition to two chromosomal homologs CsrA and CsrR [137]. Ectopic expression of CsrT caused pleiotropic effects, such as reduced transfer efficiency of ICE-βox, resistance to hydrogen peroxide, infection in macrophage, reduced motility, and flagellar assembly.

Hfq has been intensively studied in chromosomal RNA networks, but the host factor for Qβ bacteriophage should directly play an important role in the function of other mobile genetic elements. Hfq is implicated in ColE1 plasmid replication in addition to Rom [138]. Moreover, Hfq regulates F plasmid transfer through destabilization of the *traY* and *traM* mRNAs [139]. It still needs to study in detail how Hfq regulates these crucial events of the two model plasmids.

Bacterial regulatory RNAs encoded on mobile genetic elements need to be transcribed in the new recipient cells, but they might take a shortcut by direct transfer even across the kingdom. Bacterial membrane vesicles (MVs) are known to contain proteins, DNA, and RNA [140]. MVs released from *E. coli*, *Vibrio cholerae*, and *Porphyromonas gingivalis* cells contain a variety of RNAs and can be internalized into human epithelial cells [141–144]. Moreover, a particular tRNA fragment packaged into *Pseudomonas aeruginosa* MVs functions to suppress immune response in human epithelial cells [145]. Further studies are necessary to clarify how the directly transferred RNAs actually function in the new recipient cells.

Acknowledgments This chapter is dedicated to the memory of Dr. Jun-ichi Tomizawa who passed away on 26th Jan. 2017. MM is supported by Tomizawa Jun-ichi and Keiko Fund of Molecular Biology Society of Japan for Young Scientists and MEXT LEADER program. Research in Miyakoshi group is supported by JSPS KAKENHI Grant (JP15H06528 and JP16H06190).

References

1. Jacob F, Monod J. Genetic regulatory mechanisms in the synthesis of proteins. J Mol Biol. 1961;3:318–56.
2. Arraiano CM, Andrade JM, Domingues S, Guinote IB, Malecki M, Matos RG, Moreira RN, Pobre V, Reis FP, Saramago M, et al. The critical role of RNA processing and degradation in the control of gene expression. FEMS Microbiol Rev. 2010;34:883–923.
3. Mohanty BK, Kushner SR. Regulation of mRNA decay in bacteria. Annu Rev Microbiol. 2016;70:25–44.
4. Hui MP, Foley PL, Belasco JG. Messenger RNA degradation in bacterial cells. Annu Rev Genet. 2014;48:537–59.
5. Shine J, Dalgarno L. The 3′-terminal sequence of Escherichia coli 16S ribosomal RNA: complementarity to nonsense triplets and ribosome binding sites. Proc Natl Acad Sci U S A. 1974;71:1342–6.
6. Kohler R, Mooney RA, Mills DJ, Landick R, Cramer P. Architecture of a transcribing-translating expressome. Science. 2017;356:194–7.
7. De Lay N, Schu DJ, Gottesman S. Bacterial small RNA-based negative regulation: Hfq and its accomplices. J Biol Chem. 2013;288:7996–8003.
8. Frohlich KS, Vogel J. Activation of gene expression by small RNA. Curr Opin Microbiol. 2009;12:674–82.
9. Papenfort K, Vanderpool CK. Target activation by regulatory RNAs in bacteria. FEMS Microbiol Rev. 2015;39:362–78.
10. Repoila F, Darfeuille F. Small regulatory non-coding RNAs in bacteria: physiology and mechanistic aspects. Biol Cell. 2009;101:117–31.
11. Brantl S. Regulatory mechanisms employed by cis-encoded antisense RNAs. Curr Opin Microbiol. 2007;10:102–9.
12. Wagner EG, Altuvia S, Romby P. Antisense RNAs in bacteria and their genetic elements. Adv Genet. 2002;46:361–98.
13. Gottesman S, Storz G. Bacterial small RNA regulators: versatile roles and rapidly evolving variations. Cold Spring Harb Perspect Biol. 2011;3:a003798.
14. Vogel J, Luisi BF. Hfq and its constellation of RNA. Nat Rev Microbiol. 2011;9:578–89.
15. Wagner EG, Romby P. Small RNAs in bacteria and archaea: who they are, what they do, and how they do it. *Adv Genet*. 2015;90:133–208.
16. Gimpel M, Brantl S. Dual-function small regulatory RNAs in bacteria. Mol Microbiol. 2017;103:387–97.
17. Vanderpool CK, Balasubramanian D, Lloyd CR. Dual-function RNA regulators in bacteria. Biochimie. 2011;93:1943–9.
18. Bouloc P, Repoila F. Fresh layers of RNA-mediated regulation in Gram-positive bacteria. Curr Opin Microbiol. 2016;30:30–5.
19. De Lay NR, Garsin DA. The unmasking of 'junk' RNA reveals novel sRNAs: from processed RNA fragments to marooned riboswitches. Curr Opin Microbiol. 2016;30:16–21.
20. Miyakoshi M, Chao Y, Vogel J. Regulatory small RNAs from the 3′ regions of bacterial mRNAs. Curr Opin Microbiol. 2015;24:132–9.
21. Tomizawa J, Itoh T, Selzer G, Som T. Inhibition of ColE1 RNA primer formation by a plasmid-specified small RNA. Proc Natl Acad Sci U S A. 1981;78:1421–5.
22. Weaver KE. Emerging plasmid-encoded antisense RNA regulated systems. Curr Opin Microbiol. 2007;10:110–6.
23. Saramago M, Barria C, Arraiano CM, Domingues S. Ribonucleases, antisense RNAs and the control of bacterial plasmids. Plasmid. 2015;78:26–36.
24. Smirnov A, Forstner KU, Holmqvist E, Otto A, Gunster R, Becher D, Reinhardt R, Vogel J. Grad-seq guides the discovery of ProQ as a major small RNA-binding protein. Proc Natl Acad Sci U S A. 2016;113:11591–6.

25. Brantl S, Jahn N. sRNAs in bacterial type I and type III toxin-antitoxin systems. FEMS Microbiol Rev. 2015;39:413–27.
26. Gerdes K, Wagner EG. RNA antitoxins. Curr Opin Microbiol. 2007;10:117–24.
27. Harms A, Brodersen DE, Mitarai N, Gerdes K. Toxins, targets, and triggers: an overview of toxin-antitoxin biology. Mol Cell. 2018;70(5):768–84.
28. Guo Y, Quiroga C, Chen Q, McAnulty MJ, Benedik MJ, Wood TK, Wang X. RalR (a DNase) and RalA (a small RNA) form a type I toxin-antitoxin system in Escherichia coli. Nucleic Acids Res. 2014;42:6448–62.
29. Tomizawa J, Som T. Control of ColE1 plasmid replication: enhancement of binding of RNA I to the primer transcript by the Rom protein. Cell. 1984;38:871–8.
30. Yanisch-Perron C, Vieira J, Messing J. Improved M13 phage cloning vectors and host strains: nucleotide sequences of the M13mp18 and pUC19 vectors. Gene. 1985;33:103–19.
31. Arutyunov D, Frost LS. F conjugation: back to the beginning. Plasmid. 2013;70:18–32.
32. Frost LS, Ippen-Ihler K, Skurray RA. Analysis of the sequence and gene products of the transfer region of the F sex factor. Microbiol Rev. 1994;58:162–210.
33. Penfold SS, Simon J, Frost LS. Regulation of the expression of the traM gene of the F sex factor of Escherichia coli. Mol Microbiol. 1996;20:549–58.
34. Finlay BB, Frost LS, Paranchych W, Willetts NS. Nucleotide sequences of five IncF plasmid finP alleles. J Bacteriol. 1986;167:754–7.
35. Cheah KC, Skurray R. The F plasmid carries an IS3 insertion within finO. J Gen Microbiol. 1986;132:3269–75.
36. Yoshioka Y, Ohtsubo H, Ohtsubo E. Repressor gene finO in plasmids R100 and F: constitutive transfer of plasmid F is caused by insertion of IS3 into F finO. J Bacteriol. 1987;169:619–23.
37. Jerome LJ, van Biesen T, Frost LS. Degradation of FinP antisense RNA from F-like plasmids: the RNA-binding protein, FinO, protects FinP from ribonuclease E. J Mol Biol. 1999;285:1457–73.
38. Lee SH, Frost LS, Paranchych W. FinOP repression of the F plasmid involves extension of the half-life of FinP antisense RNA by FinO. Mol Gen Genet. 1992;235:131–9.
39. Koraimann G, Teferle K, Markolin G, Woger W, Hogenauer G. The FinOP repressor system of plasmid R1: analysis of the antisense RNA control of traJ expression and conjugative DNA transfer. Mol Microbiol. 1996;21:811–21.
40. van Biesen T, Frost LS. The FinO protein of IncF plasmids binds FinP antisense RNA and its target, traJ mRNA, and promotes duplex formation. Mol Microbiol. 1994;14:427–36.
41. van Biesen T, Soderbom F, Wagner EG, Frost LS. Structural and functional analyses of the FinP antisense RNA regulatory system of the F conjugative plasmid. Mol Microbiol. 1993;10:35–43.
42. Franze de Fernandez MT, Eoyang L, August JT. Factor fraction required for the synthesis of bacteriophage Qbeta-RNA. Nature. 1968;219:588–90.
43. Schuppli D, Georgijevic J, Weber H. Synergism of mutations in bacteriophage Qbeta RNA affecting host factor dependence of Qbeta replicase. J Mol Biol. 2000;295:149–54.
44. Chao Y, Vogel J. The role of Hfq in bacterial pathogens. Curr Opin Microbiol. 2010;13:24–33.
45. Sobrero P, Valverde C. The bacterial protein Hfq: much more than a mere RNA-binding factor. Crit Rev Microbiol. 2012;38:276–99.
46. Otaka H, Ishikawa H, Morita T, Aiba H. PolyU tail of rho-independent terminator of bacterial small RNAs is essential for Hfq action. Proc Natl Acad Sci U S A. 2011;108:13059–64.
47. Sauer E, Weichenrieder O. Structural basis for RNA 3′-end recognition by Hfq. Proc Natl Acad Sci U S A. 2011;108:13065–70.
48. Link TM, Valentin-Hansen P, Brennan RG. Structure of Escherichia coli Hfq bound to polyriboadenylate RNA. Proc Natl Acad Sci U S A. 2009;106:19292–7.

49. Mikulecky PJ, Kaw MK, Brescia CC, Takach JC, Sledjeski DD, Feig AL. Escherichia coli Hfq has distinct interaction surfaces for DsrA, rpoS and poly(A) RNAs. Nat Struct Mol Biol. 2004;11:1206–14.
50. Schumacher MA, Pearson RF, Moller T, Valentin-Hansen P, Brennan RG. Structures of the pleiotropic translational regulator Hfq and an Hfq-RNA complex: a bacterial Sm-like protein. EMBO J. 2002;21:3546–56.
51. Sauer E, Schmidt S, Weichenrieder O. Small RNA binding to the lateral surface of Hfq hexamers and structural rearrangements upon mRNA target recognition. Proc Natl Acad Sci U S A. 2012;109:9396–401.
52. Zhang A, Schu DJ, Tjaden BC, Storz G, Gottesman S. Mutations in interaction surfaces differentially impact E. coli Hfq association with small RNAs and their mRNA targets. J Mol Biol. 2013;425:3678–97.
53. Zhang A, Wassarman KM, Rosenow C, Tjaden BC, Storz G, Gottesman S. Global analysis of small RNA and mRNA targets of Hfq. Mol Microbiol. 2003;50:1111–24.
54. Bilusic I, Popitsch N, Rescheneder P, Schroeder R, Lybecker M. Revisiting the coding potential of the E. coli genome through Hfq co-immunoprecipitation. RNA Biol. 2014;11:641–54.
55. Chao Y, Papenfort K, Reinhardt R, Sharma CM, Vogel J. An atlas of Hfq-bound transcripts reveals 3′ UTRs as a genomic reservoir of regulatory small RNAs. EMBO J. 2012;31:4005–19.
56. Sittka A, Lucchini S, Papenfort K, Sharma CM, Rolle K, Binnewies TT, Hinton JC, Vogel J. Deep sequencing analysis of small noncoding RNA and mRNA targets of the global post-transcriptional regulator, Hfq. PLoS Genet. 2008;4:e1000163.
57. Tree JJ, Granneman S, McAteer SP, Tollervey D, Gally DL. Identification of bacteriophage-encoded anti-sRNAs in pathogenic Escherichia coli. Mol Cell. 2014;55:199–213.
58. Holmqvist E, Wright PR, Li L, Bischler T, Barquist L, Reinhardt R, Backofen R, Vogel J. Global RNA recognition patterns of post-transcriptional regulators Hfq and CsrA revealed by UV crosslinking in vivo. EMBO J. 2016;35:991–1011.
59. Gorski SA, Vogel J, Doudna JA. RNA-based recognition and targeting: sowing the seeds of specificity. Nat Rev Mol Cell Biol. 2017;18:215–28.
60. Morita T, Nishino R, Aiba H. Role of terminator hairpin in biogenesis of functional Hfq-binding sRNAs. RNA. 2017;23(9):1419–31.
61. Morita T, Ueda M, Kubo K, Aiba H. Insights into transcription termination of Hfq-binding sRNAs of Escherichia coli and characterization of readthrough products. RNA. 2015;21:1490–501.
62. Updegrove TB, Zhang A, Storz G. Hfq: the flexible RNA matchmaker. Curr Opin Microbiol. 2016;30:133–8.
63. Melamed S, Peer A, Faigenbaum-Romm R, Gatt YE, Reiss N, Bar A, Altuvia Y, Argaman L, Margalit H. Global mapping of small RNA-target interactions in bacteria. Mol Cell. 2016;63:884–97.
64. Lalaouna D, Carrier MC, Semsey S, Brouard JS, Wang J, Wade JT, Masse E. A 3′ external transcribed spacer in a tRNA transcript acts as a sponge for small RNAs to prevent transcriptional noise. Mol Cell. 2015;58:393–405.
65. Miyakoshi M, Chao Y, Vogel J. Cross talk between ABC transporter mRNAs via a target mRNA-derived sponge of the GcvB small RNA. EMBO J. 2015;34:1478–92.
66. Vakulskas CA, Potts AH, Babitzke P, Ahmer BM, Romeo T. Regulation of bacterial virulence by Csr (Rsm) systems. Microbiol Mol Biol Rev. 2015;79:193–224.
67. Yakhnin AV, Baker CS, Vakulskas CA, Yakhnin H, Berezin I, Romeo T, Babitzke P. CsrA activates flhDC expression by protecting flhDC mRNA from RNase E-mediated cleavage. Mol Microbiol. 2013;87:851–66.
68. Figueroa-Bossi N, Schwartz A, Guillemardet B, D'Heygere F, Bossi L, Boudvillain M. RNA remodeling by bacterial global regulator CsrA promotes Rho-dependent transcription termination. Genes Dev. 2014;28:1239–51.
69. Liu MY, Gui G, Wei B, Preston JF 3rd, Oakford L, Yuksel U, Giedroc DP, Romeo T. The RNA molecule CsrB binds to the global regulatory protein CsrA and antagonizes its activity in Escherichia coli. J Biol Chem. 1997;272:17502–10.

70. Weilbacher T, Suzuki K, Dubey AK, Wang X, Gudapaty S, Morozov I, Baker CS, Georgellis D, Babitzke P, Romeo T. A novel sRNA component of the carbon storage regulatory system of Escherichia coli. Mol Microbiol. 2003;48:657–70.
71. Potts AH, Vakulskas CA, Pannuri A, Yakhnin H, Babitzke P, Romeo T. Global role of the bacterial post-transcriptional regulator CsrA revealed by integrated transcriptomics. Nat Commun. 2017;8:1596.
72. Mukherjee S, Yakhnin H, Kysela D, Sokoloski J, Babitzke P, Kearns DB. CsrA-FliW interaction governs flagellin homeostasis and a checkpoint on flagellar morphogenesis in Bacillus subtilis. Mol Microbiol. 2011;82:447–61.
73. Dugar G, Svensson SL, Bischler T, Waldchen S, Reinhardt R, Sauer M, Sharma CM. The CsrA-FliW network controls polar localization of the dual-function flagellin mRNA in Campylobacter jejuni. Nat Commun. 2016;7:11667.
74. Mark Glover JN, Chaulk SG, Edwards RA, Arthur D, Lu J, Frost LS. The FinO family of bacterial RNA chaperones. Plasmid. 2015;78:79–87.
75. Olejniczak M, Storz G. ProQ/FinO-domain proteins: another ubiquitous family of RNA matchmakers? Mol Microbiol. 2017;104:905–15.
76. Milner JL, Wood JM. Insertion proQ220::Tn5 alters regulation of proline porter II, a transporter of proline and glycine betaine in Escherichia coli. J Bacteriol. 1989;171:947–51.
77. Stalmach ME, Grothe S, Wood JM. Two proline porters in Escherichia coli K-12. J Bacteriol. 1983;156:481–6.
78. Kerr CH, Culham DE, Marom D, Wood JM. Salinity-dependent impacts of ProQ, Prc, and Spr deficiencies on Escherichia coli cell structure. J Bacteriol. 2014;196:1286–96.
79. Chaulk SG, Smith Frieday MN, Arthur DC, Culham DE, Edwards RA, Soo P, Frost LS, Keates RA, Glover JN, Wood JM. ProQ is an RNA chaperone that controls ProP levels in Escherichia coli. Biochemistry. 2011;50:3095–106.
80. Sexton JA, Vogel JP. Regulation of hypercompetence in Legionella pneumophila. J Bacteriol. 2004;186:3814–25.
81. Attaiech L, Boughammoura A, Brochier-Armanet C, Allatif O, Peillard-Fiorente F, Edwards RA, Omar AR, MacMillan AM, Glover M, Charpentier X. Silencing of natural transformation by an RNA chaperone and a multitarget small RNA. Proc Natl Acad Sci U S A. 2016;113:8813–8.
82. Smirnov A, Wang C, Drewry LL, Vogel J. Molecular mechanism of mRNA repression in trans by a ProQ-dependent small RNA. EMBO J. 2017;36:1029–45.
83. Holmqvist E, Li L, Bischler T, Barquist L, Vogel J. Global maps of ProQ binding in vivo reveal target recognition via RNA structure and stability control at mRNA 3′ ends. Mol Cell. 2018;70:971–982 e976.
84. Chaulk S, Lu J, Tan K, Arthur DC, Edwards RA, Frost LS, Joachimiak A, Glover JN. N. meningitidis 1681 is a member of the FinO family of RNA chaperones. RNA Biol. 2010;7:812–9.
85. Ghetu AF, Gubbins MJ, Frost LS, Glover JN. Crystal structure of the bacterial conjugation repressor finO. Nat Struct Biol. 2000;7:565–9.
86. Gonzalez GM, Hardwick SW, Maslen SL, Skehel JM, Holmqvist E, Vogel J, Bateman A, Luisi BF, Broadhurst RW. Structure of the Escherichia coli ProQ RNA-binding protein. RNA. 2017;23:696–711.
87. Fabrega A, Vila J. Salmonella enterica serovar Typhimurium skills to succeed in the host: virulence and regulation. Clin Microbiol Rev. 2013;26:308–41.
88. Marcus SL, Brumell JH, Pfeifer CG, Finlay BB. Salmonella pathogenicity islands: big virulence in small packages. Microbes Infect. 2000;2:145–56.
89. Porwollik S, McClelland M. Lateral gene transfer in Salmonella. Microbes Infect. 2003;5:977–89.
90. Kroger C, Colgan A, Srikumar S, Handler K, Sivasankaran SK, Hammarlof DL, Canals R, Grissom JE, Conway T, Hokamp K, et al. An infection-relevant transcriptomic compendium for Salmonella enterica Serovar Typhimurium. Cell Host Microbe. 2013;14:683–95.

91. Srikumar S, Kroger C, Hebrard M, Colgan A, Owen SV, Sivasankaran SK, Cameron AD, Hokamp K, Hinton JC. RNA-seq brings new insights to the intra-macrophage transcriptome of Salmonella Typhimurium. PLoS Pathog. 2015;11:e1005262.
92. Ellermeier JR, Slauch JM. Adaptation to the host environment: regulation of the SPI1 type III secretion system in Salmonella enterica serovar Typhimurium. Curr Opin Microbiol. 2007;10:24–9.
93. Fass E, Groisman EA. Control of Salmonella pathogenicity island-2 gene expression. Curr Opin Microbiol. 2009;12:199–204.
94. Bustamante VH, Martinez LC, Santana FJ, Knodler LA, Steele-Mortimer O, Puente JL. HilD-mediated transcriptional cross-talk between SPI-1 and SPI-2. Proc Natl Acad Sci U S A. 2008;105:14591–6.
95. Lucchini S, Rowley G, Goldberg MD, Hurd D, Harrison M, Hinton JC. H-NS mediates the silencing of laterally acquired genes in bacteria. PLoS Pathog. 2006;2:e81.
96. Navarre WW, Porwollik S, Wang Y, McClelland M, Rosen H, Libby SJ, Fang FC. Selective silencing of foreign DNA with low GC content by the H-NS protein in Salmonella. Science. 2006;313:236–8.
97. Dorman CJ. H-NS, the genome sentinel. Nat Rev Microbiol. 2007;5:157–61.
98. Azam MS, Vanderpool CK. Translational regulation by bacterial small RNAs via an unusual Hfq-dependent mechanism. Nucleic Acids Res. 2018;46:2585–99.
99. Chen J, Gottesman S. Hfq links translation repression to stress-induced mutagenesis in E. coli. Genes Dev. 2017;31(13):1382–95.
100. Desnoyers G, Masse E. Noncanonical repression of translation initiation through small RNA recruitment of the RNA chaperone Hfq. Genes Dev. 2012;26:726–39.
101. Prevost K, Salvail H, Desnoyers G, Jacques JF, Phaneuf E, Masse E. The small RNA RyhB activates the translation of shiA mRNA encoding a permease of shikimate, a compound involved in siderophore synthesis. Mol Microbiol. 2007;64:1260–73.
102. Salvail H, Caron MP, Belanger J, Masse E. Antagonistic functions between the RNA chaperone Hfq and an sRNA regulate sensitivity to the antibiotic colicin. EMBO J. 2013;32:2764–78.
103. Pfeiffer V, Sittka A, Tomer R, Tedin K, Brinkmann V, Vogel J. A small non-coding RNA of the invasion gene island (SPI-1) represses outer membrane protein synthesis from the Salmonella core genome. Mol Microbiol. 2007;66:1174–91.
104. Ipinza F, Collao B, Monsalva D, Bustamante VH, Luraschi R, Alegria-Arcos M, Almonacid DE, Aguayo D, Calderon IL, Gil F, et al. Participation of the Salmonella OmpD porin in the infection of RAW264.7 macrophages and BALB/c mice. PLoS One. 2014;9:e111062.
105. Choi E, Han Y, Cho YJ, Nam D, Lee EJ. A trans-acting leader RNA from a Salmonella virulence gene. Proc Natl Acad Sci U S A. 2017;114:10232–7.
106. Padalon-Brauch G, Hershberg R, Elgrably-Weiss M, Baruch K, Rosenshine I, Margalit H, Altuvia S. Small RNAs encoded within genetic islands of Salmonella typhimurium show host-induced expression and role in virulence. Nucleic Acids Res. 2008;36:1913–27.
107. Hershko-Shalev T, Odenheimer-Bergman A, Elgrably-Weiss M, Ben-Zvi T, Govindarajan S, Seri H, Papenfort K, Vogel J, Altuvia S. Gifsy-1 prophage IsrK with dual function as small and messenger RNA modulates vital bacterial machineries. PLoS Genet. 2016;12:e1005975.
108. Calderon IL, Morales EH, Collao B, Calderon PF, Chahuan CA, Acuna LG, Gil F, Saavedra CP. Role of Salmonella Typhimurium small RNAs RyhB-1 and RyhB-2 in the oxidative stress response. Res Microbiol. 2014;165:30–40.
109. Calderon PF, Morales EH, Acuna LG, Fuentes DN, Gil F, Porwollik S, McClelland M, Saavedra CP, Calderon IL. The small RNA RyhB homologs from Salmonella typhimurium participate in the response to S-nitrosoglutathione-induced stress. Biochem Biophys Res Commun. 2014;450:641–5.
110. Kim JN, Kwon YM. Identification of target transcripts regulated by small RNA RyhB homologs in Salmonella: RyhB-2 regulates motility phenotype. Microbiol Res. 2013;168:621–9.
111. Ortega AD, Gonzalo-Asensio J, Garcia-del Portillo F. Dynamics of Salmonella small RNA expression in non-growing bacteria located inside eukaryotic cells. RNA Biol. 2012;9:469–88.

112. Gong H, Vu GP, Bai Y, Chan E, Wu R, Yang E, Liu F, Lu S. A Salmonella small non-coding RNA facilitates bacterial invasion and intracellular replication by modulating the expression of virulence factors. PLoS Pathog. 2011;7:e1002120.
113. Lee YH, Kim S, Helmann JD, Kim BH, Park YK. RaoN, a small RNA encoded within Salmonella pathogenicity island-11, confers resistance to macrophage-induced stress. Microbiology. 2013;159:1366–78.
114. Smith C, Stringer AM, Mao C, Palumbo MJ, Wade JT. Mapping the regulatory network for Salmonella enterica serovar Typhimurium invasion. MBio. 2016;7:e01024–16.
115. Colgan AM, Kroger C, Diard M, Hardt WD, Puente JL, Sivasankaran SK, Hokamp K, Hinton JC. The Impact of 18 ancestral and horizontally-acquired Regulatory proteins upon the transcriptome and sRNA landscape of Salmonella enterica serovar Typhimurium. PLoS Genet. 2016;12:e1006258.
116. Sharma CM, Darfeuille F, Plantinga TH, Vogel J. A small RNA regulates multiple ABC transporter mRNAs by targeting C/A-rich elements inside and upstream of ribosome-binding sites. Genes Dev. 2007;21:2804–17.
117. Sharma CM, Papenfort K, Pernitzsch SR, Mollenkopf HJ, Hinton JC, Vogel J. Pervasive post-transcriptional control of genes involved in amino acid metabolism by the Hfq-dependent GcvB small RNA. Mol Microbiol. 2011;81:1144–65.
118. Westermann AJ, Forstner KU, Amman F, Barquist L, Chao Y, Schulte LN, Muller L, Reinhardt R, Stadler PF, Vogel J. Dual RNA-seq unveils noncoding RNA functions in host-pathogen interactions. Nature. 2016;529:496–501.
119. Martinez LC, Yakhnin H, Camacho MI, Georgellis D, Babitzke P, Puente JL, Bustamante VH. Integration of a complex regulatory cascade involving the SirA/BarA and Csr global regulatory systems that controls expression of the Salmonella SPI-1 and SPI-2 virulence regulons through HilD. Mol Microbiol. 2011;80:1637–56.
120. Lopez-Garrido J, Puerta-Fernandez E, Casadesus J. A eukaryotic-like 3′ untranslated region in Salmonella enterica hilD mRNA. Nucleic Acids Res. 2014;42:5894–906.
121. El Mouali Y, Gaviria-Cantin T, Sanchez-Romero MA, Gibert M, Westermann AJ, Vogel J, Balsalobre C. CRP-cAMP mediates silencing of Salmonella virulence at the post-transcriptional level. PLoS Genet. 2018;14:e1007401.
122. Papenfort K, Podkaminski D, Hinton JC, Vogel J. The ancestral SgrS RNA discriminates horizontally acquired Salmonella mRNAs through a single G-U wobble pair. Proc Natl Acad Sci U S A. 2012;109:E757–64.
123. Ellis MJ, Trussler RS, Haniford DB. A cis-encoded sRNA, Hfq and mRNA secondary structure act independently to suppress IS200 transposition. Nucleic Acids Res. 2015;43:6511–27.
124. Ellis MJ, Trussler RS, Charles O, Haniford DB. A transposon-derived small RNA regulates gene expression in Salmonella Typhimurium. Nucleic Acids Res. 2017;45:5470–86.
125. Gonzalo-Asensio J, Ortega AD, Rico-Perez G, Pucciarelli MG, Garcia-Del Portillo F. A novel antisense RNA from the Salmonella virulence plasmid pSLT expressed by non-growing bacteria inside eukaryotic cells. PLoS One. 2013;8:e77939.
126. Serna A, Espinosa E, Camacho EM, Casadesus J. Regulation of bacterial conjugation in microaerobiosis by host-encoded functions ArcAB and sdhABCD. Genetics. 2010;184:947–58.
127. Camacho EM, Casadesus J. Conjugal transfer of the virulence plasmid of Salmonella enterica is regulated by the leucine-responsive regulatory protein and DNA adenine methylation. Mol Microbiol. 2002;44:1589–98.
128. Camacho EM, Casadesus J. Regulation of traJ transcription in the Salmonella virulence plasmid by strand-specific DNA adenine hemimethylation. Mol Microbiol. 2005;57:1700–18.
129. Camacho EM, Serna A, Casadesus J. Regulation of conjugal transfer by Lrp and Dam methylation in plasmid R100. Int Microbiol. 2005;8:279–85.
130. Papenfort K, Espinosa E, Casadesus J, Vogel J. Small RNA-based feedforward loop with AND-gate logic regulates extrachromosomal DNA transfer in Salmonella. Proc Natl Acad Sci U S A. 2015;112:E4772–81.

131. Sterzenbach T, Nguyen KT, Nuccio SP, Winter MG, Vakulskas CA, Clegg S, Romeo T, Baumler AJ. A novel CsrA titration mechanism regulates fimbrial gene expression in Salmonella typhimurium. EMBO J. 2013;32:2872–83.
132. Frohlich KS, Papenfort K. Interplay of regulatory RNAs and mobile genetic elements in enteric pathogens. Mol Microbiol. 2016;101:701–13.
133. Papenfort K, Corcoran CP, Gupta SK, Miyakoshi M, Heidrich N, Chao Y, Fröhlich KS, Ziebuhr W, Böhm A, Vogel J. Regulatory mechanisms of special significance: role of small RNAs in virulence regulation. In: Regulation of bacterial virulence. Washington, DC: American Society of Microbiology; 2013. p. 493–527.
134. Doyle M, Fookes M, Ivens A, Mangan MW, Wain J, Dorman CJ. An H-NS-like stealth protein aids horizontal DNA transmission in bacteria. Science. 2007;315:251–2.
135. Vrentas C, Ghirlando R, Keefer A, Hu Z, Tomczak A, Gittis AG, Murthi A, Garboczi DN, Gottesman S, Leppla SH. Hfqs in Bacillus anthracis: Role of protein sequence variation in the structure and function of proteins in the Hfq family. Protein Sci. 2015;24:1808–19.
136. Keefer AB, Asare ĒK, Pomerantsev AP, Moayeri M, Martens C, Porcella SF, Gottesman S, Leppla SH, Vrentas CE. In vivo characterization of an Hfq protein encoded by the Bacillus anthracis virulence plasmid pXO1. BMC Microbiol. 2017;17:63.
137. Abbott ZD, Flynn KJ, Byrne BG, Mukherjee S, Kearns DB, Swanson MS. csrT represents a new class of csrA-like regulatory genes associated with integrative conjugative elements of Legionella pneumophila. J Bacteriol. 2016;198:553–64.
138. Cech GM, Pakula B, Kamrowska D, Wegrzyn G, Arluison V, Szalewska-Palasz A. Hfq protein deficiency in Escherichia coli affects ColE1-like but not lambda plasmid DNA replication. Plasmid. 2014;73:10–5.
139. Will WR, Frost LS. Hfq is a regulator of F-plasmid TraJ and TraM synthesis in Escherichia coli. J Bacteriol. 2006;188:124–31.
140. Domingues S, Nielsen KM. Membrane vesicles and horizontal gene transfer in prokaryotes. Curr Opin Microbiol. 2017;38:16–21.
141. Blenkiron C, Simonov D, Muthukaruppan A, Tsai P, Dauros P, Green S, Hong J, Print CG, Swift S, Phillips AR. Uropathogenic Escherichia coli releases extracellular vesicles that are associated with RNA. PLoS One. 2016;11:e0160440.
142. Ghosal A, Upadhyaya BB, Fritz JV, Heintz-Buschart A, Desai MS, Yusuf D, Huang D, Baumuratov A, Wang K, Galas D, et al. The extracellular RNA complement of Escherichia coli. Microbiology. 2015;*4*(2):252–66.
143. Ho MH, Chen CH, Goodwin JS, Wang BY, Xie H. Functional advantages of Porphyromonas gingivalis vesicles. PLoS One. 2015;10:e0123448.
144. Sjostrom AE, Sandblad L, Uhlin BE, Wai SN. Membrane vesicle-mediated release of bacterial RNA. Sci Rep. 2015;5:15329.
145. Koeppen K, Hampton TH, Jarek M, Scharfe M, Gerber SA, Mielcarz DW, Demers EG, Dolben EL, Hammond JH, Hogan DA, et al. A novel mechanism of host-pathogen interaction through sRNA in bacterial outer membrane vesicles. PLoS Pathog. 2016;12:e1005672.

Bacteria–Virus Interactions

5

Takashi Yoshida, Daichi Morimoto, and Shigeko Kimura

Abstract

Viruses that infect bacteria (bacteriophages) are ubiquitous and abundant in the environment. In marine ecosystems, viruses may infect as much as 20–40% of the bacterial population. Therefore, viruses play important roles in global biogeochemical cycle. Viruses also affect the genome evolution and fitness of their host microorganisms through horizontal gene transfer. On the other hand, viruses may also acquire metabolic genes from their hosts and temporally manipulate host metabolism during infection through the expression of specialized "auxiliary metabolic genes." In addition, host-selective viral infection is believed to maintain bacterial diversity via frequency-dependent selection, whereby virus infection reduces bacterial species that have become relatively abundant in a population and thereby allows coexistence of multiple host–virus combinations. Further, continual bacteria–virus interactions drive reciprocal evolution of bacterial defense and viral counterdefense mechanisms (coevolution), resulting in increased diversity. Thus, examination of bacteria–virus interactions is essential for understanding microbial evolution and diversity.

Keywords

Lytic virus · Temperate virus · Provirus · Auxiliary metabolic gene · Kill-the-winner · Coevolution

T. Yoshida (✉) · D. Morimoto · S. Kimura
Marine Microbiology, Graduate School of Agriculture, Kyoto University, Kyoto, Japan
e-mail: yoshiten@kais.kyoto-u.ac.jp; dai@kais.kyoto-u.ac.jp; shigeko@kais.kyoto-u.ac.jp

H. Nishida, T. Oshima (eds.), *DNA Traffic in the Environment*,
https://doi.org/10.1007/978-981-13-3411-5_5

5.1 Introduction

Viruses that infect microorganisms are ubiquitous and abundant in almost all environments [1]. They inject their genomes into their host microorganisms for their propagation, ultimately resulting in host cell lysis (lytic cycle). Therefore, viruses play important roles in the movement of nutrients and organic carbon from host cells to dissolved and particulate organic matter pools ("viral shunt"). Thus, virus-mediated microbial lysis is widely recognized as affecting marine biogeochemical cycles [1–3]. Although bacteriophage (or phage) is the name commonly used to describe viruses that infect and propagate within bacteria (and sometimes includes archaeal tailed viruses), we will refer to bacteriophages (or phages) as viruses in this chapter according to their description by the International Committee on Taxonomy of Viruses (ICTV) of the Bacterial and Archaeal Viruses Subcommittee [4].

Viruses affect the genome evolution and fitness of their host microorganisms through horizontal gene transfer (HGT). They may also acquire metabolic genes from their hosts and can temporally manipulate host metabolism during infection through the expression of specialized viral genes called "auxiliary metabolic genes" (AMGs) [5–7]. In addition to a lytic cycle, temperate viruses undergo a lysogenic cycle that results in vertical propagation within host lineages through integration into the bacterial genome as proviruses or plasmids. These proviruses often cause phenotypic changes in their hosts (lysogenic conversion) [8, 9] and transfer a neighboring segment of the host genome to other hosts during the excision process (specialized transduction; process 4 in Fig. 5.1) [9–12]. Partial host genome segments can also be incorrectly packaged into the capsid and subsequently transferred to other hosts (generalized transduction; process 1 in Fig. 5.1) [9, 13].

In addition to mediating HGT, viruses affect the diversity and dynamics of host populations and communities. Selective viral infection prevents exponential increases in dominant microbial species (populations), which enables the coexistence of multiple competing species ("kill-the-winner" model) [14, 15]. At the same time, continual host–virus interactions promote the emergence of host antiviral defense systems, as well as counterdefense systems in viruses (coevolution), which also generates genetic diversity [16–18].

Thus, studying bacteria–virus interactions is essential for understanding microbial evolution and diversity. Here, we review the effects of viruses on microbial evolution and diversity and examine coevolutionary dynamics arising through bacteria–virus interactions.

5.2 Temperate Virus–Host Interactions

During the lysogenic cycle of temperate viruses (proviruses), most viral genes are repressed to maintain normal cell viability [19]. Proviruses and their hosts have a shared evolutionary history because viral DNA replication is completely synchronized with host DNA replication, which results in increases in both provirus and host populations [20]. Accordingly, proviruses protect the host from various

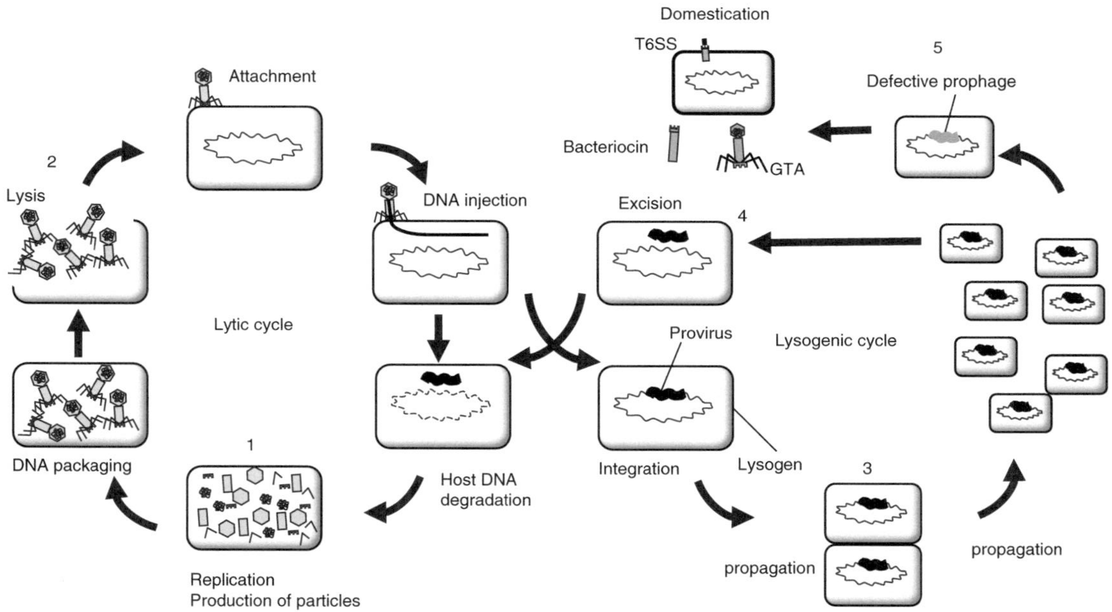

Fig. 5.1 Schematic illustration of viral and bacterial evolution. Lytic viruses only undergo a lytic cycle, while temperate viruses undergo either lytic or lysogenic cycles. Viruses affect the genetic diversity of both their host and themselves through the processes shown in steps 1–5. (*1*) General transduction. Partial host genome segments can be incorrectly packaged into the capsid and transferred to other hosts. (*2*) Release of plasmid. Some lytic viruses can increase transformation efficiency of plasmids from host cells after viral lysis. (*3*) Lysogenic conversion. Proviruses often cause phenotypic changes of their hosts. (*4*) Specialized transduction. Proviruses can transfer a neighboring segment of the host genome to other hosts during the excision process. (*5*) Domestication of viruses. Proviruses are rapidly inactivated because intact proviruses are likely to kill the host in the lytic cycle

opponents to ensure the lysogenic state, often by inducing phenotypic changes in their hosts in a process known as lysogenic conversion [19, 21] (process 3 in Fig. 5.1). Of these, changes that protect against "superinfection," during which infected cells are also invaded by other viruses, are most intensively studied. This "superinfection exclusion" can be achieved via a wide variety of mechanisms [22–26]. Most cases involve alterations to the cell surface or other cell envelope components. For example, provirus-encoded proteins block the adsorption of superinfecting viruses to the cell surface, which is the first step in the infection cycle [25, 26]. Similarly, the viral genome injection step is a common target for provirus-mediated superinfection exclusion [22–24]. A number of provirus-encoded proteins also affect bacterial pathogenicity. Among these, outer membrane protein Lom and lipoprotein Bor increase the ability of lysogens to adhere to human epithelial cells or improve the survival rate of the lysogens in animal serum [27–29]. In *Neisseria meningitidis*, provirus-encoded protein TspB, which is present on the bacterial cell surface and binds human IgG, leads to the formation of large host cell aggregates in a biofilm. Thus, proviruses can protect bacteria from host immune responses [30].

Proviruses are also a source of virulence genes and many pathogenic bacteria depend on provirus-encoded proteins for their toxin production. Virus-borne toxin genes (one gene in some cases) can turn harmless bacteria into dangerous pathogens and facilitate the emergence of novel pathogens [31–33]. Famously, *Escherichia coli* O157:H7 acquired the ability to produce a Shiga-like toxin after being lysogenized by several viruses [32]. Similarly, *Vibrio cholerae* was converted from a harmless water-dwelling bacterium into a significant human pathogen upon the acquisition of virus VPIΦ, which encodes the toxin co-regulated pilus [31, 33].

Temperate viruses are also known to encode accessory genes that increase host bacterial fitness under certain conditions. For example, P1, P2, or Mu lysogenic *E. coli* strains can proliferate more rapidly than nonlysogenic strains during aerobic growth under glucose-limited conditions. The increased fitness of the lysogens correlates with higher metabolic activity compared with nonlysogens under glucose exhaustion [34]. Thus, these proviruses confer an evolutionarily significant reproductive growth advantage to their host *E. coli* strains [34]. In *Shewanella oneidensis* MR-1, which is capable of forming highly structured surface-attached communities, extracellular DNA serves as a structural component in all stages of biofilm formation under static and hydrodynamic conditions [35]. Extracellular DNA, like many other crucial biofilm-promoting factors, is released during provirus LambdaSo-, MuSo1-, and MuSo2-mediated cell lysis [35]. Proviruses also assist with antibiotic tolerance. *E. coli* K-12 contains nine cryptic proviruses, CP4-6, DLP12, e14, rac, Qin, CP4-44, CPS-53, CPZ-55, and CP4-57, which have lost the ability to produce infective phage particles [36]. These cryptic proviruses contribute to increased resistance to sublethal concentrations of quinolone and β-lactam antibiotics, primarily through the production of proteins that inhibit cell division [36]. In addition, these proviruses help their host to withstand osmotic, oxidative, and acid stresses, as well as increasing growth and influencing biofilm formation [36]. Similarly, a *Listeria monocytogenes* provirus provides its host with regulatory switches for escaping from macrophage phagosomes. In the DNA uptake competence (Com) system of *L. monocytogenes*, the Com master activator gene, *comK*, is interrupted by a provirus [37]. The Com system is regulated by the formation of a functional *comK* gene via provirus excision. The provirus is specifically excised during intracellular growth, primarily within phagosomes, yet, in contrast to this classical induction, progeny virions are not produced. Thus, the provirus serves as a genetic switch to modulate the virulence of its bacterial host during the course of infection [37].

The induction of proviruses can counteract or delay the colonization of nonlysogens. Cocultures of *Salmonella* strains with and without proviruses undergo rapid bacterial composition changes as a result of virus-mediated killing of sensitive bacteria and lysogenic conversion of survivors [38]. Thus, spontaneous provirus induction in a few lysogenic cells enhances the competitive fitness of the lysogenic population as a whole, setting up a selection regime that forces maintenance and spread of viral DNA [38].

Although proviruses possess various mechanisms for superinfection exclusion, as described above, the chimeric nature of viruses may suggest that superinfections have rarely occurred in lysogens, and that proviruses also serve as a genetic pool for viruses. For example, the genome of *Staphylococcus pasteuri* virus SpaA1 consists of three genomic regions derived from apparently different origins that form a mosaic pattern. The three genomic regions consist of partial provirus sequences from *Bacillus thuringiensis kurstaki str. T03a001* (genes involved in DNA replication/transcription, and cell entry and exit), *Bacillus cereus* AH676 (additional regulatory and recombination genes), and the majority of the MZTP02 virus genome [39]. One possible mechanism for the mosaicism of viral genomes is virus-encoded homologous recombination enzymes (recombinases). A previous study showed that compared with RecA, viral Rad52-like recombinases have relaxed fidelity for nucleotide recognition sites in vivo, leading to gene exchanges among proviruses [40].

5.3 Domestication of Proviruses by Hosts

A study by Bobay et al. pointed out the strong selection for mutations leading to provirus inactivation because intact proviruses are likely to kill the host upon induction of the lytic cycle [20]. Comparative genomics using *E. coli* and *Salmonella enterica* proviruses revealed that the size distribution of provirus elements is bimodal, with small elements, which are rapidly inactivated proviruses ("domestication"; process 5 in Fig. 5.1), and large elements, which are intact proviruses [20, 41]. Accordingly, a pervasive pattern of systematic counterselection for nonsynonymous mutations in provirus genes is observed. Importantly, such patterns of purifying selection are observed not only in accessory regions but also in core viral genes, such as those encoding structural and lysis components [20]. This suggests that bacterial hosts utilize these components as functional units [20]. In this context, partly degraded prophage elements are assumed to have adaptive functions in the host, as outlined below.

R-type and F-type bacteriocins are typically composed of domesticated tail and lysis genes from myoviruses and siphoviruses, respectively, and kill surrounding bacterial cells lacking the appropriate cognate immunity genes [20]. Notably, the largest orthologous provirus family, which is related to P2-like viruses (Myoviridae), seems to have been stabilized in many *Salmonella* strains because they lack integrases for viral DNA packaging. The putative domestication of this provirus might even predate the split between *Escherichia* and *Salmonella*, because a very similar small *Salmonella* provirus also lacks an integrase at the same position in two *E. coli* strains. Although most provirus families do not so easily fit the description of R-type or F-type bacteriocins, they do act as killer virus particles. Thus, defective viruses can easily give rise to killer virus particles [20]. Similarly, *S. enterica* serovar Typhimurium produces pore-forming bacteriocins, known as group A and B colicins, which are released in the gut environment to kill its competitors. Typically, colicins are co-expressed with cognate lysis genes in only a fraction of cells, which results in bacterial lysis and colicin release into the environment [42].

Gene transfer agents (GTAs) are also derived from defective viruses [43]. GTAs contain a random piece of the host cell genome but contain insufficient DNA to encode the protein components required to generate viral particles themselves. Therefore, GTAs cannot transfer a complete set of structural genes to a recipient cell. It is important to note that these agents are distinct from generalized transducing viruses [43]. All known GTAs have tailed-virus structures and are presumably released into the environment by lysis of the producing cell. Released GTA particles may transfer DNA from the producing cell to a recipient cell. In contrast, transducing viral particles are produced during the replication of a virus inside a host cell, and only the occasional viral particle contains host DNA in place of viral DNA. The amount of host DNA in the transducing particles is equivalent to the normal viral genome size [43].

Type VI secretion systems (T6SSs) are the most widespread of all bacterial secretion systems [44]. T6SSs are structurally related to the cell-puncturing device of tailed viruses and are predicted to function as contractile injection machineries that perforate eukaryotic and prokaryotic target membranes for effector delivery. The activity of T6SSs can play an important role in virulence by modifying the eukaryotic host cytoskeleton through actin crosslinking [45]. As such, they are considered efficient weaponry in interbacterial warfare and provide a fitness advantage by hydrolyzing the cell walls of opponent bacteria. Therefore, T6SSs might enable pathogens to outcompete commensal bacteria and aid in host colonization [45]. Indeed, in *V. cholerae*, protein secretion by the T6SS requires the action of a dynamic intracellular tubular component that is structurally and functionally homologous to contractile viral tail sheath protein [46].

5.4 Lytic Virus–Host Interactions

To date, most of the cultured marine microbial viruses display lytic traits, both physiologically and genetically. These viruses often contain multiple "host-like genes," which have more recently been termed "auxiliary metabolic genes" (AMGs)

[5, 6]. AMGs involved in photosynthesis and the pentose phosphate pathway were discovered in cyanobacterial viruses over a decade ago [47–50], while recent large-scale marine metagenomic studies have revealed a wide variety of AMGs involved in many different cellular functions, including amino acid and carbohydrate metabolism, energy production, and iron–sulfur cluster assembly and modification, in addition to photosynthesis [6, 51, 52]. AMGs can largely be divided into two classes based on their function: (1) Class I AMGs, which encode proteins whose metabolic functions are assigned in the pathway database of the Kyoto Encyclopedia of Genes and Genomes database (KEGG PATHWAY), and (2) Class II AMGs, which encode proteins that are not listed in KEGG PATHWAY but have a supplemental role in metabolism through membrane transport or iron-cluster assembly [6, 52, 53]. Genes involved in DNA replication, repair, and nucleotide synthesis are commonly present in viral genomes but are not considered AMGs [54].

AMGs are acquired from the bacterial host through partial transduction events and play critical roles in augmenting biochemical processes related to viral propagation in host metabolisms [5, 6]. For example, photosynthetic organisms possess multiple mechanisms to avoid damage to photosynthetic machinery from excessive light energy (photoinhibition) [55]. D1, encoded by *psbA*, is a subunit of a heterodimer in the reaction center of photosystem II (PSII) [56]. Damaged D1 is rapidly removed and replaced with newly synthesized protein, thereby preventing photoinhibition [57]. *psbA* is an AMG in marine cyanobacterial viruses and is expressed during infection because photosynthetic activity is sustained even after the decline of the host's photosynthesis gene expression [58]. In contrast, freshwater cyanobacterial viruses often possess an AMG-like *nblA* gene, encoding NblA, which plays a central role in the degradation of phycobilisomes, the major light-harvesting complexes of PSII in cyanobacteria [7, 59, 60]. Cyanobacteria that inhabit the surface of a freshwater lake might benefit from phycobilisome degradation rather than PSII repair; thus, the presence of viruses carrying these AMGs would increase bacterial fitness [7]. A metagenomics analysis also revealed that viruses in different environments harbor niche-specific AMGs (e.g., photic zone AMGs for iron–sulfur cluster modulation for virus production, and aphotic zone AMGs for high-pressure deep-sea survival) [54].

Another study showed that increases in the NADPH/NADP ratio in marine *Synechococcus* species are accompanied during viral infection by a decrease in Calvin cycle activity and increases in pentose phosphate pathway and light reaction activity [61]. In this way, cyanobacterial viruses maintain host photosynthesis activity and redirect carbon flux from the Calvin cycle to the pentose phosphate pathway using AMGs. This relationship is another example of the ecological importance of virus–host interactions. As described above, up to 40% of microorganisms are lysed by viral infection every day in marine environments, thereby contributing to the marine geochemical cycle [1]. However, marine viruses may further contribute to this cycle via AMG-mediated manipulation of host metabolism during infection [5].

Marine viruses can also serve as a potential genetic pool while shaping the evolution of their hosts. For example, phylogenetic analyses provide evidence that viruses drive the exchange and reshuffling of *psbA* genes between *Synechococcus* and *Prochlorococcus* strains [62, 63]. In addition, a recent study shows that lytic viruses can promote HGT via plasmid transformation [64] (process 2 in Fig. 5.1). Prokaryotes often carry plasmids that confer physiological benefits (e.g., antibiotic resistance) (see Chaps. 1 and 6). Two lytic *E. coli* viruses, referred to as "superspreaders," can increase the transformation efficiency of evolutionally distinct plasmids from host cells after viral lysis. These viruses lack hydrolytic endonucleases, which would usually be involved in the degradation of host chromosomal DNA during infection, and are therefore hypothesized to use this simple mechanism for the superspreader phenotype [64].

5.5 Ecological Advantages of Natural Host–Virus Interactions

We have so far described various HGT events mediated by lysogenic and lytic viruses that are advantageous to their bacterial hosts (Processes 1–5 in Fig. 5.1). Finally, we will discuss the contribution of host–virus interactions to the ecological dynamics of both viral and host populations and the advantages of these interactions. One ecological model for the population dynamics of bacteria–virus interactions is "kill-the-winner" dynamics, which is analogous to the classical Lotka–Volterra model used to explain predator–prey population dynamics [14, 15]. In this model, viruses infect host species that have become relatively abundant (the winner) as a result of differences in substrate affinity between the coexisting bacterial species, which results in the maintenance of competing bacterial species by oscillation of bacterial host species over time [14].

Meanwhile, metagenomic approaches have shown that several genomic regions (identified by sequence alignment analysis) will be underrepresented in all of the genomes of one bacterial species represented in a metagenome [18, 65–70]. These regions are referred to as metagenomic islands (MGIs) [69]. These MGI regions can show diversity even within a single bacterial species (closely related lineages), leading to the question of whether this diversity is important for the ecology and environmental adaptation of the closely related lineages. The concept of a pan-genome is used to describe increasing gene pool diversity ascribed to one bacterial species as the number of sequenced strains increases [69, 71]. Pan-genomes include shared genes, called core genes, and noncore genes, called accessory genes, among strains. Diversification of MGIs indicates increases in the number of accessory genes in the pan-genome. To date, several studies indicate that genes encoding the O-chain of lipopolysaccharide [72], exopolysaccharide biosynthesis pathway proteins, proteins involved in sugar modification of extracellular structures, pili, and flagellar components [18], and giant proteins [73] that are possibly MGIs encoding extracellular proteins are potential viral recognition sites [69]. Additionally, restriction-modification and CRISPR systems that are involved in viral resistance are also found in MGIs [74, 75] (see Chap. 3). These findings show that genes found in MGIs play a role in avoidance of viral infection, suggesting that viral selection significantly contributes to the diversification of MGIs through the extension of pan-genomes. Also, these similar genomic arrangements were found between closely related bacterial strains like staphylococcal pathogenicity islands, containing superantigen genes, and other mobile elements [76].

Continual interactions between bacteria and viruses drive the evolution of bacterial defense mechanisms as well as viral counterdefense mechanisms, thereby generating diversity in both (coevolution) [17]. Based on the diversity of MGIs among closely related lineages and metagenomes, the constant-diversity (CD) dynamics model was proposed to explain microbial diversity among closely related microbial lineages in similar environments [18]. In the CD dynamics model, bacteria compete with each other and lytic viruses have an equal chance of infecting any cell within a population of a single prokaryotic species in an idealized natural environment with dissolved organic nutrients. A large amount of diversity in host viral sensitivity types is required to avoid catastrophic lysis by each virus. Host receptor or sensitivity types are each recognized by a different virus type. The number of viruses that target the receptor of fitter host lineages would increase, meaning that the targeted lineage would eventually be replaced by other host lineages. In this way, a constant high level of diversity among closely related lineages would be maintained [18]. For example, we analyzed the leader-end of the subtype I-D CRISPR system in *Microcystis* isolates [77, 78] along with the tail sheath gene (*g91*) of a *Microcystis* Ma-LMM01-type virus [79] in natural populations and found that the generation and maintenance of diversity in both the *Microcystis* and viral populations is partly driven by CD dynamics.

In addition to the models of bacteria–virus interactions based on lytic viruses described above, a model has recently been proposed to describe the contribution of temperate viruses to population dynamics [80]. Basically, bacterial viruses are believed to control host abundance in a frequency-dependent manner, as described above, and viral abundance is typically tenfold higher than that of prokaryotes [80]. However, two viral metagenomics studies have shown that viral particles are relatively less abundant at high microbial densities ($>10^6$ colony-forming units/mL) in the environment [80, 81]. However, a coral reef viral metagenomics analysis showed that the relative frequency of genes corresponding to temperate viral genomes increased with microbial density [81]. These observations can be explained by a theory known as the Piggyback-the-Winner model [81], which assumes that lysogeny is favored at high cell densities because lysogenized viruses can replicate quickly in a fast-growing hosts. Meanwhile, this model appears to be contradicted by laboratory experiments that show that lysogeny is often favored under poor growth conditions. However, a mechanism called the "arbitrium system" was recently described for the coordination of lysis-lysogeny decisions in SPβ-group *Bacillus* viruses [82]. These viruses produce a six-residue peptide during infection of the host cell and then release the peptide into the surrounding environment. During subsequent infections, progeny viruses measure the concentration of this peptide and lysogenize if the concentration is sufficiently high [82].

It is possible that many viral-mediated HGT events are yet to be uncovered. Most of the known HGT events have been observed in cultured viral–host systems. Therefore, metagenomic approaches will be a useful for the detection of HGT events in natural populations. At the same time, theoretical models describing viral–host interactions (e.g., a merged model of CD dynamics and the Piggyback-the-Winner model) are needed to understand where and when virus-mediated HGT events occur at high frequencies and how the results of these HGT events contribute to the evolution of viruses and their hosts.

References

1. Suttle CA. Marine viruses – major players in the global ecosystem. Nat Rev Microbiol. 2007;5:801–12. https://doi.org/10.1038/nrmicro1750.
2. Fuhrman JA. Marine viruses and their biogeochemical and ecological effects. Nature. 1999;399:541–8. https://doi.org/10.1038/21119.
3. Suttle CA. Viruses in the sea. Nature. 2005;437:356–61. https://doi.org/10.1038/nature04160.
4. Krupovic M, Dutilh BE, Adriaenssens EM, Wittmann J, Vogensen FK, Sullivan MB, et al. Taxonomy of prokaryotic viruses: update from the ICTV bacterial and archaeal viruses subcommittee. Arch Virol. 2016;161:1095–9. https://doi.org/10.1007/s00705-015-2728-0.
5. Breitbart M, Thompson L, Suttle C, Sullivan M. Exploring the vast diversity of marine viruses. Oceanography. 2007;20:135–9. https://doi.org/10.5670/oceanog.2007.58.
6. Hurwitz BL, U'Ren JM. Viral metabolic reprogramming in marine ecosystems. Curr Opin Microbiol. 2016;31:161–8. https://doi.org/10.1016/j.mib.2016.04.002.
7. Yoshida T, Nagasaki K, Takashima Y, Shirai Y, Tomaru Y, Takao Y, et al. Ma-LMM01 infecting toxic *Microcystis aeruginosa* illuminates diverse cyanophage genome strategies. J Bacteriol. 2008;190:1762–72. https://doi.org/10.1128/JB.01534-07.
8. Canchaya C, Proux C, Fournous G, Bruttin A, Brussow H. Prophage genomics. Microbiol Mol Biol Rev. 2003;67:238–76. https://doi.org/10.1128/MMBR.67.2.238-276.2003.
9. Touchon M, Moura de Sousa JA, Rocha EP. Embracing the enemy: the diversification of microbial gene repertoires by phage-mediated horizontal gene transfer. Curr Opin Microbiol. 2017;38:66–73. https://doi.org/10.1016/j.mib.2017.04.010.
10. Campos J, Martínez E, Marrero K, Silva Y, Rodríguez BL, Suzarte E, et al. Novel type of specialized transduction for CTXφ or its satellite phage RS1 mediated by filamentous phage VGJφ in *Vibrio cholerae*. J Bacteriol. 2003;185:7231–40. https://doi.org/10.1128/JB.185.24.7231-7240.2003.
11. Fernandes RM, de Lencastre H, Archer LJ. Specialized transduction in *Bacillus subtilis* by the phages IG1, IG3, and IG4, Arch Virol. 1989;105:137–40. Available at: http://www.ncbi.nlm.nih.gov/pubmed/2497716.
12. Gottesman ME, Yarmolinsky MB. Integration-negative mutants of bacteriophage lambda. J Mol Biol. 1968;31:487–505. https://doi.org/10.1016/0022-2836(68)90423-3.
13. Thierauf A, Perez G, Maloy S. Generalized transduction. Methods Mol Biol. 2009;501:267–86. https://doi.org/10.1007/978-1-60327-164-6_23.
14. Thingstad TF. Elements of a theory for the mechanisms controlling abundance, diversity, and biogeochemical role of lytic bacterial viruses in aquatic systems. Limnol Oceanogr. 2000;45:1320–8. https://doi.org/10.4319/lo.2000.45.6.1320.
15. Thingstad TF, Lignell R. Theoretical models for the control of bacterial growth rate, abundance, diversity and carbon demand. Aquat Microb Ecol. 1997;13:19–27. https://doi.org/10.3354/ame013019.
16. Kimura S, Uehara M, Morimoto D, Yamanaka M, Sako Y, Yoshida T. Incomplete selective sweeps of *Microcystis* population detected by the leader-end CRISPR fragment analysis in a natural pond. Front Microbiol. 2018;9:425. https://doi.org/10.3389/fmicb.2018.00425.
17. Koonin EV, Makarova KS, Wolf YI. Evolutionary genomics of defense systems in archaea and bacteria. Annu Rev Microbiol. 2017;71:233–61. https://doi.org/10.1146/annurev-micro-090816-093830.

18. Rodriguez-Valera F, Martin-Cuadrado AB, Rodriguez-Brito B, Pašić L, Thingstad TF, Rohwer F, et al. Explaining microbial population genomics through phage predation. Nat Rev Microbiol. 2009;7:828–36. https://doi.org/10.1038/nrmicro2235.
19. Bondy-Denomy J, Davidson AR. When a virus is not a parasite: the beneficial effects of prophages on bacterial fitness. J Microbiol. 2014;52:235–42. https://doi.org/10.1007/s12275-014-4083-3.
20. Bobay L-M, Touchon M, Rocha EPC. Pervasive domestication of defective prophages by bacteria. Proc Natl Acad Sci. 2014;111:12127–32. https://doi.org/10.1073/pnas.1405336111.
21. Obeng N, Pratama AA, van Elsas JD. The significance of mutualistic phages for bacterial ecology and evolution. Trends Microbiol. 2016;24:440–9. https://doi.org/10.1016/j.tim.2015.12.009.
22. Cumby N, Edwards AM, Davidson AR, Maxwell KL. The bacteriophage HK97 gp15 moron element encodes a novel superinfection exclusion protein. J Bacteriol. 2012;194:5012–9. https://doi.org/10.1128/JB.00843-12.
23. Hofer B, Ruge M, Dreiseikelmann B. The superinfection exclusion gene (sieA) of bacteriophage P22: identification and overexpression of the gene and localization of the gene product. J Bacteriol. 1995;177:3080–6. https://doi.org/10.1128/jb.177.11.3080-3086.1995.
24. Nesper J, Blaß J, Fountoulakis M, Reidl J. Characterization of the major control region of *Vibrio* cholerae bacteriophage K139: immunity, exclusion, and integration. J Bacteriol. 1999;181:2902–13. Available at: http://www.ncbi.nlm.nih.gov/pubmed/10217785.
25. Newton GJ, Daniels C, Burrows LL, Kropinski AM, Clarke AJ, Lam JS. Three-component-mediated serotype conversion in *Pseudomonas aeruginosa* by bacteriophage D3. Mol Microbiol. 2001;39:1237–47. https://doi.org/10.1046/j.1365-2958.2001.02311.x.
26. Vostrov AA, Vostrukhina OA, Svarchevsky AN, Rybchin VN. Proteins responsible for lysogenic conversion caused by coliphages N15 and φ80 are highly homologous. J Bacteriol. 1996;178:1484–6. https://doi.org/10.1128/jb.178.5.1484-1486.1996.
27. Barondess JJ, Beckwith J. A bacterial virulence determinant encoded by lysogenic coliphage λ. J Colloid Interface Sci. 1990;346:871–4. https://doi.org/10.1038/346871a0.
28. Pacheco SV, González OG, Contreras GLP. The Lom gene of bacteriophage λ is involved in Escherichia coli K12 adhesion to human buccal epithelial cells. FEMS Microbiol Lett. 1997;156:129–32. https://doi.org/10.1111/j.1574-6968.1997.tb12717.x.
29. Vaca-Pacheco S, Paniagua-Contreras GL, García-González O, de la Garza M. The clinically isolated FIZ15 bacteriophage causes lysogenic conversion in *Pseudomonas aeruginosa* PAO1. Curr Microbiol. 1999;38:239–43. https://doi.org/10.1007/PL00006794.
30. Müller MG, Ing JY, Cheng MK-W, Flitter BA, Moe GR. Identification of a phage-encoded immunoglobulin-binding protein from invasive *Neisseria meningitidis*. J Immunol. 2013;191:3287–96. https://doi.org/10.4049/jimmunol.1301153.
31. Karaolis DKR, Somara S, Maneval DR, Johnson JA, Kaper JB. A bacteriophage encoding a pathogenicity island, a type-IV pilus and a phage receptor in cholera bacteria. Nature. 1999;399:375–9. https://doi.org/10.1038/20715.
32. O'Brien AD, Newland JW, Miller SF, Holmes RK, Smith HW, Formal SB. Shiga-like toxin-converting Phages from *Escherichia coli* strains that cause hemorrhagic colitis or infantile Diarrhea. Science. 1983;226:694–6. https://doi.org/10.1126/science.6387911.
33. Waldor MK, Mekalanos JJ. Lysogenic conversion by a filamentous phage encoding cholera toxin. Science. 1996;272:1910–4. https://doi.org/10.1126/science.272.5270.1910.
34. Edlin G, Lin L, Bitner R. Reproductive fitness of P1, P2, and mu lysogens of *Escherichia coli.* J Virol. 1977;21:560–4. https://doi.org/10.1007/978-1-59745-501-5_1.
35. Gödeke J, Paul K, Lassak J, Thormann KM. Phage-induced lysis enhances biofilm formation in *Shewanella oneidensis* MR-1. ISME J. 2011;5:613–26. https://doi.org/10.1038/ismej.2010.153.
36. Wang X, Kim Y, Ma Q, Hong SH, Pokusaeva K, Sturino JM, et al. Cryptic prophages help bacteria cope with adverse environments. Nat Commun. 2010;1:147. https://doi.org/10.1038/ncomms1146.

37. Rabinovich L, Sigal N, Borovok I, Nir-Paz R, Herskovits AA. Prophage excision activates listeria competence genes that promote phagosomal escape and virulence. Cell. 2012;150:792–802. https://doi.org/10.1016/j.cell.2012.06.036.
38. Bossi L, Fuentes JA, Mora G, Figueroa-bossi N. Prophage contribution to bacterial population dynamics. J Bacteriol. 2003;185:6467–71. https://doi.org/10.1128/JB.185.21.6467.
39. Swanson MM, Reavy B, Makarova KS, Cock PJ, Hopkins DW, Torrance L, et al. Novel bacteriophages containing a genome of another bacteriophage within their genomes. PLoS One. 2012;7:e40683. https://doi.org/10.1371/journal.pone.0040683.
40. De Paepe M, Hutinet G, Son O, Amarir-Bouhram J, Schbath S, Petit MA. Temperate phages acquire DNA from defective prophages by relaxed homologous recombination: the role of Rad52-like recombinases. PLoS Genet. 2014;10:e1004181. https://doi.org/10.1371/journal.pgen.1004181.
41. Casjens S. Prophages and bacterial genomics: what have we learned so far? Mol Microbiol. 2003;49:277–300. https://doi.org/10.1046/j.1365-2958.2003.03580.x.
42. Argov T, Azulay G, Pasechnek A, Stadnyuk O, Ran-Sapir S, Borovok I, et al. Temperate bacteriophages as regulators of host behavior. Curr Opin Microbiol. 2017;38:81–7. https://doi.org/10.1016/j.mib.2017.05.002.
43. Lang AS, Zhaxybayeva O, Beatty JT. Gene transfer agents: phage-like elements of genetic exchange. Nat Rev Microbiol. 2012;10:472–82. https://doi.org/10.1038/nrmicro2802.
44. Bingle LE, Bailey CM, Pallen MJ. Type VI secretion: a beginner's guide. Curr Opin Microbiol. 2008;11:3–8. https://doi.org/10.1016/j.mib.2008.01.006.
45. Kapitein N, Mogk A. Deadly syringes: type VI secretion system activities in pathogenicity and interbacterial competition. Curr Opin Microbiol. 2013;16:52–8. https://doi.org/10.1016/j.mib.2012.11.009.
46. Basler M, Pilhofer M, Henderson PD, Jensen JG, Mekalanos J. Type VI secretion requires a dynamic contractile phage tail-like structure. Nature. 2012;483:182–6. https://doi.org/10.1109/TMI.2012.2196707.Separate.
47. Mann NH, Clokie MRJ, Millard A, Cook A, Wilson WH, Wheatley PJ, et al. The genome of S-PM2, a "photosynthetic" T4-type bacteriophage that infects marine *Synechococcus* strains. J Bacteriol. 2005;187:3188–200. https://doi.org/10.1128/JB.187.9.3188-3200.2005.
48. Mann NH, Cook A, Bailey S, Clokie M, Amanullah A, Azam N, et al. Bacterial photosynthesis genes in a virus. Nature. 2003;424:741. https://doi.org/10.1038/424741a.
49. Sullivan MB, Coleman ML, Weigele P, Rohwer F, Chisholm SW. Three *Prochlorococcus* cyanophage genomes: signature features and ecological interpretations. PLoS Biol. 2005;3:e144. https://doi.org/10.1371/journal.pbio.0030144.
50. Sullivan MB, Lindell D, Lee JA, Thompson LR, Bielawski JP, Chisholm SW. Prevalence and evolution of core photosystem II genes in marine cyanobacterial viruses and their hosts. PLoS Biol. 2006;4:e234. https://doi.org/10.1371/journal.pbio.0040234.
51. Enav H, Mandel-Gutfreund Y, Béjà O. Comparative metagenomic analyses reveal viral-induced shifts of host metabolism towards nucleotide biosynthesis. Microbiome. 2014;2:9. https://doi.org/10.1186/2049-2618-2-9.
52. Hurwitz BL, Brum JR, Sullivan MB. Depth-stratified functional and taxonomic niche specialization in the "core" and "flexible" Pacific Ocean Virome. ISME J. 2015;9:472–84. https://doi.org/10.1038/ismej.2014.143.
53. Brum JR, Sullivan MB. Rising to the challenge: accelerated pace of discovery transforms marine virology. Nat Rev Microbiol. 2015;13:147–59. https://doi.org/10.1038/nrmicro3404.
54. Reichardt N, Duncan SH, Young P, Belenguer A, McWilliam Leitch C, Scott KP, et al. Phylogenetic distribution of three pathways for propionate production within the human gut microbiota. ISME J. 2014;8:1323–35. https://doi.org/10.1038/ismej.2014.14.
55. Powles SB. Photoinhibition of photosynthesis induced by visible light. Annu Rev Plant Physiol. 1984;35:15–44. https://doi.org/10.1146/annurev.pp.35.060184.000311.
56. Mellis A. Photosystem-II damage and repair cycle in chloroplasts: what modulates the rate of photodamage in vivo? Trends Plant Sci. 1999;4:130–5. https://doi.org/10.1016/S1360-1385(99)01387-4.

57. Aro EM, Virgin I, Andersson B. Photoinhibition of photosystem II. Inactivation, protein damage and turnover. BBA-Bioenergetics. 1993;1143:113–34. https://doi.org/10.1016/0005-2728(93)90134-2.
58. Lindell D, Jaffe JD, Johnson ZI, Church GM, Chisholm SW. Photosynthesis genes in marine viruses yield proteins during host infection. Nature. 2005;438:86–9. https://doi.org/10.1038/nature04111.
59. Gao E-B, Gui J-F, Zhang Q-Y. A novel cyanophage with a cyanobacterial nonbleaching protein A gene in the genome. J Virol. 2012;86:236–45. https://doi.org/10.1128/JVI.06282-11.
60. Yoshida-Takashima Y, Yoshida M, Ogata H, Nagasaki K, Hiroishi S, Yoshida T. Cyanophage infection in the bloom-forming cyanobacteria *Microcystis aeruginosa* in surface freshwater. Microbes Environ. 2012;27:350–5. https://doi.org/10.1264/jsme2.ME12037.
61. Thompson LR, Zeng Q, Kelly L, Huang KH, Singer AU, Stubbe J, et al. Phage auxiliary metabolic genes and the redirection of cyanobacterial host carbon metabolism. Proc Natl Acad Sci. 2011;108:E757–64. https://doi.org/10.1073/pnas.1102164108.
62. Millard A, Clokie MRJ, Shub DA, Mann NH. Genetic organization of the *psbAD* region in phages infecting marine *Synechococcus* strains. Proc Natl Acad Sci. 2004;101:11007–12. https://doi.org/10.1073/pnas.0401478101.
63. Zeidner G, Bielawski JP, Shmoish M, Scanlan DJ, Sabehi G, Béjà O. Potential photosynthesis gene recombination between *Prochlorococcus* and *Synechococcus* via viral intermediates. Environ Microbiol. 2005;7:1505–13. https://doi.org/10.1111/j.1462-2920.2005.00833.x.
64. Keen EC, Bliskovsky VV, Malagon F, Baker JD, Prince JS, Klaus JS, et al. Novel "Superspreader" bacteriophages promote horizontal gene transfer by transformation. mBio. 2017;8:e02115–6. https://doi.org/10.1128/mBio.02115-16.
65. Coleman ML, Chisholm SW. Code and context: *Prochlorococcus* as a model for cross-scale biology. Trends Microbiol. 2007;15:398–407. https://doi.org/10.1016/j.tim.2007.07.001.
66. Cuadros-Orellana S, Martin-Cuadrado AB, Legault B, D'Auria G, Zhaxybayeva O, Papke RT, et al. Genomic plasticity in prokaryotes: the case of the square haloarchaeon. ISME J. 2007;1:235–45. https://doi.org/10.1038/ismej.2007.35.
67. Frias-Lopez J, Shi Y, Tyson GW, Coleman ML, Schuster SC, Chisholm SW, et al. Microbial community gene expression in ocean surface waters. Proc Natl Acad Sci. 2008;105:3805–10. https://doi.org/10.1073/pnas.0708897105.
68. Kettler GC, Martiny AC, Huang K, Zucker J, Coleman ML, Rodrigue S, et al. Patterns and implications of gene gain and loss in the evolution of *Prochlorococcus*. PLoS Genet. 2007;3:e231. https://doi.org/10.1371/journal.pgen.0030231.
69. Rodriguez-Valera F, Ussery DW. Is the pan-genome also a pan-selectome? F1000Research. 2012;1:16. https://doi.org/10.3410/f1000research.1-16.v1.
70. Wilhelm LJ, Tripp HJ, Givan SA, Smith DP, Giovannoni SJ. Natural variation in SAR11 marine bacterioplankton genomes inferred from metagenomic data. Biol Direct. 2007;2:27. https://doi.org/10.1186/1745-6150-2-27.
71. Tettelin H, Masignani V, Cieslewicz MJ, Donati C, Medini D, Ward NL, et al. Genome analysis of multiple pathogenic isolates of *Streptococcus agalactiae*: implications for the microbial "pan-genome". Proc Natl Acad Sci. 2005;102:13950–5 https://doi.org/10.1073/pnas.0506758102.
72. Sharma RS, Mishra V, Mohmmed A, Babu CR. Phage specificity and lipopolysaccarides of stem- and root-nodulating bacteria (*Azorhizobium caulinodans, Sinorhizobium* spp., and *Rhizobium* spp.) of *Sesbania* spp. Arch Microbiol. 2008;189:411–8. https://doi.org/10.1007/s00203-007-0322-x.
73. Reva O, Tümmler B. Think big – Giant genes in bacteria. Environ Microbiol. 2008;10:768–77. https://doi.org/10.1111/j.1462-2920.2007.01500.x.
74. Sorek R, Kunin V, Hugenholtz P. CRISPR – a widespread system that provides acquired resistance against phages in bacteria and archaea. Nat Rev Microbiol. 2008;6:181–6. https://doi.org/10.1038/nrmicro1793.

75. Wilmes P, Simmons SL, Denef VJ, Banfield JF. The dynamic genetic repertoire of microbial communities. FEMS Microbiol Rev. 2009;33:109–32. https://doi.org/10.1111/j.1574-6976.2008.00144.x.
76. Chen J, Novick RP. Phage-mediated Intergeneric transfer of toxin genes. Science. 2009;323:139–41. https://doi.org/10.1126/science.1164783.
77. Kuno S, Sako Y, Yoshida T. Diversification of CRISPR within coexisting genotypes in a natural population of the bloom-forming cyanobacterium *Microcystis aeruginosa*. Microbiology. 2014;160:903–16. https://doi.org/10.1099/mic.0.073494-0.
78. Kuno S, Yoshida T, Kaneko T, Sako Y. Intricate interactions between the bloom-forming cyanobacterium *Microcystis aeruginosa* and foreign genetic elements, revealed by diversified clustered regularly interspaced short palindromic repeat (CRISPR) signatures. Appl Environ Microbiol. 2012;78:5353–60. https://doi.org/10.1128/AEM.00626-12.
79. Kimura S, Sako Y, Yoshida T. Rapid *Microcystis* cyanophage gene diversification revealed by long- and short-term genetic analyses of the tail sheath gene in a natural pond. Appl Environ Microbiol. 2013;79:2789–95. https://doi.org/10.1128/AEM.03751-12.
80. Wigington CH, Sonderegger DL, Brussaard CPD, Buchan A, Finke JF, Fuhrman JA, et al. Re-examining of the relationship between marine virus and microbial cell abundances. Nat Microbiol. 2016;1:15024. https://doi.org/10.1038/nmicrobiol.2015.24.
81. Knowles B, Silveira CB, Bailey BA, Barott K, Cantu VA, Cobian-Guëmes AG, et al. Lytic to temperate switching of viral communities. Nature. 2016;531:466–70. https://doi.org/10.1038/nature17193.
82. Erez Z, Steinberger-Levy I, Shamir M, Doron S, Stokar-Avihail A, Peleg Y, et al. Communication between viruses guides lysis-lysogeny decisions. Nature. 2017;541:488–93. https://doi.org/10.1038/nature21049.

Plasmids and Their Hosts

6

Masaki Shintani and Haruo Suzuki

Abstract

Plasmids have a key role in the rapid evolution and adaptation of their hosts by conferring new phenotypes upon them. It is therefore important to understand the relationships between plasmids and their host organisms and "who can carry which." Here, factors that determine and affect the host ranges of plasmids are reviewed, including features of replication, maintenance, conjugative transfer, and the effects on host fitness caused by plasmid carriage. Recent trials to identify the unknown hosts of plasmids found in natural environments are also discussed.

Abbreviations

FACS	Fluorescence activated cell sorter
FP	Fluorescence protein
GC	Guanine and cytosine
H-NS	Histone-like nucleoid-structuring

M. Shintani (✉)
Department of Engineering, Graduate School of Integrated Science and Technology, Shizuoka University, Hamamatsu, Japan

Department of Bioscience, Graduate School of Science and Technology, Shizuoka University, Hamamatsu, Japan
e-mail: shintani.masaki@shizuoka.ac.jp

H. Suzuki
Faculty of Environment and Information Studies, Keio University, Fujisawa, Kanagawa, Japan

Institute for Advanced Biosciences, Keio University, Tsuruoka, Yamagata, Japan
e-mail: haruo@sfc.keio.ac.jp

H. Nishida, T. Oshima (eds.), *DNA Traffic in the Environment*,
https://doi.org/10.1007/978-981-13-3411-5_6

MDA	Multiple displacement amplification
MOB	Mobility
MPF	Mating pair formation
NAP	Nucleoid-associated protein
oriT	Origin of transfer
oriV	Origin of vegetative replication
PBRT	PCR-based replicon typing systems
pMLST	Plasmid multilocus sequence typing
RCR	Rolling-circle replication
T4CP	Type IV coupling protein
T4SS	Type IV secretion system
TA	Toxin-antitoxin

6.1 Introduction

Plasmids are circular- or linear-extrachromosomal replicons found in the microorganisms of *Bacteria*, *Archaea*, and *Eukaryota* [1]. Plasmids are not only vertically inherited from parent cells to daughter cells but can also be horizontally transferred between cells by conjugation and natural transformation [2]. Conjugation is one of the most effective mechanisms to spread genetic elements among bacteria, and plasmids are thus important "vehicles" for facilitating rapid evolution without mutations in the host genome and for adapting to new environments. While plasmids are important genetic tools for microbial engineering, as cloning vectors for biotechnology, some plasmids can mediate horizontal gene transfer (HGT), which spreads antibiotic resistance, virulence, and other traits among different bacteria in microbial communities. The World Health Organization (WHO) has stated that "antibiotic resistance is one of the biggest threats to global health, food security, and development today" [3].

One of the most important aspects of plasmids is "host range," i.e., which plasmid can be hosted by which microbe. The host range of plasmids can be defined by either plasmid replication or conjugation. The so-called broad-host-range (BHR) plasmids can be hosted (replicated and transferred) by phylogenetically distant organisms, while narrow-host-range (NHR) plasmids can be hosted by closely related organisms (e.g., those belonging to the same species), although the definition of broadness or narrowness is still controversial. For example, BHR plasmid pB10 can be replicated in, and transferred to, bacteria belonging to different classes of *Proteobacteria.* Other examples of BHR plasmids and their host ranges are recently summarized in [4]. Various factors have been found to determine or affect the host range of a plasmid, including factors affecting the replication, maintenance, and/or conjugation of the plasmid, as well as factors affecting the host chromosome. Recently, several other factors have been reported to regulate host fitness during plasmid maintenance in the host cells. In this chapter, we focus on bacterial plasmids and features that could affect the fate of plasmids in the host candidates.

6.2 Plasmid Function

In this section, we summarize factors that determine plasmid host ranges, including plasmid genes for replication, maintenance, and conjugative transfer, and nucleoid-associated proteins (NAPs).

6.2.1 Replication and Maintenance

In plasmids, DNA replication initiates at a specific site known as the origin of vegetative replication (*oriV*). The well-known replication systems of circular plasmids include theta-type replication, strand displacement-type replication, and rolling-circle replication. Many theta-type replicating plasmids, whose lagging strand is synthesized discontinuously producing replication intermediates that look like the Greek letter "theta," contain repeated DNA sequences, or iterons, to which the replication initiation protein binds [reviewed in [5]]. The protein bound to the iteron sequences opens the double-stranded DNA with host factors (DnaA or PriA) and recruits the DNA polymerase of the host cell [6, 7]. ColE1-family plasmids also use theta-type replication systems strictly controlled by an antisense RNA [8, 9]. Representative plasmids with strand displacement-type replication are incompatibility (discussed later) group Q (IncQ) plasmids, which encode a helicase RepA, a specific primase RepB, and replication initiation protein RepC [10]. Plasmids with this system can be continuously replicated, including the lagging strand [10]. This system is independent of host factors for its replication initiation, enabling the host range of the plasmid to be broad [10]. The other major plasmid replication system is the rolling-circle replication (RCR) mechanism, present in many small multicopy plasmids [11, 12]. In any system, the replication of plasmids is dependent on several molecules in the host cell including DnaA, DNA polymerase, RNA polymerase, RNase, ribosomes, helicase, nucleotides, and ATP. Because the chromosome and plasmid(s) in the same host share the host's replication system, plasmids with similar replication initiation systems should have similar host ranges.

Two major maintenance systems of plasmids in the host cells are partition (*par*) systems and toxin-antitoxin (TA) systems. The former involves actively delivering low-copy-number plasmids from parental cells to daughter cells [recently reviewed in [13]]. In brief, the *par* system is composed of two proteins, ATPase or GTPase and DNA-binding protein, which requires one *cis* site for its binding. For the TA systems encoded on the plasmid, one of the two gene products (stable toxin) can kill or stop growth of cells without the other gene product (unstable antitoxin) [14] (see Chap. 3). These systems contribute to the fate of plasmids in the host cells.

Incompatibility (Inc) is one of the classical methods of plasmid classification, based on the Inc test, to assess whether two different plasmids can be propagated stably in the same host cell line. If the two plasmids share similar replication and/or *par* systems, either of the plasmids will be unstable in the host cell line [15]. Inc groups have been independently classified in hosts of three different taxonomic groups; there are 27 Inc groups in the family *Enterobacteriaceae*, 14 Inc groups in the genus *Pseudomonas*, and approximately 18 Inc groups in the genus *Staphylococcus* (Table 6.1) [73, 75–77, 82, 83]. Several Inc groups of *Pseudomonas* are identical to those in *Enterobacteriaceae*, such as IncP-1 (equivalent to IncP), IncP-3 (equivalent to IncA and/or IncC) [84], IncP-4 (equivalent to IncQ) [85], and IncP-6 (equivalent to IncG/U) [86] (Table 6.1). Recently, it was reported that *parMRC* partitioning systems in plasmids of *Clostridium perfringens* determined their compatibility, even though they had almost identical replication initiation protein genes [87, 88]. The Inc test can yield biologically relevant information, but recent classification of Inc groups is usually based on their similarity to genes involved in replication and/or partition. Rozwandowicz et al. showed that IncK plasmids formed two distinct subclusters, IncK1 and IncK2, based on the presence or absence of accessory genes, and that the plasmids were incompatible with each other within the subcluster, but were compatible between subclusters, using a traditional Inc test [89]. It was shown that IncA and IncC plasmids are compatible and thus they proposed that they should not be referred to as "IncA/C" [90]. These findings showed that even plasmids with homologous replication initiation proteins could be compatible in Inc tests. In-depth comparisons of similar plasmids based on their nucleotide sequences and experimental Inc tests will provide more information about co-occurrence of two different plasmids in the same cell and their host ranges.

6.2.2 Conjugation

Conjugation is an important mechanism for horizontally transferring plasmid DNA between different organisms. Below, we focus on the conjugative transfer of plasmids found in both Gram-negative and Gram-positive bacteria.

Self-transmissible plasmids in Gram-negative bacteria generally carry complete sets of the genes required for transfer, i.e., the origin of transfer (*oriT*), relaxase protein, type IV coupling protein (T4CP), and type IV secretion system (T4SS). Garcillán-Barcia et al. [91, 92] and Smillie et al. [78] classified the self-transmissible, or mobilizable, plasmids in the GenBank database into six mobility (MOB) types (MOB_C, MOB_F, MOB_H, MOB_P, MOB_Q, and MOB_V) and four classes of mating pair formation (MPF; MPF_F, MPF_G, MPF_I, and MPF_T) based on all-against-all BLASTP analysis followed by Markov clustering (MCL), to identify and classify homologous proteins of relaxases (for MOB), T4CPs, and T4SSs (for MPF). Mobilizable plasmids are non-self-transmissible because they have only MOB, or MOB and T4SS, but could be transferred by other self-transmissible plasmids, such as helper plasmids with T4SS and MPF. Conjugation is also affected by the type of sex pili, one of the features of MPF (rigid or flexible), and whether the preference in mating conditions (solid surface or liquid environment) between donor and recipient cells is different [93, 94]. It should be noted that the combination of Inc groups, MOB types, and MPF classes could be important for the host range of plasmids (Table 6.1).

Gram-positive bacteria transfer plasmids by two methods. First, a single strand of plasmid DNA is transported via a T4SS-like plasmid in Gram-negative bacteria, which seems to be widely used as means for transferring plasmids in Gram-positive

Table 6.1 List of plasmids in different incompatibility groups

Incompatibility[a]	Representative plasmid[b]	Accession number	MOB[c]	MPF[c]	Notes[d]	References
Inc groups in *Enterobacteriaceae*						
IncA and/or IncC (IncP-3)	RA1	NC_012885	MOB_H	MPF_F	Theta	[16, 17]
IncB/O	R3521	GU256641	MOB_P	MPF_I	Theta	[18, 19]
IncD	R711b	NA	NA	NA		[20]
IncF	F	AP001918	MOB_F	MPF_F	Theta	[21]
IncG/U (=IncP-6)	Rms149	AJ877225	MOB_P	–	Theta	[22]
IncH	R27	AF250878	MOB_H	MPF_F	Theta	[23]
IncI	R64	AP005147	MOB_P	MPF_I	Theta	[24]
IncJ	R391(integrative and conjugative element)	AY090559	MOB_H	MPF_F	Rolling circle	[25–28]
IncK	R387	MCTC_50022[e]	MOB_P	MPF_I	Theta	[19]
IncL/M	pEL60 pKOI-34	NC_005246 AB715422	MOB_P	MPF_I	Theta	[29–31]
IncN	N3	NC_015599	MOB_F	MPF_T	Theta	[32, 33]
IncP (=IncP-1)	RK2	BN000925	MOB_P	MPF_T	Theta	[34]
IncQ (=IncP-4)	RSF1010	M28829	MOB_Q	–	Strand displacement	[35]
IncR	pKP1780	JX424614	–	–	Theta	[36]
IncS (=IncHI2)	R478	BX664015	MOB_H	MPF_F	Theta	[37]
IncT	Rts1	AP004237	MOB_H	MPF_F	Theta	[38, 39]
IncV	R753	NCTC_50521 (planned)[e]	NA	NA	NA	[40]
IncW	R388	NC_028464	MOB_F	MPF_T	Theta	[41]
IncX	R6K	LT827129	MOB_P	MPF_T	Theta	[42]
IncY	P1 (prophage), pMCR-1-P3	AF234172 (phage P1 *mod749*::IS*5* *c*1-100) AF234173 (prophage P1 *mod1902*::IS*5* *c*1-100 *rev-6 dmt*ΔMB) KX880944	–	–	–	[43]
IncZ	pEI545	M93064 (partial)	NA	NA	NA	[19]

(continued)

Table 6.1 (continued)

Incompatibility[a]	Representative plasmid[b]	Accession number	MOB[c]	MPF[c]	Notes[d]	References
Inc groups in *Pseudomonas*						
IncP-1 (=IncP)	RK2	BN000925	MOB_P	MPF_T	Theta	[34]
IncP-2	pOZ176	KC543497	–	–	Theta	[44]
IncP-3 (=IncA and/ or IncC)	RA1	NC_012885	MOB_H	MPF_F	Theta	[16, 17]
IncP-4 (=IncQ)	RSF1010	M28829	MOB_Q	–	Strand displacement	[35]
IncP-5	Rms163	NA	NA	NA		[45, 46]
IncP-6 (=IncG/ IncU)	Rms149	AJ877225	MOB_P	–	Theta	[22]
IncP-7	pCAR1	AB088420/AB474758	MOB_H	MPF_F	Theta	[47, 48]
IncP-8	FP2	NA	NA	NA	NA	[49]
IncP-9	pWW0 NAH7	AJ344068/AB237655	MOB_F	MPF_T	Theta	[50, 51]
IncP-10	R91	X54695	NA	NA	NA	[52]
IncP-11	pMG39	NA	NA	NA	NA	[53]
IncP-12	R716	NA	NA	NA	NA	[54, 55]
IncP-13	pMG25	NA	NA	NA	NA	[56]
IncP-14	pBS222	NA	NA	NA	NA	[56]
Inc groups in *Staphylococcus*						
Inc1	pSK1	GU565967	–	–	Theta Rep type: unique	[57]
Inc2	pII145	NA	NA	NA	–	[58]
Inc3	pT181	J01764	–	–	RCR Rep type:7	[59]

Inc4	pC221	X02529	MOB_P	–	RCR Rep type:7	[60]
Inc5	pS194	X06627	–	–	RCR Rep type:7	[61]
Inc6	pK545	NA	NA	NA	Theta -	[58]
Inc7	pUB101	AY373761	–	–	RCR Rep type: 19	[62, 63]
Inc8	pC194	V01277	–	–	RCR Rep type: 13	[64]
Inc9	pUB112	NA	NA	NA	RCR Rep type: 7	[65, 66]
Inc10	pC223	AY355285	MOB_P	–	RCR Rep type: 7b	[67]
Inc11	pE194	V01278	MOB_V	–	RCR Rep type: unique	[68]
Inc12	pE1764	NA	NA	–	–	[69]
Inc13	pUB110	M19465	MOB_V	–	RCR Rep type: 22	[70, 71]
Inc14	pCW7	J03323 (rep only)	NA	NA	Rep type: 7	[72]
Inc15	pWBG637	KX086582	–	–	–	[73]
Inc18	pIP501	Deposited separately	MOB_v	Unidentified MPF	Theta Rep type: 1	[74]

[a]Several Inc groups are identical, e.g., IncA = IncC, which are also identical with IncP-3 in the Inc groups of *Pseudomonas* plasmids
[b]Representative plasmids are listed based on [75–77]
[c]Classification of MOB classes and MPF types is based on [78, 79]. "–" indicates that the genes involved in conjugation have not been detected, whereas "NA" indicates that the nucleotide sequences of the plasmid are not available
[d]Rep type of Inc1~18 is based on Jensen et al. and Lozano et al. [80, 81]
[e]Accession numbers in Wellcome Sanger institute (http://www.sanger.ac.uk/resources/downloads/plasmids/)

bacteria [95]. pIP501 (found in *Streptococcus agalactiae*), the broadest transfer host range plasmid in Gram-positive bacteria, contains 15 genes for T4SS [96, 97]. Second, plasmids found in the order *Actinomycetales* have conjugative systems that function in a similar manner to the segregation of chromosomal DNA during bacterial cell division and sporulation [reviewed in [98]]. The translocation of double-stranded DNA to the recipient cell is mediated by an FtsK-homologous protein, which is known as ATP-dependent DNA translocase [95, 98, 99].

Multiple regulation systems for conjugation have been found both on plasmids and host chromosomes [reviewed in [100]]. Several plasmids including F, R100, R27, and RA3 encode their own transcriptional regulators (repressors) [75, 101–106], probably because conjugation is an energetically costly process. Several elements encoded on host chromosomes also participate in the conjugation of plasmids, for example, the transfer of *Rhizobium leguminosarum* plasmid pSym is induced by homoserine lactone [107]; plasmid F is influenced by the extracellular response element CpxA [108]; and plasmid R100 is regulated by host-encoded regulators, such as Dam methylase and Lrp protein [109, 110].

All conjugative plasmids contain at least one entry exclusion systems; when the cells already contain conjugative plasmids, they become inefficient recipients. This system could limit an excess transfer of plasmids, which can kill the recipients in a process known as lethal zygosis, and is almost essential for conjugative plasmids, though the physiological importance of this is still unclear [reviewed in detail by [111]]. These systems are negative factors for conjugations, at least in laboratory conditions, and are important for the host range of plasmids.

Plasmids have been classified based on the similarity between the nucleotide sequences of genes (or the amino-acid sequences of proteins) involved in replication and conjugation, including PCR-based replicon typing systems (PBRT) [112] and MOB typing [113–115]. Plasmids of IncA and/or IncC, IncHI1, IncHI2, IncI, and IncN groups have been subtyped by plasmid multilocus sequence typing (pMLST) using specific genes in plasmids as targets for PCR [116]. These classifications are effective but not sufficient in predicting their host ranges. This is because plasmid conjugation and its efficiency can vary, depending on many other factors including cell density, growth rate, nutrient availability, temperature, and high-salt stress [93, 94, 117–119]. Recently, divalent cations (Ca^{2+} and Mg^{2+}) have been found to increase the conjugation efficiency of several Inc plasmids, especially the IncP-7 plasmid, pCAR1 [120, 121]. It is also affected by combinations of donor and recipient strains [122]. Sakuda et al. [120] found that conjugation frequencies of plasmids pCAR1 and NAH7 were similar in mating with one donor strain (*Pseudomonas putida*) and one of the two recipient strains (*P. putida* and *Pseudomonas resinovorans*) (mating with one donor and one recipient). In contrast, these plasmids were transferred more frequently to *P. putida* than to *P. resinovorans* when the two recipient strains were mixed (mating with one donor and two recipients, Sakuda et al., unpublished). The results suggest that a host-specific factor(s) could affect the host range of plasmids. There are several systems to prevent conjugation of plasmids. Restriction-modification and CRISPR-Cas systems inhibit the conjugation of plasmids by cleaving DNA sequences [123] (see Chap. 3). These physiological or environmental conditions of hosts and/or recipients could also affect the host range of plasmids.

6.2.3 Nucleoid-Associated Protein (NAP)

Nucleoid-associated proteins (NAPs) are DNA-binding proteins encoded on plasmids, as well as on chromosomes, and can function as transcriptional regulators by binding to several regions of the DNA. NAPs aid chromosomal DNA compaction in bacterial cells [124] (see Chap. 1). One member of the NAPs, a histone-like nucleoid-structuring (H-NS) protein, can bind to horizontally acquired elements and repress their transcription (silencing), which can reduce the deleterious effects of harboring foreign DNAs [125]. The NAP genes are also found in plasmids, which are proposed to have "stealth" effects that minimize the fitness reduction caused by carriage of plasmids [126], because deletion (or disruption) of the NAP genes in plasmids had greater effects on the host transcriptome than did the plasmid carriage [127, 128]. Therefore, NAPs encoded both on host chromosome and plasmids can affect host fitness. Because the numbers and combinations of NAPs alter in different plasmids and host chromosomes [129], they are important to determine the host range of plasmids. In fact, NAPs could affect the stability and conjugation of plasmids. For example, the IncHI1 plasmid R27 has a gene encoding H-NS, involved in the modulation of R27 transfer by interacting with a Hha/Ymo family protein [130]. Another example is the IncP-7 plasmid pCAR1, which has three NAP genes, *pmr*, *phu*, and *pnd* [131]. Double deletion mutants of *pmr* and *phu* or *pnd* reduced its stability and lost transferability [132]. Since the binding sites for Pmr, both on pCAR1 and its host (*P. putida*) chromosome, were found to be similar to those of chromosomally encoded H-NS-like proteins, TurA and TurB [133], it could be cooperatively regulated by the H-NS proteins encoded by both the plasmid and host chromosome, though their detailed molecular mechanisms are unclear. Similarly, another H-NS-like protein, Acr2, encoded on the IncA and/or IncC plasmid, was found to negatively regulate its conjugative transfer [134, 135]. Therefore, NAPs encoded on plasmids are important factors to determine their maintenance and conjugation in different hosts.

6.3 Fitness Cost

A plasmid can bring its host potential benefits through its accessory genes, but sometimes a burden (fitness cost), reducing the host growth rate and competitiveness under no selective pressure (e.g., that exerted by antibiotics). One of the experimental approaches to investigate the molecular mechanisms of how plasmid carriage affects the host fitness is to obtain a compensated mutant from the parental plasmid-bearing host [reviewed in [136]]. San Millan et al. showed that compensating mutations on chromosomal genes encoding helicase, kinase, or the global regulator GacA could reduce the fitness cost of plasmid carriage [137, 138]. Sota et al. [139] showed that mutation to the *trfA1* gene encoding the replication initiation protein TrfA1 (TrfA1 variant) could reduce the fitness cost of the IncP-1 plasmid in an "inappropriate" or naïve host, *Shewanella oneidensis*. This reduction was not observed in *Escherichia coli* or *Cupriavidus pinatubonensis,* but in *P. putida* [140]. The tight binding of the wild-type TrfA to the host helicase DnaB caused fitness cost of host, and the TrfA1 variant had a lower affinity to the DnaB resulting in the reduction of fitness cost [140]. Stalder et al. [141] reported that three distinct patterns of evolution exist to reduce the fitness cost of plasmids: (1) mutations in *trfA1* gene, (2) acquisition of a putative toxin-antitoxin system on a transposon from a co-existing plasmid, and (3) a mutation in the *fur* gene encoding one of the global regulators in its host. These facts indicate that the co-evolution of plasmid and host increases the persistence of plasmids in their hosts.

Effects on three different host *Pseudomonas* (*P. putida*, *P. aeruginosa*, and *P. fluorescens*) by carriage of the IncP-7 plasmid pCAR1 were compared based on their phenotypes and transcriptomes [142–144]. Changes in fitness varied between different hosts [144], although detailed molecular mechanisms have not yet been elucidated. One of the most striking responses in two of the three hosts was the induction of genes on prophages by pCAR1 carriage [143, 144]. Notably, Martinez-Garcia et al. [145, 146] showed that deletions of the four prophages from *P. putida*, which harbored no plasmids, increased the growth rate, transformation efficiency, and protein expression from the plasmid vector, suggesting that prophages themselves affect host fitness. Recently, Shintani et al. (unpublished) found that the deletion of these prophages of *P. putida* could affect the fitness of hosts with the IncP-1 and IncP-7 plasmids (Shintani et al. unpublished). Thus, there could be cross talk between the two mobile genetic elements, prophages (which integrate into the host chromosome) and plasmid, although their mechanisms remain unclear.

pCAR1 is the plasmid endowing carbazole-degrading ability (converting carbazole to catechol) to its host, but the growth rates of different hosts can change in minimal medium with carbazole as the sole carbon source. The growth of *P. fluorescens* Pf0-1 with pCAR1 was significantly slower than that of *P. aeruginosa* PAO1, or *P. putida* KT2440 [48]. This is because of the toxicity of accumulated catechol, an intermediate compound of carbazole degradation, and of differences in catechol metabolism in these hosts, whose catabolism is mediated by enzymes encoded on the host chromosomes [48, 147]. Notably, DNA rearrangements were found on pCAR1 and the chromosome of its host Pf0-1 cultured in a minimal medium with carbazole as the sole carbon source, which could allow the host to avoid the accumulation of catechol [48, 147]. This indicates that fitness could vary between different host bacteria or growth conditions. The differences in fitness costs in different host cells could influence the stability of plasmids in the host cells and their host ranges.

6.4 Prediction of Plasmid Host Range

6.4.1 Prediction of Host Range Based on Bioinformatics

Novel plasmids have been found in host genomes and metagenomes, thanks to the recent revolution in nucleotide sequencing technology and in bioinformatic tools. There are now 12,015 complete sequences of plasmids in the NCBI database: 11,710 are from *Bacteria*, 192 are from *Archaea*, and the remaining 113 are from *Eukaryota* (based on the NCBI database, ftp://ftp.ncbi.nlm.nih.gov/genomes/GENOME_REPORTS/plasmids.txt, downloaded on Mar. 2018). As more than 10,000 complete sequences of plasmids are available in public databases, we can now use bioinformatic approaches for predicting plasmid host ranges based on sequence features, including plasmid size, nucleotide composition (G+C content, oligonucleotide frequencies, and codon usage), and replication strand asymmetry (asymmetric nucleotide compositions and gene proportion between leading and lagging strands of DNA replication). Intragenomic variation in nucleotide composition has been used to detect putative alien genes acquired by horizontal transfer [148, 149].

Plasmid sizes (Kb) varied widely among sequenced plasmids, ranging from 0.537 (*Xanthomonas campestris* pv. campestris str. CN14) to 5836.680 (*Pseudomonas monteilii*). There are clear differences in sizes between non-transmissible and conjugative plasmids (including mobilizable- and self-transmissible plasmids) [78, 83]. Notably, larger plasmids and conjugative plasmids frequently carried multiple NAP genes [129]. Similar NAP genes were found on both plasmids and host chromosomes and thus, their combinations might be important for the stable maintenance of plasmids within their hosts.

The relative frequency of guanine and cytosine (G+C content), calculated by the formula (G+C)/(A+T+G+C), varies widely among bacterial genomes, especially at synonymously variable third positions within codons [150, 151]. G+C content is correlated with a number of variables including genome size, aerobiosis, lifestyle, and environments [152], and can be shaped by mutation and selection [153]. The G+C content variability reflects differences in the DNA polymerase III alpha subunit [154, 155]. G+C contents also varied widely among different plasmids, ranging from 19.3% (*Eukaryota Moniliophthora roreri* plasmid pMR2) to 87.5% (*Actinobacteria Streptomyces autolyticus* plasmid). Previous studies revealed that G+C content is lower in plasmids than in hosts [156, 157], and that there is a strong correlation between the G+C content of plasmids and host chromosomes [158]. For a correlation of G+C contents between 2296 plasmids and their corresponding hosts analyzed here, the Pearson correlation coefficient was 0.97, and in 1704 (74.2%) cases, the plasmids had lower G+C contents than their hosts (Fig. 6.1). Therefore, the G+C content of plasmids could be an important indicator for predicting their host ranges.

Karlin et al. proposed that each genome has a characteristic "signature," consisting of the relative abundance of vectors of oligonucleotides such as di-, tri-, and tetra-nucleotides (i.e., *k*-mers) [159, 160]. The oligonucleotide compositions of

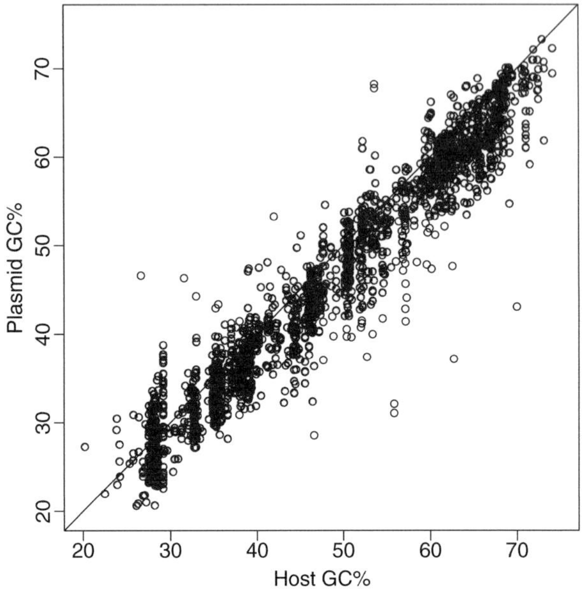

Fig. 6.1 Plot of G+C contents (GC%) between 2296 plasmids and their corresponding hosts. Each point represents a plasmid-host pair from 920 taxa. The *y*-axis indicates the G+C content of plasmids (plasmid GC%) and the *x*-axis indicates the G+C content of hosts (host GC%), retrieved from the National Center for Biotechnology Information (NCBI) genome list (ftp://ftp.ncbi.nlm.nih.gov/genomes/GENOME_REPORTS/)

DNA sequence segments are relatively constant along the genome, and those from closely related taxa tend to be more similar than those from distantly related taxa. The oligonucleotide compositions of plasmids tend to be more similar to those of their known host chromosomes than to those of other bacterial chromosomes [161, 162]. Thus, bacteria with the most similar oligonucleotide compositions may be the most probable hosts in which plasmids have evolved [163, 164].

Most amino acids can be encoded by more than one codon, with codons encoding the same amino acid called synonymous codons. Synonymous codon usage varies among genes, between different organisms, and even within a single genome. The codon usage of genes can reflect a balance between translational selection, mutational biases, and other factors [165, 166]. For example, the strength of translational selection for synonymous codon usage varies among the three replicons of *Sinorhizobium meliloti* (the class *Alphaproteobacteria*), i.e., it is much weaker in the plasmid pSymA, than in the plasmid pSymB and the chromosome [167]. In *Agrobacterium tumefaciens* (the class *Alphaproteobacteria*), the differences in codon usage between chromosomes (circular and linear) and plasmids (pAt and pTi) are larger than the differences between two chromosomes or two plasmids [168]. In *Borrelia burgdorferi* (the class *Spirochaetes*), there is a high similarity in codon usage among the cp32 family plasmids, and between the chromosomal leading strand and linear plasmid lp38 [168], as well as a significant difference in codon usage between the leading and lagging strands due to strand-specific mutational biases [169, 170]. The codon usage of genes in the largest plasmid, as well as the chromosome of *Lawsonia intracellularis* (the class *Deltaproteobacteria*), can be affected by strand-specific mutational biases [171]. Codon usage of any replicon (plasmids and chromosomes) can reflect a complex balance between host-specific mutational biases and selective pressures, resulting in varying degrees of codon-usage similarity between replicons.

Another important feature of bacterial genomes is replication strand asymmetry. In bacterial genomes, essential and highly expressed genes are preferentially located on the leading strands of DNA replication [172–175]. GC skew, defined as (C−G)/(C+G), has been used for measuring strand compositional asymmetry and for predicting the origin and terminus of replication on bacterial chromosomes and plasmids [171, 176, 177]. A measure of the strength of GC skew, quantified by the GC skew index (GCSI), detected a difference in GCSI between replicons with different types of replication machinery (e.g., GCSI between eubacteria and archaea chromosomes, and GCSI between RCR and non-RCR plasmids), and a correlation between GCSI of plasmids and their host chromosomes [178], suggesting that any replicon (plasmids and chromosomes) replicated and repaired in the same cell, has been subject to host-specific mutational biases and selective pressures, resulting in similar degrees of GC skew.

There is evidence of plasmid-mediated horizontal gene transfer, and some plasmid genes integrate into host chromosomes [179]. For example, IncP (IncP-1) plasmid sequences were detected in the chromosomes of bacteria such as *Pseudomonas* [180] and *Brucella* [181], and genes encoding the replication initiator protein TrfA were also found in bacterial chromosomes [182]. Fondi et al. [183] identified genes shared between plasmids and chromosomes as possible indications of gene transfer between them in the genus *Acinetobacter*. Network analyses of homologous DNA families shared among chromosomes and mobile elements showed that betweenness centralities are higher in plasmids than in phages, indicating that plasmids (e.g., promiscuous IncP-1 plasmid pB10), rather than viruses, are key vectors of DNA exchange between bacterial chromosomes [184]. The presence of DNA sequences shared between plasmids and chromosomes suggests that these replicons co-resided in the same hosts at some point in their history, although we cannot rule out the possibility of natural transformation, i.e., the uptake of extracellular plasmid DNA.

Bioinformatic tools have been developed to find novel plasmids in genomic and metagenomic data, which can be divided into two categories: those that reconstruct plasmids via assembly of sequencing reads (PLACNET, PlasmidSPAdes, and Recycler) [185–187], and those to identify plasmids in assembled contigs (PlasmidFinder, cBar, and PlasFlow) [188–190]. Arredondo-Alonso et al. compared the performance of four tools (PlasmidSPAdes, Recycler, cBar, and PlasmidFinder) for detecting plasmids from short read sequencing data [191]. Krawczyk et al. demonstrated that PlasFlow outperformed cBar on test data [189]. These bioinformatic tools allow us to update the range of hosts in which plasmids are found.

6.4.2 Prediction of Host Range Based on Experiments

There have been several reports for experimental detection and separation of unknown hosts of plasmids from environmental samples. Since most bacteria in natural environments have not been isolated yet [192, 193], there must be a large number of unidentified hosts of plasmids. To obtain these (potential) hosts from environmental samples, cultivation-independent methods have been developed and adopted. One of the most efficient ways is to use fluorescent protein (FP) to visualize and detect plasmid-bearing cells, as part of a method developed by Molin et al. [194–196]. They use a *lac*-like promoter, whose expression is repressed in the presence of the LacI repressor, upstream of the FP gene. The *lacI* gene is then introduced into the donor chromosome. If the plasmid is transferred from the donor cell to a recipient cell without *lacI*, then FP can be expressed, and the transconjugant cell can be observed by fluorescence microscopy. This system could be used with fluorescence microcopy and/or a fluorescence activated cell sorter (FACS) to obtain the transconjugant cells [197, 198]. After sorting a single transconjugant cell, multiple displacement amplification (MDA) can be used to amplify genomic DNA from a single bacterial cell without cultivation processes [199]. Shintani and Klümper discovered previously unknown transconjugants using these systems, FACS or MDA, and sequencing of the 16S rRNA genes of sorted transconjugants [200, 201]. These systems could be used for the detection of unidentified plasmid hosts, but still show bias in the transconjugants detected and separated from environmental samples. The biases can be caused by two major factors: (1) the fluorescence intensity of FP could be drastically different from host to host [201], probably because the expression levels of FP vary between different hosts due to its promoter or codon usage; (2) MDA may not necessarily amplify all the genomic DNA from a single cell.

Another potential method for overcoming the bias of GFP expression in different hosts in environmental samples might be the use of fluorescence in situ hybridization (FISH) for detecting plasmid DNA [201, 202], although it is still difficult to apply this method as bacterial cells are so small and also contaminated with debris and particles in environmental samples. Comparisons by Raman spectrum of cells might be efficient in detecting hosts with catabolic genes obtained via plasmids, as the host cells could incorporate a specific substrate (the target of the catabolic gene products) labeled with a stable isotope. Indeed, Huang et al. reported that they successfully identified hosts of a naphthalene-degradative plasmid by Raman microscopy [203]. Introduction of new methods to detect differences between plasmid-free and plasmid-bearing cells will enable us to identify transconjugants in natural environments and expand our knowledge of plasmid host range.

6.5 Conclusions and Remarks

The determinants for a plasmid's hosts are diverse and it is still challenging to clearly understand which can possess the plasmid. Bioinformatics will be instrumental in predicting the candidate hosts of plasmids, based on their nucleotide sequence features, with experimental data showing the prerequisite conditions required for the host cell to acquire the plasmid. Recent studies at the single-cell level have identified plasmid hosts in different environments that have the potential to provide us with important information on how plasmids spread between different bacteria in nature, including "transient" hosts that can be mediators and reservoirs of accessory genes. Nevertheless, experimental data, including from these recent studies, are still insufficient as predictors, and additional data will be required. The analyses will shed light on the mechanisms of plasmid spreading in natural environments, as well as bacterial adaptation and evolution.

Acknowledgments The writing of this chapter was partly supported by JSPS KAKENHI Grant Numbers 15H05618 and 15KK0278 (to M. S.) and research funds from Keio University Academic Development Funds for Individual Research and from the Yamagata Prefectural Government and the City of Tsuruoka (to H.S.).

References

1. Kado CI. Historical events that spawned the field of plasmid biology. Microbiol Spectr. 2014;2. https://doi.org/10.1128/microbiolspec.PLAS-0019-2013.
2. Hulter N, Ilhan J, Wein T, Kadibalban AS, Hammerschmidt K, Dagan T. An evolutionary perspective on plasmid lifestyle modes. Curr Opin Microbiol. 2017;38:74–80. https://doi.org/10.1016/j.mib.2017.05.001.
3. World Health Organization. Handle antibiotics with care: tripartite brochure, 2017. http://www.who.int/antimicrobial-resistance/OIE-FAO-WHO-AMR-LEAFLET-INFOGRAPHIC.pdf?ua=1.
4. Jain A, Srivastava P. Broad host range plasmids. FEMS Microbiol Lett. 2013;348:87–96. https://doi.org/10.1111/1574-6968.12241.
5. Konieczny I, Bury K, Wawrzycka A, Wegrzyn K. Iteron plasmids. Microbiol Spectr. 2014; 2(6). https://doi.org/10.1128/microbiolspec.PLAS-0026-2014.
6. Giraldo R, Fernandez-Tresguerres ME. Twenty years of the pPS10 replicon: insights on the molecular mechanism for the activation of DNA replication in iteron-containing bacterial plasmids. Plasmid. 2004;52:69–83. https://doi.org/10.1016/j.plasmid.2004.06.002.
7. Rakowski SA, Filutowicz M. Plasmid R6K replication control. Plasmid. 2013;69:231–42. https://doi.org/10.1016/j.plasmid.2013.02.003.
8. Cesareni G, Helmer-Citterich M, Castagnoli L. Control of ColE1 plasmid replication by antisense RNA. Trends Genet. 1991;7:230–5.
9. Polisky B. ColE1 replication control circuitry: sense from antisense. Cell. 1988;55:929–32.
10. Lilly J, Camps M. Mechanisms of theta plasmid replication. Microbiol Spectr. 2015; 3.
11. Khan SA. Plasmid rolling-circle replication: highlights of two decades of research. Plasmid. 2005;53:126–36. https://doi.org/10.1016/j.plasmid.2004.12.008.
12. Ruiz-Maso JA, Macho NC, Bordanaba-Ruiseco L, Espinosa M, Coll M, Del Solar G. Plasmid rolling-circle replication. Microbiol Spectr. 2015; 3. PLAS-0035-2014. https://doi.org/10.1128/microbiolspec.PLAS-0035-2014.
13. Baxter JC, Funnell BE. Plasmid partition mechanisms Microbiol Spectr. 2014; 2. https://doi.org/10.1128/microbiolspec.PLAS-0023-2014.
14. Harms A, Brodersen DE, Mitarai N, Gerdes K. Toxins, targets, and triggers: an overview of toxin-antitoxin biology. Mol Cell. 2018;70(5):768–84. https://doi.org/10.1016/j.molcel.2018.01.003.
15. Novick RP. Plasmid incompatibility. Microbiol Rev. 1987;51:381–95.
16. Fricke WF, et al. Comparative genomics of the IncA/C multidrug resistance plasmid family. J Bacteriol. 2009;191:4750–7. https://doi.org/10.1128/JB.00189-09.
17. Llanes C, Gabant P, Couturier M, Bayer L, Plesiat P. Molecular analysis of the replication elements of the broad-host-range RepA/C replicon. Plasmid. 1996;36:26–35. https://doi.org/10.1006/plas.1996.0028.
18. Papagiannitsis CC, Tzouvelekis LS, Kotsakis SD, Tzelepi E, Miriagou V. Sequence of pR3521, an IncB plasmid from *Escherichia coli* encoding ACC-4, SCO-1, and TEM-1 beta-lactamases. Antimicrob Agents Chemother. 2011;55:376–81. https://doi.org/10.1128/AAC.00875-10.

19. Praszkier J, Wei T, Siemering K, Pittard J. Comparative analysis of the replication regions of IncB, IncK, and IncZ plasmids. J Bacteriol. 1991;173:2393–7.
20. Coetzee JN, Bradley DE, Lecatsas G, du Toit L, Hedges RW. Bacteriophage D: an IncD group plasmid-specific phage. J Gen Microbiol. 1985;131:3375–83. https://doi.org/10.1099/00221287-131-12-3375.
21. Manwaring NP, Skurray RA, Firth N. Nucleotide sequence of the F plasmid leading region. Plasmid. 1999;41:219–25. https://doi.org/10.1006/plas.1999.1390.
22. Haines AS, Jones K, Cheung M, Thomas CM. The IncP-6 plasmid Rms149 consists of a small mobilizable backbone with multiple large insertions. J Bacteriol. 2005;187:4728–38. https://doi.org/10.1128/JB.187.14.4728-4738.2005.
23. Sherburne CK, et al. The complete DNA sequence and analysis of R27, a large IncHI plasmid from *Salmonella typhi* that is temperature sensitive for transfer. Nucleic Acids Res. 2000;28:2177–86.
24. Sampei G, Furuya N, Tachibana K, Saitou Y, Suzuki T, Mizobuchi K, Komano T. Complete genome sequence of the incompatibility group I1 plasmid R64. Plasmid. 2010;64:92–103. https://doi.org/10.1016/j.plasmid.2010.05.005.
25. Bellanger X, Payot S, Leblond-Bourget N, Guedon G. Conjugative and mobilizable genomic islands in bacteria: evolution and diversity. FEMS Microbiol Rev. 2014;38:720–60. https://doi.org/10.1111/1574-6976.12058.
26. Carraro N, Poulin D, Burrus V. Replication and active partition of integrative and conjugative elements (ICEs) of the SXT/R391 family: the line between ICEs and conjugative plasmids is getting thinner. PLoS Genet. 2015;11:e1005298. https://doi.org/10.1371/journal.pgen.1005298.
27. Guglielmini J, Quintais L, Garcillán-Barcia MP, de la Cruz F, Rocha EP. The repertoire of ICE in prokaryotes underscores the unity, diversity, and ubiquity of conjugation. PLoS Genet. 2011;7:e1002222. https://doi.org/10.1371/journal.pgen.1002222.
28. Murphy DB, Pembroke JT. Monitoring of chromosomal insertions of the IncJ elements R391 and R997 in *Escherichia coli* K-12. FEMS Microbiol Lett. 1999;174:355–61.
29. Borrell L, Yang J, Pittard AJ, Praszkier J. Interaction of initiator proteins with the origin of replication of an IncL/M plasmid. Plasmid. 2006;56:88–101. https://doi.org/10.1016/j.plasmid.2006.04.002.
30. Foster GC, McGhee GC, Jones AL, Sundin GW. Nucleotide sequences, genetic organization, and distribution of pEU30 and pEL60 from *Erwinia amylovora*. Appl Environ Microbiol. 2004;70:7539–44. https://doi.org/10.1128/AEM.70.12.7539-7544.2004.
31. Shimada N, et al. Complete nucleotide sequence of pKOI-34, an IncL/M plasmid carrying $bla_{\mathrm{IMP\text{-}34}}$ in *Klebsiella oxytoca* isolated in Japan. Antimicrob Agents Chemother. 2016;60:3156–62. https://doi.org/10.1128/AAC.02507-15.
32. Zoueva OP, Iyer VN, Matula TI, Kozlowski M. Analysis of pCU1 replication origins: dependence of *oriS* on the plasmid-encoded replication initiation protein RepA. Plasmid. 2003;49:152–9. https://doi.org/10.1016/s0147-619x(02)00151-8.
33. Humphrey B, et al. Fitness of *Escherichia coli* strains carrying expressed and partially silent IncN and IncP1 plasmids. BMC Microbiol. 2012;12:53. https://doi.org/10.1186/1471-2180-12-53.
34. Pansegrau W, et al. Complete nucleotide sequence of Birmingham IncP alpha plasmids. Compilation and comparative analysis. J Mol Biol. 1994;239:623–63. https://doi.org/10.1006/jmbi.1994.1404.
35. Scholz P, Haring V, Wittmann-Liebold B, Ashman K, Bagdasarian M, Scherzinger E. Complete nucleotide sequence and gene organization of the broad-host-range plasmid RSF1010. Gene. 1989;75:271–88.
36. Papagiannitsis CC, Miriagou V, Giakkoupi P, Tzouvelekis LS, Vatopoulos AC. Characterization of pKP1780, a novel IncR plasmid from the emerging *Klebsiella pneumoniae* ST147, encoding the VIM-1 metallo-beta-lactamase. J Antimicrob Chemother. 2013;68:2259–62. https://doi.org/10.1093/jac/dkt196.

37. Gilmour MW, Thomson NR, Sanders M, Parkhill J, Taylor DE. The complete nucleotide sequence of the resistance plasmid R478: defining the backbone components of incompatibility group H conjugative plasmids through comparative genomics. Plasmid. 2004;52:182–202. https://doi.org/10.1016/j.plasmid.2004.06.006.
38. Kamio Y, Tabuchi A, Itoh Y, Katagiri H, Terawaki Y. Complete nucleotide sequence of mini-Rts1 and its copy mutant. J Bacteriol. 1984;158:307–12.
39. Murata T, et al. Complete nucleotide sequence of plasmid Rts1: implications for evolution of large plasmid genomes. J Bacteriol. 2002;184:3194–202. https://doi.org/10.1128/jb.184.12.3194-3202.2002.
40. Hedges RW. R factors from *Proteus mirabilis* and *P. vulgaris*. J Gen Microbiol. 1975;87:301–11. https://doi.org/10.1099/00221287-87-2-301.
41. Revilla C, et al. Different pathways to acquiring resistance genes illustrated by the recent evolution of IncW plasmids. Antimicrob Agents Chemother. 2008;52:1472–80. https://doi.org/10.1128/AAC.00982-07.
42. Kelley WL, Bastia D. Conformational changes induced by integration host factor at origin gamma of R6K and copy number control. J Biol Chem. 1991;266:15924–37.
43. Lobocka MB, et al. Genome of bacteriophage P1. J Bacteriol. 2004;186:7032–68. https://doi.org/10.1128/JB.186.21.7032-7068.2004.
44. Xiong J, Alexander DC, Ma JH, Deraspe M, Low DE, Jamieson FB, Roy PH. Complete sequence of pOZ176, a 500-kilobase IncP-2 plasmid encoding IMP-9-mediated carbapenem resistance, from outbreak isolate *Pseudomonas aeruginosa* 96. Antimicrob Agents Chemother. 2013;57:3775–82. https://doi.org/10.1128/AAC.00423-13.
45. Iyobe S, Hasuda K, Fuse A, Mitsuhashi S. Demonstration of R factors from *Pseudomonas aeruginosa*. Antimicrob Agents Chemother. 1974;5:547–52.
46. Sagai H, Hasuda K, Iyobe S, Bryan LE, Holloway BW, Mitsuhashi S. Classification of R plasmids by incompatibility in *Pseudomonas aeruginosa*. Antimicrob Agents Chemother. 1976;10:573–8.
47. Maeda K, Nojiri H, Shintani M, Yoshida T, Habe H, Omori T. Complete nucleotide sequence of carbazole/dioxin-degrading plasmid pCAR1 in *Pseudomonas resinovorans* strain CA10 indicates its mosaicity and the presence of large catabolic transposon Tn*4676*. J Mol Biol. 2003;326:21–33. https://doi.org/10.1016/S0022-2836(02)01400-6.
48. Takahashi Y, Shintani M, Li L, Yamane H, Nojiri H. Carbazole-degradative IncP-7 plasmid pCAR1.2 is structurally unstable in *Pseudomonas fluorescens* Pf0-1, which accumulates catechol, the intermediate of the carbazole degradation pathway. Appl Environ Microbiol. 2009;75:3920–9. https://doi.org/10.1128/aem.02373-08.
49. Finger J, Krishnapillai V. Host range, entry exclusion, and incompatibility of *Pseudomonas aeruginosa* FP plasmids. Plasmid. 1980;3:332–42.
50. Greated A, Lambertsen L, Williams PA, Thomas CM. Complete sequence of the IncP-9 TOL plasmid pWW0 from *Pseudomonas putida*. Environ Microbiol. 2002;4:856–71.
51. Sota M, et al. Genomic and functional analysis of the IncP-9 naphthalene-catabolic plasmid NAH7 and its transposon Tn*4655* suggests catabolic gene spread by a tyrosine recombinase. J Bacteriol. 2006;188:4057–67. https://doi.org/10.1128/JB.00185-06.
52. Cain D, Holloway BW. Prime plasmids derived from the IncP-10 plasmid R91-5 in *Pseudomonas putida*. FEMS Microbiol Lett. 1984;24:97–101. https://doi.org/10.1111/j.1574-6968.1984.tb01252.x.
53. Medeiros AA, Cohenford M, Jacoby GA. Five novel plasmid-determined beta-lactamases. Antimicrob Agents Chemother. 1985;27:715–9.
54. Bradley DE. Specification of the conjugative pili and surface mating systems of *Pseudomonas* plasmids. J Gen Microbiol. 1983;129:2545–56. https://doi.org/10.1099/00221287-129-8-2545.
55. Bryan LE, Semaka SD, Van den Elzen HM, Kinnear JE, Whitehouse RL. Characteristics of R931 and other *Pseudomonas aeruginosa* R factors. Antimicrob Agents Chemother. 1973;3:625–37.

56. Boronin AM. Diversity of *Pseudomonas* plasmids: to what extent? FEMS Microbiol Lett. 1992;100:461–7. https://doi.org/10.1016/0378-1097(92)90246-K.
57. Kwong SM, Lim R, Lebard RJ, Skurray RA, Firth N. Analysis of the pSK1 replicon, a prototype from the staphylococcal multiresistance plasmid family. Microbiology. 2008;154:3084–94. https://doi.org/10.1099/mic.0.2008/017418-0.
58. Ruby C, Novick RP. Plasmid interactions in *Staphylococcus aureus*: nonadditivity of compatible plasmid DNA pools. Proc Natl Acad Sci U S A. 1975;72:5031–5.
59. Khan SA, Novick RP. Complete nucleotide sequence of pT181, a tetracycline-resistance plasmid from *Staphylococcus aureus*. Plasmid. 1983;10:251–9. https://doi.org/10.1016/0147-619X(83)90039-2.
60. Shaw WV, Brenner DG, LeGrice SF, Skinner SE, Hawkins AR. Chloramphenicol acetyltransferase gene of staphylococcal plasmid pC221. Nucleotide sequence analysis and expression studies. FEBS Lett. 1985;179:101–6.
61. Projan SJ, Moghazeh S, Novick RP. Nucleotide sequence of pS194, a streptomycin-resistance plasmid from *Staphylococcus aureus*. Nucleic Acids Res. 1988;16:2179–87.
62. Boe L, Gros MF, te Riele H, Ehrlich SD, Gruss A. Replication origins of single-stranded-DNA plasmid pUB110. J Bacteriol. 1989;171:3366–72.
63. O'Brien FG, Price C, Grubb WB, Gustafson JE. Genetic characterization of the fusidic acid and cadmium resistance determinants of *Staphylococcus aureus* plasmid pUB101. J Antimicrob Chemother. 2002;50:313–21.
64. Horinouchi S, Weisblum B. Nucleotide sequence and functional map of pC194, a plasmid that specifies inducible chloramphenicol resistance. J Bacteriol. 1982a;150:815–25.
65. Ehret M, Matzura H. Replication control of the *Staphylococcus aureus* chloramphenicol resistance plasmids pC223 and pUB112 in *Bacillus subtilis*. Nucleic Acids Res. 1988;16:2045–62.
66. Projan SJ, Novick R. Comparative analysis of five related staphylococcal plasmids. Plasmid. 1988;19:203–21. https://doi.org/10.1016/0147-619X(88)90039-X.
67. Smith MC, Thomas CD. An accessory protein is required for relaxosome formation by small staphylococcal plasmids. J Bacteriol. 2004;186:3363–73. https://doi.org/10.1128/JB.186.11.3363-3373.2004.
68. Horinouchi S, Weisblum B. Nucleotide sequence and functional map of pE194, a plasmid that specifies inducible resistance to macrolide, lincosamide, and streptogramin type B antibodies. J Bacteriol. 1982b;150:804–14.
69. Iordănescu S, Surdeanu M. New incompatibility groups for *Staphylococcus aureus* plasmids. Plasmid. 1980;4:256–60. https://doi.org/10.1016/0147-619X(80)90064-5.
70. McKenzie T, Hoshino T, Tanaka T, Sueoka N. The nucleotide sequence of pUB110: some salient features in relation to replication and its regulation. Plasmid. 1986;15:93–103.
71. McKenzie T, Hoshino T, Tanaka T, Sueoka N. Correction. A revision of the nucleotide sequence and functional map of pUB110. Plasmid. 1987;17:83–5.
72. Balson DF, Shaw WV. Nucleotide sequence of the *rep* gene of staphylococcal plasmid pCW7. Plasmid. 1990;24:74–80.
73. Udo EE, Grubb WB. A new incompatibility group plasmid in *Staphylococcus aureus*. FEMS Microbiol Lett. 1991;62:33–6.
74. Thompson JK, Collins MA. Completed sequence of plasmid pIP501 and origin of spontaneous deletion derivatives. Plasmid. 2003;50:28–35. https://doi.org/10.1016/s0147-619x(03)00042-8.
75. Lawley T, Frost LS, Wilkins BM. Bacterial conjugation in gram-negative bacteria. In: Funnell BE, Phillips GJ, editors. Plasmid biology. Washington DC: ASM press; 2004. p. 203–26. https://doi.org/10.1128/9781555817732.ch9.
76. Thomas CM, Haines AS. Plasmids of the genus *Pseudomonas*. In: Ramos J-L, editor. *Pseudomonas*: volume 1 genomics, life style and molecular architecture. Boston: Springer US; 2004. p. 197–231. https://doi.org/10.1007/978-1-4419-9086-0_7.
77. Taylor DE, Gibreel A, Tracz DM, Lawley TD. Antibiotic resistance plasmids. In: Funnell BE, Phillips GJ, editors. Plasmid biology. Washington DC: ASM press; 2004. p. 473–92. https://doi.org/10.1128/9781555817732.ch23.

78. Smillie C, Garcillán-Barcia MP, Francia MV, Rocha EP, de la Cruz F. Mobility of plasmids. Microbiol Mol Biol Rev. 2010;74:434–52. https://doi.org/10.1128/MMBR.00020-10.
79. Guglielmini J, Neron B, Abby SS, et al. Key components of the eight classes of type IV secretion systems involved in bacterial conjugation or protein secretion. Nucleic Acids Res. 2014;42:5715–27.
80. Jensen LB, Garcia-Migura L, Valenzuela AJ, Løhr M, Hasman H, Aarestrup FM. A classification system for plasmids from enterococci and other gram-positive bacteria. J Microbiol Methods. 2010;80:25–43. https://doi.org/10.1016/j.mimet.2009.10.012.
81. Lozano C, García-Migura L, Aspiroz C, Zarazaga M, Torres C, Aarestrup FM. Expansion of a plasmid classification system for gram-positive bacteria and determination of the diversity of plasmids in *Staphylococcus aureus* strains of human, animal, and food origins. Appl Environ Microbiol. 2012;78:5948–55. https://doi.org/10.1128/AEM.00870-12.
82. Carattoli A. Resistance plasmid families in *Enterobacteriaceae*. Antimicrob Agents Chemother. 2009;53:2227–38. https://doi.org/10.1128/AAC.01707-08.
83. Shintani M, Sanchez ZK, Kimbara K. Genomics of microbial plasmids: classification and identification based on replication and transfer systems and host taxonomy. Front Microbiol. 2015a;6:242. https://doi.org/10.3389/fmicb.2015.00242.
84. Johnson TJ, Lang KS. IncA/C plasmids: an emerging threat to human and animal health? Mob Genet Elem. 2012;2:55–8. https://doi.org/10.4161/mge.19626.
85. Loftie-Eaton W, Rawlings DE. Diversity, biology and evolution of IncQ-family plasmids. Plasmid. 2012;67:15–34. https://doi.org/10.1016/j.plasmid.2011.10.001.
86. Haines AS, Cheung M, Thomas CM. Evidence that IncG (IncP-6) and IncU plasmids form a single incompatibility group. Plasmid. 2006;55:210–5. https://doi.org/10.1016/j.plasmid.2005.11.003.
87. Adams V, Watts TD, Bulach DM, Lyras D, Rood JI. Plasmid partitioning systems of conjugative plasmids from *Clostridium perfringens*. Plasmid. 2015;80:90–6. https://doi.org/10.1016/j.plasmid.2015.04.004.
88. Watts TD, Johanesen PA, Lyras D, Rood JI, Adams V. Evidence that compatibility of closely related replicons in *Clostridium perfringens* depends on linkage to *parMRC*-like partitioning systems of different subfamilies. Plasmid. 2017;91:68–75. https://doi.org/10.1016/j.plasmid.2017.03.008.
89. Rozwandowicz M, et al. Plasmids of distinct IncK lineages show compatible phenotypes. Antimicrob Agents Chemother. 2017;61:AAC–01954. https://doi.org/10.1128/AAC.01954-16.
90. Ambrose SJ, Harmer CJ, Hall RM. Compatibility and entry exclusion of IncA and IncC plasmids revisited: IncA and IncC plasmids are compatible. Plasmid. 2018;96–97:7–12. https://doi.org/10.1016/j.plasmid.2018.02.002.
91. Garcillán-Barcia MP, Alvarado A, de la Cruz F. Identification of bacterial plasmids based on mobility and plasmid population biology. FEMS Microbiol Rev. 2011;35:936–56. https://doi.org/10.1111/j.1574-6976.2011.00291.x.
92. Garcillán-Barcia MP, Francia MV, de la Cruz F. The diversity of conjugative relaxases and its application in plasmid classification. FEMS Microbiol Rev. 2009;33:657–87.
93. Bradley DE, Taylor DE, Cohen DR. Specification of surface mating systems among conjugative drug resistance plasmids in *Escherichia coli* K-12. J Bacteriol. 1980;143:1466–70.
94. Nakazawa S, et al. Different transferability of incompatibility (Inc) P-7 plasmid pCAR1 and IncP-1 plasmid pBP136 in stirring liquid conditions. PLoS One. 2017;12:e0186248. https://doi.org/10.1371/journal.pone.0186248.
95. Goessweiner-Mohr N, Arends K, Keller W, Grohmann E. Conjugative type IV secretion systems in gram-positive bacteria. Plasmid. 2013;70:289–302. https://doi.org/10.1016/j.plasmid.2013.09.005.
96. Kurenbach B, Bohn C, Prabhu J, Abudukerim M, Szewzyk U, Grohmann E. Intergeneric transfer of the *Enterococcus faecalis* plasmid pIP501 to *Escherichia coli* and *Streptomyces lividans* and sequence analysis of its tra region. Plasmid. 2003;50:86–93.

97. Kurenbach B, et al. The TraA relaxase autoregulates the putative type IV secretion-like system encoded by the broad-host-range *Streptococcus agalactiae* plasmid pIP501. Microbiology. 2006;152:637–45. https://doi.org/10.1099/mic.0.28468-0.
98. Thoma L, Muth G. Conjugative DNA-transfer in *Streptomyces*, a mycelial organism. Plasmid. 2016;87-88:1–9. https://doi.org/10.1016/j.plasmid.2016.09.004.
99. Thoma L, Dobrowinski H, Finger C, Guezguez J, Linke D, Sepulveda E, Muth G. A multiprotein DNA translocation complex directs intramycelial plasmid spreading during *Streptomyces* conjugation. MBio. 2015;6:e02559–14. https://doi.org/10.1128/mBio.02559-14.
100. Banuelos-Vazquez LA, Torres Tejerizo G, Brom S. Regulation of conjugative transfer of plasmids and integrative conjugative elements. Plasmid. 2017;91:82–9. https://doi.org/10.1016/j.plasmid.2017.04.002.
101. Gibert M, Juarez A, Madrid C, Balsalobre C. New insights in the role of HtdA in the regulation of R27 conjugation. Plasmid. 2013;70:61–8. https://doi.org/10.1016/j.plasmid.2013.01.009.
102. Gibert M, Paytubi S, Beltran S, Juarez A, Balsalobre C, Madrid C. Growth phase-dependent control of R27 conjugation is mediated by the interplay between the plasmid-encoded regulatory circuit TrhR/TrhY-HtdA and the cAMP regulon. Environ Microbiol. 2016;18:5277–87. https://doi.org/10.1111/1462-2920.13579.
103. Kulinska A, Godziszewska J, Wojciechowska A, Ludwiczak M, Jagura-Burdzy G. Global transcriptional regulation of backbone genes in broad-host-range plasmid RA3 from the IncU group involves segregation protein KorB (ParB family). Appl Environ Microbiol. 2016;82:2320–35. https://doi.org/10.1128/AEM.03541-15.
104. Ludwiczak M, Dolowy P, Markowska A, Szarlak J, Kulinska A, Jagura-Burdzy G. Global transcriptional regulator KorC coordinates expression of three backbone modules of the broad-host-range RA3 plasmid from IncU incompatibility group. Plasmid. 2013;70:131–45. https://doi.org/10.1016/j.plasmid.2013.03.007.
105. Sandt CH, Herson DS. Mobilization of the genetically engineered plasmid pHSV106 from *Escherichia coli* HB101(pHSV106) to *Enterobacter cloacae* in drinking water. Appl Environ Microbiol. 1991;57:194–200.
106. Taki K, Abo T, Ohtsubo E. Regulatory mechanisms in expression of the *traY-I* operon of sex factor plasmid R100: involvement of *traJ* and *traY* gene products. Genes Cells. 1998;3:331–45.
107. Danino VE, Wilkinson A, Edwards A, et al. Recipient-induced transfer of the symbiotic plasmid pRL1JI in *Rhizobium leguminosarum* bv. viciae is regulated by a quorum-sensing relay. Mol Microbiol. 2003;50:511–25.
108. Gubbins MJ, Lau I, Will WR, et al. The positive regulator, TraJ, of the *Escherichia coli* F plasmid is unstable in a cpxA* background. J Bacterial. 2002;184:5781–8.
109. Camacho EM, Casadesus J. Conjugal transfer of the virulence plasmid of *Salmonella enterica* is regulated by the leucine-responsive regulatory protein and DNA adenine methylation. Mol Microbiol. 2002;44:1589–98.
110. Camacho EM, Serna A, Casadesus J. Regulation of conjugal transfer by Lrp and dam methylation in plasmid R100. Int Microbiol. 2005;8:279–85.
111. Garcillan-Barcia MP, de la Cruz F. Why is entry exclusion an essential feature of conjugative plasmids? Plasmid. 2008;60:1–18. https://doi.org/10.1016/j.plasmid.2008.03.002.
112. Carattoli A. Plasmids and the spread of resistance. Int J Med Microbiol. 2013;303:298–304. https://doi.org/10.1016/j.ijmm.2013.02.001.
113. Fernandez-Lopez R, Redondo S, Garcillan-Barcia MP, de la Cruz F. Towards a taxonomy of conjugative plasmids. Curr Opin Microbiol. 2017;38:106–13. https://doi.org/10.1016/j.mib.2017.05.005.
114. Garcillan-Barcia MP, de la Cruz F. Ordering the bestiary of genetic elements transmissible by conjugation. Mobile Genet Elem. 2013;3:e24263. https://doi.org/10.4161/mge.24263.
115. Orlek A, et al. Ordering the mob: insights into replicon and MOB typing schemes from analysis of a curated dataset of publicly available plasmids. Plasmid. 2017;91:42–52. https://doi.org/10.1016/j.plasmid.2017.03.002.

116. Jolley KA, Maiden MC. BIGSdb: scalable analysis of bacterial genome variation at the population level. BMC Bioinformatics. 2010;11:595. https://doi.org/10.1186/1471-2105-11-595.
117. Beuls E, Modrie P, Deserranno C, Mahillon J. High-salt stress conditions increase the pAW63 transfer frequency in *Bacillus thuringiensis*. Appl Environ Microbiol. 2012;78:7128–31. https://doi.org/10.1128/AEM.01105-12.
118. Schuurmans JM, van Hijum SA, Piet JR, Handel N, Smelt J, Brul S, ter Kuile BH. Effect of growth rate and selection pressure on rates of transfer of an antibiotic resistance plasmid between *E. coli* strains. Plasmid. 2014;72:1–8. https://doi.org/10.1016/j.plasmid.2014.01.002.
119. Verma T, Ramteke PW, Garg SK. Effect of ecological factors on conjugal transfer of chromium-resistant plasmid in *Escherichia coli* isolated from tannery effluent. Appl Biochem Biotechnol. 2002;102–103:5–20.
120. Sakuda A, et al. Divalent cations increase the conjugation efficiency of the incompatibility P-7 group plasmid pCAR1 among different *Pseudomonas* hosts. Microbiology. 2018;164:20–7. https://doi.org/10.1099/mic.0.000583.
121. Shintani M, Matsui K, Takemura T, Yamane H, Nojiri H. Behavior of the IncP-7 carbazole-degradative plasmid pCAR1 in artificial environmental samples. Appl Microbiol Biotechnol. 2008b;80:485–97. https://doi.org/10.1007/s00253-008-1564-5.
122. Shintani M, Fukushima N, Tezuka M, Yamane H, Nojiri H. Conjugative transfer of the IncP-7 carbazole degradative plasmid, pCAR1, in river water samples. Biotechnol Lett. 2008a;30:117–22. https://doi.org/10.1007/s10529-007-9519-y.
123. Johnston C, Martin B, Polard P, Claverys JP. Postreplication targeting of transformants by bacterial immune systems? Trends Microbiol. 2013;21:516–21. https://doi.org/10.1016/j.tim.2013.08.002.
124. Dillon SC, Dorman CJ. Bacterial nucleoid-associated proteins, nucleoid structure and gene expression. Nat Rev Microbiol. 2010;8:185–95. https://doi.org/10.1038/nrmicro2261.
125. Ali SS, Xia B, Liu J, Navarre WW. Silencing of foreign DNA in bacteria. Curr Opin Microbiol. 2012;15:175–81. https://doi.org/10.1016/j.mib.2011.12.014.
126. Dorman CJ. H-NS-like nucleoid-associated proteins, mobile genetic elements and horizontal gene transfer in bacteria. Plasmid. 2014;75:1–11. https://doi.org/10.1016/j.plasmid.2014.06.004.
127. Doyle M, Fookes M, Ivens A, Mangan MW, Wain J, Dorman CJ. An H-NS-like stealth protein aids horizontal DNA transmission in bacteria. Science. 2007;315:251–2. https://doi.org/10.1126/science.1137550.
128. Yun CS, et al. Pmr, a histone-like protein H1 (H-NS) family protein encoded by the IncP-7 plasmid pCAR1, is a key global regulator that alters host function. J Bacteriol. 2010;192:4720–31. https://doi.org/10.1128/jb.00591-10.
129. Shintani M, Suzuki-Minakuchi C, Nojiri H. Nucleoid-associated proteins encoded on plasmids: occurrence and mode of function. Plasmid. 2015b;80:32–44. https://doi.org/10.1016/j.plasmid.2015.04.008.
130. Forns N, Banos RC, Balsalobre C, Juarez A, Madrid C. Temperature-dependent conjugative transfer of R27: role of chromosome- and plasmid-encoded Hha and H-NS proteins. J Bacteriol. 2005;187:3950–9. https://doi.org/10.1128/JB.187.12.3950-3959.2005.
131. Nojiri H. Structural and molecular genetic analyses of the bacterial carbazole degradation system. Biosci Biotechnol Biochem. 2012;76:1–18. https://doi.org/10.1271/bbb.110620.
132. Suzuki-Minakuchi C, et al. Effects of three different nucleoid-associated proteins encoded on IncP-7 plasmid pCAR1 on host *Pseudomonas putida* KT2440. Appl Environ Microbiol. 2015;81:2869–80. https://doi.org/10.1128/AEM.00023-15.
133. Yun CS, et al. MvaT family proteins encoded on IncP-7 plasmid pCAR1 and the host chromosome regulate the host transcriptome cooperatively but differently. Appl Environ Microbiol. 2016;82:832–42. https://doi.org/10.1128/AEM.03071-15.
134. Fernandez-Alarcon C, Singer RS, Johnson TJ. Comparative genomics of multidrug resistance-encoding IncA/C plasmids from commensal and pathogenic *Escherichia coli* from multiple animal sources. PLoS One. 2011;6:e23415. https://doi.org/10.1371/journal.pone.0023415.

135. Lang KS, Johnson TJ. Characterization of Acr2, an H-NS-like protein encoded on A/C2-type plasmids. Plasmid. 2016;87-88:17–27. https://doi.org/10.1016/j.plasmid.2016.07.004.
136. San Millan A, MacLean RC. Fitness costs of plasmids: a limit to plasmid transmission. Microbiol Spectr. 2017; 5. https://doi.org/10.1128/microbiolspec.MTBP-0016-2017.
137. San Millan A, Pena-Miller R, Toll-Riera M, Halbert ZV, McLean AR, Cooper BS, MacLean RC. Positive selection and compensatory adaptation interact to stabilize non-transmissible plasmids. Nat Commun. 2014;5:5208. https://doi.org/10.1038/ncomms6208.
138. San Millan A, Toll-Riera M, Qi Q, MacLean RC. Interactions between horizontally acquired genes create a fitness cost in *Pseudomonas aeruginosa*. Nat Commun. 2015;6:6845. https://doi.org/10.1038/ncomms7845.
139. Sota M, Yano H, Hughes JM, Daughdrill GW, Abdo Z, Forney LJ, Top EM. Shifts in the host range of a promiscuous plasmid through parallel evolution of its replication initiation protein. ISME J. 2010;4:1568–80. https://doi.org/10.1038/ismej.2010.72.
140. Yano H, et al. Evolved plasmid-host interactions reduce plasmid interference cost. Mol Microbiol. 2016;101:743–56. https://doi.org/10.1111/mmi.13407.
141. Stalder T, Rogers LM, Renfrow C, Yano H, Smith Z, Top EM. Emerging patterns of plasmid-host coevolution that stabilize antibiotic resistance. Sci Rep. 2017;7:4853. https://doi.org/10.1038/s41598-017-04662-0.
142. Miyakoshi M, Shintani M, Terabayashi T, Kai S, Yamane H, Nojiri H. Transcriptome analysis of *Pseudomonas putida* KT2440 harboring the completely sequenced IncP-7 plasmid pCAR1. J Bacteriol. 2007;189:6849–60. https://doi.org/10.1128/jb.00684-07.
143. Shintani M, et al. Response of the *Pseudomonas* host chromosomal transcriptome to carriage of the IncP-7 plasmid pCAR1. Environ Microbiol. 2010;12:1413–26. https://doi.org/10.1111/j.1462-2920.2009.02110.x.
144. Takahashi Y, et al. Modulation of primary cell function of host *Pseudomonas* bacteria by the conjugative plasmid pCAR1. Environ Microbiol. 2015;17:134–55. https://doi.org/10.1111/1462-2920.12515.
145. Martinez-Garcia E, Jatsenko T, Kivisaar M, de Lorenzo V. Freeing *Pseudomonas putida* KT2440 of its proviral load strengthens endurance to environmental stresses. Environ Microbiol. 2015;17:76–90. https://doi.org/10.1111/1462-2920.12492.
146. Martinez-Garcia E, Nikel PI, Aparicio T, de Lorenzo V. *Pseudomonas* 2.0: genetic upgrading of *P. putida* KT2440 as an enhanced host for heterologous gene expression. Microb Cell Factories. 2014;13:159. https://doi.org/10.1186/s12934-014-0159-3.
147. Shintani M. The behavior of mobile genetic elements (MGEs) in different environments. Biosci Biotechnol Biochem. 2017;81:854–62. https://doi.org/10.1080/09168451.2016.1270743.
148. Becq J, Churlaud C, Deschavanne P. A benchmark of parametric methods for horizontal transfers detection. PLoS One. 2010;5:e9989. https://doi.org/10.1371/journal.pone.0009989.
149. Karlin S. Detecting anomalous gene clusters and pathogenicity islands in diverse bacterial genomes. Trends Microbiol. 2001;9:335–43.
150. Muto A, Osawa S. The guanine and cytosine content of genomic DNA and bacterial evolution. Proc Natl Acad Sci U S A. 1987;84:166–9.
151. Sharp PM, Bailes E, Grocock RJ, Peden JF, Sockett RE. Variation in the strength of selected codon usage bias among bacteria. Nucleic Acids Res. 2005;33:1141–53. https://doi.org/10.1093/nar/gki242.
152. Dutta C, Paul S. Microbial lifestyle and genome signatures. Curr Genomics. 2012;13:153–62. https://doi.org/10.2174/138920212799860698.
153. Hildebrand F, Meyer A, Eyre-Walker A. Evidence of selection upon genomic GC-content in bacteria. PLoS Genet. 2010;6:e1001107. https://doi.org/10.1371/journal.pgen.1001107.
154. Wu H, Zhang Z, Hu S, Yu J. On the molecular mechanism of GC content variation among eubacterial genomes. Biol Direct. 2012;7:2. https://doi.org/10.1186/1745-6150-7-2.
155. Zhao X, Zhang Z, Yan J, Yu J. GC content variability of eubacteria is governed by the pol III alpha subunit. Biochem Biophys Res Commun. 2007;356:20–5. https://doi.org/10.1016/j.bbrc.2007.02.109.

156. Rocha EP, Danchin A. Base composition bias might result from competition for metabolic resources. Trends Genet. 2002;18:291–4. https://doi.org/10.1016/S0168-9525(02)02690-2.
157. van Passel MW, Bart A, Luyf AC, van Kampen AH, van der Ende A. Compositional discordance between prokaryotic plasmids and host chromosomes. BMC Genomics. 2006;7:26. https://doi.org/10.1186/1471-2164-7-26.
158. Nishida H. Comparative analyses of base compositions, DNA sizes, and dinucleotide frequency profiles in archaeal and bacterial chromosomes and plasmids. Int J Evol Biol. 2012;2012:342482. https://doi.org/10.1155/2012/342482.
159. Karlin S, Campbell AM, Mrazek J. Comparative DNA analysis across diverse genomes. Annu Rev Genet. 1998;32:185–225. https://doi.org/10.1146/annurev.genet.32.1.185.
160. Mrazek J. Phylogenetic signals in DNA composition: limitations and prospects. Mol Biol Evol. 2009;26:1163–9. https://doi.org/10.1093/molbev/msp032.
161. Campbell A, Mrazek J, Karlin S. Genome signature comparisons among prokaryote, plasmid, and mitochondrial DNA. Proc Natl Acad Sci U S A. 1999;96:9184–9.
162. Suzuki H, Sota M, Brown C, Top E. Using mahalanobis distance to compare genomic signatures between bacterial plasmids and chromosomes. Nucleic Acids Res. 2008;36:e147. https://doi.org/10.1093/nar/gkn753.
163. Norberg P, Bergstrom M, Jethava V, Dubhashi D, Hermansson M. The IncP-1 plasmid backbone adapts to different host bacterial species and evolves through homologous recombination. Nat Commun. 2011;2:268. https://doi.org/10.1038/ncomms1267.
164. Suzuki H, Yano H, Brown CJ, Top EM. Predicting plasmid promiscuity based on genomic signature. J Bacteriol. 2010;192:6045–55. https://doi.org/10.1128/Jb.00277-10.
165. Behura SK, Severson DW. Codon usage bias: causative factors, quantification methods and genome-wide patterns: with emphasis on insect genomes. Biol Rev Camb Philos Soc. 2013;88:49–61. https://doi.org/10.1111/j.1469-185X.2012.00242.x.
166. Ermolaeva MD. Synonymous codon usage in bacteria. Curr Issues Mol Biol. 2001;3:91–7.
167. Peixoto L, Zavala A, Romero H, Musto H. The strength of translational selection for codon usage varies in the three replicons of *Sinorhizobium meliloti*. Gene. 2003;320:109–16.
168. Davis JJ, Olsen GJ. Modal codon usage: assessing the typical codon usage of a genome. Mol Biol Evol. 2010;27:800–10. https://doi.org/10.1093/molbev/msp281.
169. Lafay B, Lloyd AT, McLean MJ, Devine KM, Sharp PM, Wolfe KH. Proteome composition and codon usage in spirochaetes: species-specific and DNA strand-specific mutational biases. Nucleic Acids Res. 1999;27:1642–9.
170. McInerney JO. Replicational and transcriptional selection on codon usage in *Borrelia burgdorferi*. Proc Natl Acad Sci U S A. 1998;95:10698–703.
171. Guo FB, Yuan JB. Codon usages of genes on chromosome, and surprisingly, genes in plasmid are primarily affected by strand-specific mutational biases in *Lawsonia intracellularis*. DNA Res. 2009;16:91–104. https://doi.org/10.1093/dnares/dsp001.
172. French S. Consequences of replication fork movement through transcription units *in vivo*. Science. 1992;258:1362–5.
173. Rocha EP, Danchin A. Essentiality, not expressiveness, drives gene-strand bias in bacteria. Nat Genet. 2003;34:377–8. https://doi.org/10.1038/ng1209.
174. Srivatsan A, Tehranchi A, MacAlpine DM, Wang JD. Co-orientation of replication and transcription preserves genome integrity. PLoS Genet. 2010;6:e1000810. https://doi.org/10.1371/journal.pgen.1000810.
175. Zheng WX, Luo CS, Deng YY, Guo FB. Essentiality drives the orientation bias of bacterial genes in a continuous manner. Sci Rep. 2015;5:16431. https://doi.org/10.1038/srep16431.
176. Arakawa K, Tomita M. Measures of compositional strand bias related to replication machinery and its applications. Curr Genomics. 2012;13:4–15. https://doi.org/10.2174/138920212799034749.
177. Sernova NV, Gelfand MS. Identification of replication origins in prokaryotic genomes. Brief Bioinform. 2008;9:376–91. https://doi.org/10.1093/bib/bbn031.

178. Arakawa K, Suzuki H, Tomita M. Quantitative analysis of replication-related mutation and selection pressures in bacterial chromosomes and plasmids using generalised GC skew index. BMC Genomics. 2009;10:640. https://doi.org/10.1186/1471-2164-10-640.
179. Harrison E, Brockhurst MA. Plasmid-mediated horizontal gene transfer is a coevolutionary process. Trends Microbiol. 2012;20:262–7. https://doi.org/10.1016/j.tim.2012.04.003.
180. Chiu CM, Thomas CM. Evidence for past integration of IncP-1 plasmids into bacterial chromosomes. FEMS Microbiol Lett. 2004;241:163–9. https://doi.org/10.1016/j.femsle.2004.10.016.
181. Lavigne JP, Vergunst AC, Bourg G, O'Callaghan D. The IncP island in the genome of *Brucella suis* 1330 was acquired by site-specific integration. Infect Immun. 2005;73:7779–83. https://doi.org/10.1128/IAI.73.11.7779-7783.2005.
182. Stenger DC, Lee MW. Phylogeny of replication initiator protein TrfA reveals a highly divergent clade of incompatibility group P1 plasmids. Appl Environ Microbiol. 2011;77:2522–6. https://doi.org/10.1128/AEM.02789-10.
183. Fondi M, et al. Exploring the evolutionary dynamics of plasmids: the *Acinetobacter* pan-plasmidome. BMC Evol Biol. 2010;10:59. https://doi.org/10.1186/1471-2148-10-59.
184. Halary S, Leigh JW, Cheaib B, Lopez P, Bapteste E. Network analyses structure genetic diversity in independent genetic worlds. Proc Natl Acad Sci U S A. 2010;107:127–32. https://doi.org/10.1073/pnas.0908978107.
185. Antipov D, Hartwick N, Shen M, Raiko M, Lapidus A, Pevzner PA. plasmidSPAdes: assembling plasmids from whole genome sequencing data. Bioinformatics. 2016;32:3380–7. https://doi.org/10.1093/bioinformatics/btw493.
186. Lanza VF, de Toro M, Garcillan-Barcia MP, Mora A, Blanco J, Coque TM, de la Cruz F. Plasmid flux in *Escherichia coli* ST131 sublineages, analyzed by plasmid constellation network (PLACNET), a new method for plasmid reconstruction from whole genome sequences. PLoS Genet. 2014;10:e1004766. https://doi.org/10.1371/journal.pgen.1004766.
187. Rozov R, Brown Kav A, Bogumil D, Shterzer N, Halperin E, Mizrahi I, Shamir R. Recycler: an algorithm for detecting plasmids from *de novo* assembly graphs. Bioinformatics. 2017;33:475–82. https://doi.org/10.1093/bioinformatics/btw651.
188. Carattoli A, et al. *In silico* detection and typing of plasmids using PlasmidFinder and plasmid multilocus sequence typing. Antimicrob Agents Chemother. 2014;58:3895–903. https://doi.org/10.1128/AAC.02412-14.
189. Krawczyk PS, Lipinski L, Dziembowski A. PlasFlow: predicting plasmid sequences in metagenomic data using genome signatures. Nucleic Acids Res. 2018;46(6):e35. https://doi.org/10.1093/nar/gkx1321.
190. Zhou F, Xu Y. cBar: a computer program to distinguish plasmid-derived from chromosome-derived sequence fragments in metagenomics data. Bioinformatics. 2010;26:2051–2. https://doi.org/10.1093/bioinformatics/btq299.
191. Arredondo-Alonso S, Willems RJ, van Schaik W, Schurch AC. On the (im)possibility of reconstructing plasmids from whole-genome short-read sequencing data. Microb Genomics. 2017;3:e000128. https://doi.org/10.1099/mgen.0.000128.
192. Kyrpides NC, et al. Genomic encyclopedia of bacteria and archaea: sequencing a myriad of type strains. PLoS Biol. 2014;12:e1001920. https://doi.org/10.1371/journal.pbio.1001920.
193. Stewart EJ. Growing unculturable bacteria. J Bacteriol. 2012;194:4151–60. https://doi.org/10.1128/JB.00345-12.
194. Christensen B, Sternberg C, Andersen J, Eberl L, Moller S, Givskov M, Molin S. Establishment of new genetic traits in a microbial biofilm community. Appl Environ Microbiol. 1998;64:2247–55.
195. Christensen B, Sternberg C, Molin S. Bacterial plasmid conjugation on semi-solid surfaces monitored with the green fluorescent protein (GFP) from *Aequorea victoria* as a marker. Gene. 1996;173:59–65. https://doi.org/10.1016/0378-1119(95)00707-5.
196. Haagensen JA, Hansen SK, Johansen T, Molin S. *In situ* detection of horizontal transfer of mobile genetic elements. FEMS Microbiol Ecol. 2002;42:261–8. https://doi.org/10.1111/j.1574-6941.2002.tb01016.x.

197. Musovic S, Dechesne A, Sørensen J, Smets BF. Novel assay to assess permissiveness of a soil microbial community toward receipt of mobile genetic elements. Appl Environ Microbiol. 2010;76:4813–8. https://doi.org/10.1128/AEM.02713-09.
198. Musovic S, Oregaard G, Kroer N, Sørensen S. Cultivation-independent examination of horizontal transfer and host range of an IncP-1 plasmid among gram-positive and gram-negative bacteria indigenous to the barley rhizosphere. Appl Environ Microbiol. 2006;72:6687–92. https://doi.org/10.1128/AEM.00013-06.
199. Hutchison CA 3rd, Smith HO, Pfannkoch C, Venter JC. Cell-free cloning using phi29 DNA polymerase. Proc Natl Acad Sci U S A. 2005;102:17332–6. https://doi.org/10.1073/pnas.0508809102.
200. Klumper U, Riber L, Dechesne A, Sannazzarro A, Hansen LH, Sorensen SJ, Smets BF. Broad host range plasmids can invade an unexpectedly diverse fraction of a soil bacterial community. ISME J. 2015;9:934–45. https://doi.org/10.1038/ismej.2014.191.
201. Shintani M, et al. Single-cell analyses revealed transfer ranges of IncP-1, IncP-7, and IncP-9 plasmids in a soil bacterial community. Appl Environ Microbiol. 2014;80:138–45. https://doi.org/10.1128/AEM.02571-13.
202. Niki H, Hiraga S. Subcellular distribution of actively partitioning F plasmid during the cell division cycle in *E. coli*. Cell. 1997;90:951–7.
203. Huang WE, Stoecker K, Griffiths R, Newbold L, Daims H, Whiteley AS, Wagner M. Raman-FISH: combining stable-isotope Raman spectroscopy and fluorescence *in situ* hybridization for the single cell analysis of identity and function. Environ Microbiol. 2007;9:1878–89. https://doi.org/10.1111/j.1462-2920.2007.01352.x.

Overlooked Broad-Host-Range Vector Particles in the Environment

7

Hiroshi X. Chiura

Abstract

Horizontal gene transfer (HGT) has been proven to have a much stronger impact than mutation in the microbial world. Evidence of HGT is currently expressed as comparative sequence homology, whereas a substance with residing transfer initiative remains ambiguous. The simultaneous transfer of multiple genes could have more profound impacts than single-gene exchange on the evolutionary processes, but such data are still insufficient. Although three novel modes of HGT mechanisms have come to light—gene transfer agents (GTAs), membrane vesicles (MVs), and intercellular nanotubes (Ins)—the classic concepts of transformation, transduction, and conjugation must not be ignored.

In addition to the previously mentioned HGT phenomena, the author proposes another brand-new concept: broad-host-range vector particles (VPs) in the natural virus-like particle (VLP) assemblage. VPs are capable of transferring chromosomal genes, plasmids, and cytoplasmic material toward recipients with broad phylogenetic ranges, although the VP-production responsible gene(s) are unidentified to date. VPs exhibit similar morphological characteristics to double bilayered MVs and tailless virions. Accordingly, confidently discriminating between VPs, bona fide tailless virions, and MVs is impossible, even with electron microscopy; hence, the environmental VLP fraction should be inevitably reconsidered as a "continuous vector variety" composed of virus–VP–MV.

VPs can be characterized explicitly by the following: (1) low recipient lethality of ~10% irrespective of particle ultraviolet treatment; (2) transfer of the host chromosomal fragments (dsDNA, ~400 kb) toward broad phylogenic recipient range (Archaea-Bacteria-Eukarya), accompanied by a high generalized trans-

Expertise: Molecular microbial genetics and microbial ecology of HGT media.

H. X. Chiura (✉)
Laboratory of Molecular and Cellular Biology, Department of Bioregulation and Biointeraction, Graduate School of Agriculture, Tokyo University of Agriculture and Technology, Tokyo, Japan

H. Nishida, T. Oshima (eds.), *DNA Traffic in the Environment*,
https://doi.org/10.1007/978-981-13-3411-5_7

duction frequency up to 1.16×10^{-2} CFU/VP; (3) progeny VP production from the VP-mediated transductant, in a phenomenon referred to as "serial transduction"; (4) spontaneous budding production that starts when the host attains the stationary phase without host cell lysis; (5) various discrete size distribution of progeny particles; (6) VPs that can only be revealed by electron microscopy by its budding from the regenerated VP-mediated transductant. The yield of a VP is strictly regulated by 3 ± 2 particles per cell, whereas introduction of the *rel* mutation would induce overproduction to bring forward a period of production to the host log-phase, which implies that a VP is an attribute of the host cell. VPs can transfer cytosolic substances, such as plasmid and protein, along with host chromosomal fragments; furthermore, both single genes and multiple genes for the host's thermo- and psychro-tolerance enhance the host's environmental adaptation.

Keywords
Broad-host range transfer · Vector particle (VP) · Serial transduction · Virus-like particle or vesicle (VLP or VLV) · Outer membrane vesicle (OMV) · Membrane vesicle (MV)

7.1 Why Do We Need to Propose New Concepts of HGT Mediators?

The currently recognized horizontal gene transfer (HGT) vectors are summarized in Table 7.1. Gene transfer agents (GTAs) [17], DNA transfer by membrane vesicles (MVs) [30, 34], and intercellular nanotubes [19] have come to light as novel modes of HGT in bacteria (http://biobabel.wordpress.com/2012/01/24/novel-modes-of-lateral-gene-transfer-in-bacteria/) in addition to the three classic mechanisms: (1) transformation (in which naked DNA is taken up from the environment), (2) transduction (by which bacteriophages facilitate gene transfer by packaging host DNA as well as their own), and (3) conjugation (when plasmids encode a pilus by which they can be transferred from cell to cell) [49].

Some transduction-like unique mechanisms to mediate HGT between individual cells of the same bacterial population, termed GTAs (the specific names for individual GTAs are given by the abbreviation of genera and species of the original host as the prefix in principle), that are accomplished by a phage-like particle have already been reported for the α-proteobacteria *Bartonella bacilliformis, Bartonella grahamii, Bartonella henselae, Bartonella vinsonii, Rhodobacter capsulatus* (RcGTA), the δ-proteobacterium *Desulfovibrio desulfuricans* (Dd1), the spirochete *Brachyspira hyodysenteriae* (VSH-1), and the archaeon *Methanococcus voltae* (VTA) [50–54]. None of the packages have more than 14 kb of DNA, ranging in size from 4.4 to 13.6 kb; all of them take the form of small bacteriophages, although no commitment of a virus-like gene to the transportation of host DNA has been reported [50, 55]. Four known GTAs resemble "constitutive" generalized transducing phages

Table 7.1 Classic and novel HGT mediators

Mediator	Broad-host-range vector particle (VP)	Conjugative plasmid (Cp)	Gene transfer agent (GTA)	Intercellular nanotubes (IN)	Membrane vesicle (MV)	Transducing particle (TP)
Cell-to-cell contact	Not required	The conjugative plasmid harboring cell	GTA-responsible gene encoded microorganism	Living cells from the whole domain	Living cells from the whole domain	Virus (bacteriophage, phage)
Synonym	–	Integrative and conjugative elements (ICE)	–	–	Extracellular vesicle: EV, microvesicle: MV, outer membrane vesicle: OMV, virus-like particle: VLP, virus-like vesicle: VLV	–
Morphology	Sphere, from 20 to several hundred nm. Coating membrane is composed of protein, carbohydrate and lipid	CCC form DNA occasionally integrated into the host chromosome	Short-tailed icosahedral phage	Generally, tube length ranged up to 1 μm, whereas width ranged approximately from 30 to 130 nm. The relatively large size of the tubes concords with an assumption that they could easily accommodate the passage of proteins such as GFP (ca. 40 Å)	Gram-negative bacterial MVs range from 10 to 300 nm in diameter and contain OM and periplasmic constituents, including proteins, lipoproteins, phospholipids, and LPS. Gram-positive bacteria are similarly sized (50–150 nm in diameter) and are rich in membrane lipids as well as toxins (including the anthrax toxin)	Consistent with the originating virus. The most popular head size is ca. 60 nm in diameter in the marine environment

(continued)

Table 7.1 (continued)

Mediator	Broad-host-range vector particle (VP)	Conjugative plasmid (Cp)	Gene transfer agent (GTA)	Intercellular nanotubes (IN)	Membrane vesicle (MV)	Transducing particle (TP)
Cargo	Host chromosomal DNA fragment up to 400 kb so far examined, yet several segments could reside in one particle. Plasmid and cytoplasmic material, however, RNA has not been found.	DNA	Host chromosomal DNA fragment of 4–13 kb	The cytoplasmic material, host chromosomal DNA fragment, plasmid, RNA	The cytoplasmic material, host chromosomal DNA fragment, plasmid, RNA	Host chromosomal DNA fragment corresponding to ca. 10% of the virus genome that TP originating from
HGT frequency	~10^{-2}/VP	10^{-4}~10^{-3}/Cp	~10^{-2}/GTA	Uncertain	Uncertain	~10^{-7}/TP
Accessible recipient range	The whole domain	Mating pair formation (MPF) types of eight delimitate the host range	The same class	Uncertain	The whole domain	Consistent with the originating virus

Production scheme	The electron dense body (EDS), composed of the chromosomal fragment and proteinaceous material, is formed in the centre of the cytoplasm, transferred to the vicinity of the cell membrane, and finally released by budding without host cell disruption. Infected recipient cell acquires VP production of the same biological traits, termed as 'serial transduction'. VP production is ca. 3 VPs/cell per generation	The mechanism is the same for plasmids: First, the relaxase nicks the DNA at *oriT*, the nucleoprotein filament to be transferred to the recipient cell and then, replication in the original and novel hosts followed by	GTA production responsible gene (ca. 15 kb) encoded host cell produces GTA like the lytic virus, whereas the cargo is host chromosomal fragment. Mature GTA is released by the host cell lysis	N/A	Most models of MV release, studied primarily in *Proteobacteria*, propose an "either-or" involvement of proteins or LPS in this process. However, it is likely that MV formation results from the contribution of multiple dynamic surface components and there are several proposed mechanisms of MV release	On lysogenic virus infection, the host cell replicates abundant copies of virus DNA. These copies are then packaged into newly synthesised virions. A mispackaging of small pieces of bacterial DNA into the virions instead of the virus genome makes transducing particle (TP)
Reference	[1–9, 130], this chapter	[10–15]	[16–18, 50–57]	[19, 20]	[11, 21–40]	[3, 41–48]

[43]. All four GTA package double-stranded DNA (dsDNA) resemble "constitutive" generalized transducing phages, with a tailed phage structure and package of less DNA than would be expected to represent a complete tailed phage genome (except for Dd1). None of these agents appear to be particles with lytic activity. GTAs seem most likely to have been derived from bacteriophages that lost their ability to self-propagate [55]. Unfortunately, genetic information about the production of these particles is currently available only for RcGTA [55]. For a release of the RcGTA, cell lysis requires the genes rcc00555 and rcc00556, an endolysin and holin, respectively [56]. The production of RcGTA is regulated by host systems, including a putative histidine kinase (CckA) that is required for maximal expression of the holin gene (rcc00555) and for maturation of RcGTA to yield gene transduction-functional particles. VPs were observed to be distinct from GTAs and related entities from their susceptible recipient phylogenic range, cargo size, and budding production.

Although it has been recognized that MV production is a ubiquitous phenomenon found in many bacteria, the existence of a universal MV formation mechanism to explain it has not been clarified. The MV should not be collectively dealt with as one of the HGT mediators because it has several different functional aspects irrelevant to gene transfer. The MV is a universal cellular feature, common to the three domains of life [57]; however, the dimensions, morphology, and molecular composition of MVs in the environment are very comparable to those of some virions [58]. Despite this, molecular ecologists have not paid much attention to environmental MVs because their origin is ambiguous. Hence, the presence of MVs in natural environments has been mostly ignored; however, recent findings [59] demonstrating the abundance of bacterial MVs, comparable to that of VLPs, in marine ecosystems could change the situation. Biller et al. [59] observed an only negligible number of apparent tailed phages (or GTAs) in vesicle-rich ocean samples ranging from $\sim 10^5$ to $\sim 10^6$ MVs/mL, for which the MV isolating method was the same as that employed traditionally for the isolation of viruses. Hence, the isolation methods for viruses also result in the enrichment of MVs, so estimates of viral abundance might have been merged with MV abundance [60]. Therefore, complexity could be added to estimations of the environmental impact of either group of DNA contained in derived vesicles. Consequently, it is necessary to elucidate the releasing mechanisms of molecular contents to the environment for an understanding of the ecological impacts of MVs [61].

MV research has been performed mainly on functions and formation mechanisms. A cell under stress accumulates material, which is worsened upon damage of the regular housekeeping and stress-responsive mechanisms. The production of bacterial outer membrane vesicles (OMVs) that occurs under stress is an entirely independent, general envelope stress response that allows selective elimination of unwanted material cell by the preferential packaging of a misfolded protein. It has unknown how the released MV adheres to the cell and further fuses and transfers its contents [25]. Regarding the function of MVs, their involvement in pathogenicity has been studied; however, considering the presence of endogenous plant bacteria and intestinal bacteria, among others, the coordinated action of MVs on the host should also be considered sufficiently. Furthermore, little is known about the role of MVs in the environment. Considering that many MVs have been identified from biofilms, it is predicted that MVs are produced throughout the environment.

The MV induction mechanism differs depending on the environment; hence, MVs have different properties and functions depending on the environment being produced. Comprehensively identifying proteins contained in the MVs of *P. aeruginosa* revealed that MVs from free-living organisms contained toxins, whereas ones originating from biofilm contained a large number of siderophore proteins instead of toxins, distinct from free-living MVs. Although the role of MVs against infected hosts has been drawing attention because MVs produced by *P. aeruginosa* contain toxins such as protease, newly elucidated characteristics from the biofilm-originating MVs indicate an expanded role for environmental MVs [62].

DNA carrying MVs [38] represents only a small proportion (0.01–1%) of vesicles [22, 63], although some vesicles contain sufficient DNA to be visible after staining with SYBR fluorescent DNA dyes that are typically used to enumerate viruses. Thus, DNA is packaged heterogeneously within vesicle populations, and it appears that vesicles are likely to be a minor component of SYBR-visible particles in natural seawater compared with viruses [61]. Considering the content of the marine virus fraction that is mainly composed of bona fide viruses would be dependable, whereas employing an appropriate method to classify viruses from MVs, such as buoyant density, would be indispensable. Therefore, an argument concerning the environmental gene transfer frequency would be plausible to construct on reported VLP assemblages in the environment because the study of MV-mediated HGT is currently underway. Whether the adhesion of MVs to cells is selective or equally attached is of great interest to researchers. By elucidating this, the whole picture of MV formation, diffusion, and adhesion/fusion—that is, the whole bacterial membrane traffic—can be clarified and further application developments would be expected [37, 64].

Dubey and Ben-Yehuda [19] demonstrated the existence of tubular conduits forming between *Bacillus subtilis* cells. These nanotubes were shown to be able to mediate the exchange of proteins and non-conjugative plasmids. Nanotubes were also formed between both gram-positive *B. subtilis* and *Staphylococcus aureus*, and a thinner variety was formed between either of the gram-positive species and gram-negative *E. coli*. The plasmid transfer does not require any intrinsic plasmid elements, and a given cell can be either donor or recipient. Dubey and Ben-Yehuda suggested that the formation of "syncytium-like synergistic consortia" mediated by nanotube connections underlies many of the traits displayed by biofilms. How cargo is transported through the nanotubes remains unanswered. Whether the transport is active and requires energy or is passive and prompted by diffusion is also unknown. Both mechanisms possibly coexist, and utilization would depend on the delivered cargo. In eukaryotic cells, nanotubes are frequently associated with cytoskeletal and motor proteins, implying a role for active transport [65].

"Trans-kingdom gene transfer" has been reported [66–69], although the Ti plasmid is the sole molecular device found to be practically acting in situ to date [70, 71]. Bacterial cells generally restrict their acceptance of foreign genetic materials [72] by the restriction-modification systems, which interfere with flexible gene transfer between species of different genera [72–77]. Gene transfer is costly to the bacterium; the benefits provided by the acquisition of advantageous genes also act as a repair mechanism [78]. Studies of the genetic structure of bacterial populations have clearly revealed that accessory element (phages, plasmids, transposons, and insertion sequences) transfer has occurred during their recent evolutionary history [75, 79–81] and further conferred that some accessory elements have been transferred across much greater phylogenic differences [82].

Recently discovered Clustered Regularly Interspaced Short Palindromic Repeats (CRISPR) found in 40% of eubacterial and 90% of archaeal sequences prokaryotes [83] act as a kind of acquired immune system for plasmids and phages [84, 85]. A CRISPR-Cas system being considered to limit HGT [48]. Bacteria can acquire an entire chromosomal CRISPR-Cas system through transduction [48, 86, 87], whereas the protective effect from phage infection increases the HGT of phage-sensitive members of mixed populations [88]. Consequently, the overall ability of CRISPR-Cas promotes HGT through transduction with broader implications for microbial evolution [48, 88]; it is also involved in virus budding [89] in eukaryotes and archaea.

The host range of an HGT mediator is defined as the breadth of organisms capable of infecting and expressing its function, with limits on the host cell ascribed to the mediator, host, or environmental characteristics. Reflective of the mediator's discipline, there is no unified definition for a "broad-host-range" mediator because the host of the mediator regulates the physicochemical interaction with the mediator to infect, replicate, and maintain [90]. Therefore, the author empirically defines the a broad-host-range VP as one capable of infecting, transferring, and budding reproduction in the recipient cell in excess of the phylogenetic family level from the original host.

In addition to the HGT phenomena, the author has proposed another brand-new concept of broad-host-range VPs, which are composed in part of natural virus assemblages. Viruses have been considered as the most abundant biological entities in the living world and the reservoir of most genetic diversity in the sea, which amounts to ~10^5 to ~10^9 virus-like particles/mL [91]. The barrier hindering the acceptance of virus-mediated gene transfer resides in the concept that most viruses have a narrow host-range [92–94], which reduces the likelihood of extensive gene transfer within a mixed bacterial population. It is also believed that the host range of bacterial viruses (bacteriophages) is restricted to specific bacterial strains or closely related species [95, 96]. A given type of virus usually has a restricted range of hosts—often a single species; however, some viruses infect only certain subspecies, whereas others may infect more than one related species or even genus [92]. Such viruses with extended host range exceeding one genus are called "broad-host-range viruses", which likely comprise <0.5% of the total virus population [92].

Furthermore, viruses may exhibit a broader host range in deep waters than in surface waters [97]. The supposed host specificity of viruses, the mostly unknown species diversity and composition of marine microbial assemblages [98], and the quantitative evaluation of contact between marine viruses and hosts still involve much guesswork [99]. Hence, the contribution of virus-mediated gene transfer (transduction) has been considered a factor of minor importance for genetic diversity and evolution in the natural microbial community. Furthermore, transduction is a reductive process, in that the genetic donor is killed (lysis) in the process of producing a transducing phage particle [100]. However, because viruses are released in a free form and do not necessarily require cell contact to survive, transduction may represent an ideal method for dispersing genes in the environment [101].

HGT among organisms in natural environments is already a fact, not a "theory" as Koonin expressed [102]. Much like evolution itself, extensive HGT in the microbial world is not doubted anymore, as HGT is a dominant process in microbial evolution that generally occurs at a high rate. Rapid advancements in genome sequencing technology have provided evidence of HGT. The determination and analysis of complete genome sequences have led to the suggestion that HGT dominates microbial evolution, with the rate of gene gain and loss being comparable to the rate of spontaneous point mutations and much greater than the duplication rate [103]. Moreover, the relevance of "horizontal" as applied to gene flow is validated by strong evidence of the existence of a central vertical, tree-like trend in genome evolution. Thus, the focus of research has shifted toward the how and why of HGT, although much more remains to be done than has been accomplished already in these directions. The tree of life may be unresolvable because of the extent of HGT [104, 105], and that which we consider phylogeny may be defined in large part by HGT [106]. Many of these studies, however, rely on evidence that could be generated by forces other than gene transfer, including selection, variable evolutionary rates, and biased sampling [107].

Additionally, little information on the probable donor lineage and the likely time of evolution has been provided in most proposed cases. Even when donors and recipients have been proposed, there is rarely supporting evidence regarding the absence of the genes from relatives of the recipient lineage that diverged before the transfer. For a complete picture of any proposed case of HGT, it will be essential to have information regarding the vectors, what (if any) selective forces were involved in the transfer, and the extent of amelioration [107]. It is necessary to integrally consider these HGT phenomena as one of the elements constituting the mobilome, although there are only a few examples of the findings on the HGT for these elements. However, other than VP, the evidence for HGT has been studied primarily by using amplicon sequencings of plasmid, which would be transferred. To the author's knowledge, no other system intensively accumulates findings by examining HGT in cooperation with the culture system other than VP. First, the author recognized the VP as a part of the lysogenic virus and studied it as a kind of transducing particle; hence, the following argument is mainly based upon a comparison with the virus-mediated transduction.

Biological features of the entity the author incidentally discovered are undoubtedly involved with such VLP-MV conflated assemblages with morphological similarity and residing DNA; however, they are entirely distinct from what has been reported. Hence, this novel concept is proposed for consideration as an overlooked biological information-transfer medium.

7.2 Curiosity as an Incentive to Thinking Led to Novel Findings

An attempt to screen restriction enzymes from marine isolates was offered by Prof U. Simidu, Ocean Research Institute, the University of Tokyo. Three type 2 restriction enzymes—*Fsp*I1604, *Fsp*I1611, and *Asp*MD1—were consequently obtained from *Flavobacterium* sp. I 1604, *F.* sp. I 1611, and *Alcaligenes* sp. MD1, amongst approximately 150 species of marine bacteria [108, 109]. These Gram-negative strains were collected at approximately 2000 m deep in the central part of the Indian Ocean. They showed stiff resistance to streptomycin, harboring plasmid-like DNA, ultraviolet light, and/or mitomycin-inducible phages. The spontaneous release of phage-like particles (PLPs) into the culture medium without artificial induction attracted curiosity because no host-vector system had been developed for marine lysogenic phages. For microbial genetics to establish a host-vector system to elucidate somatic gene transfer mediated by transduction, the development of auxotrophic mutants is inevitable [110].

Unfortunately, the minimal medium essential for the marine bacterial genetics study has not been available since 1991 [111]. Therefore, an ability of dichlorophenoxyacetic acid (2,4-D) utilization was used as a marker for the gene transfer of the marine bacteria instead of the somatic gene mutant. 2,4-D is an anthropogenic persistent organic compound that did not originally have a natural origin; therefore, to be useful for the gene transfer index, previous studies used a decomposition of 2,4-D [112–114]. The bacterial 2,4-D biodegradation is plasmid-dependent, with properties that have been studied as a model of gene transfer in nature [115–122].

Contrary to prediction, 18 strains out of 25 ubiquinone-10 possessing marine isolates belonging to α-proteobacteria exhibited strong 2,4-D degradation that was superior to *Escherichia coli* JMP397. *E. coli* JMP397 is a transformant [123] harboring 2,4-D degradation pJP4 (80 kb); it exhibited higher 2,4-D degradation ability than the plasmid originating from *Alcaligenes eutrophus* JMP134 [118]. Currently, *A. eutrophus* has been reclassified as *Ralstonia eutropha* JMP 134 [124].

Extraction of plasmid-like elements (PLEs) from those high 2,4-D utilization marine strains (Alc 096, 233, 252, *Agrobacterium kieliense* IAM12618 [125]: basonym *Ahrensia kielensis*: [126]) and *F.* sp. I1604 was somewhat fastidious [127] by requiring repeated phenol treatment, with traits ascribed to tight coupling of DNA content with proteinaceous material to be encapsulated into the particle, as described in Fig. 7.2. The extracted PLEs were successfully generated transformants in the minimal medium supplemented with 2,4-D as a sole carbon and energy source from the recipient *E. coli* AB1157 (F^-; *thr-1 leuB6 thi-1 lacY1 galK2 ara-14 xyl-5 mtl-1 proA2 hisG4 argE3 rpsL31 tsx-33 supE44*), with a higher frequency than authentic pJP4 (PLEs: 1.5~13.1 × 10^6 CFU/μg DNA; pJP4: 5.7 × 10^5 CFU/μg DNA); the amino acid requirements of prototrophic revertants were also generated [1]. Hence, PLE is considered to combine a 2,4-D degrading ability and somatic genes to complement the auxotroph. However, the conjugative transfer of 2,4-D assimilability from those marine strains to *E. coli* was not achieved (Chiura, Unpublished data). A report indicated the physiological situation of marine donor bacteria would become the essential key to the conjugative transfer to terrestrial bacteria because the accomplishment of the conjugation is severely affected by the ionic strength and type of ion required [140–143]. Conjugative transfer of pJP4 has reported toward strains belonging to a variety of genera of the α-, β-, γβ- and γ-classes of the Proteobacteria; however, only *Pseudomonas putida* and *Delftia* sp. strains were able to grow on 2,4-D as the sole carbon source [144]. The broad-host-range (BHR) plasmids were defined as those plasmids that can self-transfer themselves and can stably replicate and maintain in bacterial species from at least two subgroups within the Proteobacteria (e.g., between α- and β- Proteobacteria) [145, 146]. Accordingly, the definition of BHR is not uniform with the overall HGT mediators. The BHR plasmids typically have mosaic genomes that include two distinct regions [147]. The "plasmid backbone" genes encode proteins involved in the replication, maintenance, control, and conjugative transfer of the BHR plasmid. In the marine environment, the directly determined rates of HGT in the marine bacterial community were reported to be high, ranging from 2.3×10^{-6} to 2.2×10^{-4} transconjugants per recipient [148, 149]. The HGT performance of the above PLE is comparable.

Despite PLEs being susceptible to DNase I, attempts to draw restriction maps for those extracted PLEs was a wasted effort because restriction enzymes had been incapable of producing restriction fragments. As an avoidable attempt, the electron microscopic observation of the PLE samples showed an apparent existence of particles instead of free DNA strings. Those PLEs must have been intracellularly maturing virus particles extracted from the cell [1]. Accordingly, the PLE must have been an intracellular form of that spontaneously released VLP to the culture medium, which was capable of encapsulating a piece of a chromosomal gene to achieve prototrophic reversion of recipient auxotrophy as the plasmid responsible for 2,4-D degradation [150]. Hence, the object of attention changed to the spontaneously released VLPs to the culture medium.

The release of VLPs to the culture medium started when the host cells entered to the stationary phase, and after that without accompanying any reduction of the host cell abundance. Yields of the VLP at 100 h culture at 30 °C attained 6–70% of the host cell abundance, depending upon the strain [1]. An attempt to examine the capability of the broad-host-range HGT exceeding at least the family level using the above VLPs demonstrated successful gene transfer exceeding the difference of the family level [1].

Here, VLPs can potentially accomplish transduction-like gene transfer exceeding the family level (see Table 7.4), referred to as VP. A generated VP-mediated transductant is named by combining the VP donor initial, the recipient genus initial, and "trans." Therefore, an *A. kielensis* originating VP is referred to as AkVP and a generated *E. coli* transductant is named AkEtrans. The transduction frequency is defined as the fraction of the generated colony number in each selected marker plate in the total applied particles. An overall average frequency of VLP-mediated gene transfer at a multiplicity of infection (MOI) of 0.1 was estimated to be between 2.62×10^{-3} and 3.58×10^{-5} per VP [2].

AkEtrans acquired VLP production also may transfer genes; namely, AKEVP was produced. 2,4-D transduction frequency/μg DNA was estimated as 2.1×10^6 CFU for AkVP and 2.1×10^6 CFU for AkEVP. At 6.4×10^6 CFU for FsI1604VP, it is four orders of magnitude more efficient than MVs derived from *Acinetobacter baylyi* demonstrated for drug resistance transfer [63].

The VPs of *A. kielensis* (AkVP: $\rho^{25} = 1.4157$ g/cm^3; size ± SD, nm = 123 ± 2.6, $n = 106$) are especially characterized for their enveloped structure (see Fig. 7.2) with varying lengths (0.8–1.5 μm in diameter) being encapsulated one to several spherical particles (5.1 ± 2.8 particles/envelope; $n = 60$) [1]. This phenomenon is an uncommon feature for prokaryotic viruses, although some eukaryotic viruses, such as the influenza virus, have envelopes. RNA phage Ø6 of *Pseudomonas* is a particular example of an encapsulated prokaryotic virus whose particle is incorporated into the recipient cell during infection [151]. Remarkably, the generated transductants acquired VLP production other than prototrophic reversion of the amino acid requirement and 2,4-D utilization.

Transduction is a process of abortive infection. No progenies from infected viruses' transducing particles are expected, as shown in the schematic diagram in Fig. 7.1, whereas the VLPs from the above marine bacteria were not the case. Therefore, new terms should be given to express this unusual phenomenon: the "serial transduction" and "serial transducing particle" as the broad-host-range VP.

E. coli transductant (AkEtrans) generated by AkVP acquired comparable VLP production (AkEtransVP: $\rho^{25} = 1.4159$ g/cm^3; size ± SD, nm = 106 ± 4.3, $n = 54$). This again made it possible to transfer 2,4-D utilization and prototrophic reversion of amino acid auxotrophy to the recipient *E. coli,* which exhibited a comparable overall transduction frequency of $5.25 \pm 2.85 \times 10^{-4}$ CFU/particle, comparable to that of AkVP ($2.31 \pm 1.50 \times 10^{-4}$ CFU/particle). The transducing frequencies of these VPs were found to be higher by 4–7 orders of magnitude than those reported for naturally isolated transducing phages [42, 45, 47].

The distribution of particle abundance in the sedimentation profile exhibited a similar continual increase with size—an apparent difference from findings concerning the size distribution of particles showing discrete bands, which were composed of several subpopulations. The trend curve showing the relationship between the distance of the ultracentrifuge tube from the bottom and buoyant density is given as:

$$f_{\text{Bouyant density}}(d) = -0.0004d^3 + 0.0087d^2 - 0.0978d + 1.5605 (R^2 = 0.999)$$

Here, d is the distance from the bottom in cm. As an extension, particle diameter (in nm) and content as dsDNA (kbp) could be expressed as the following functions of buoyant density (ρ):

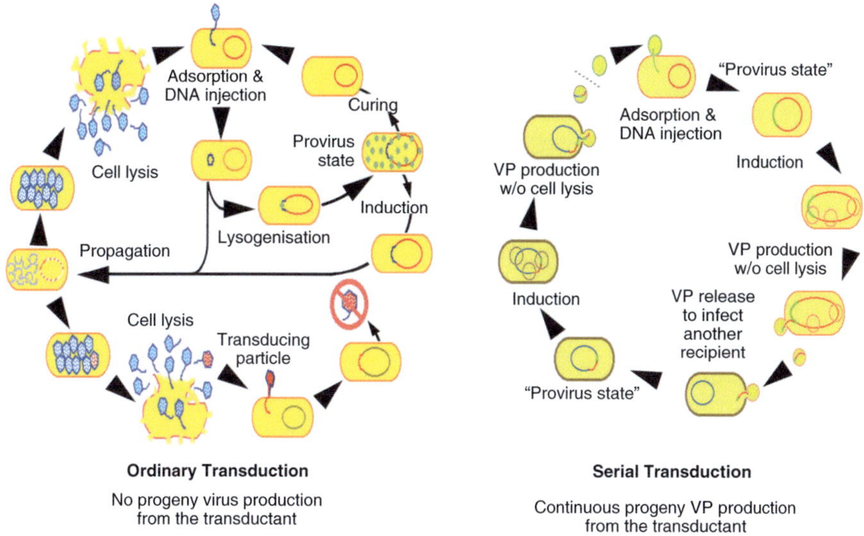

Fig. 7.1 A schematic representation of "ordinary" transduction and "serial transduction". The left panel shows the life cycle of lysogens and the production of transducing particles. Note that transducing particles would not be capable of producing progeny from the transductant. The right panel shows the serial transduction with the aid of broad-host-range vector particle (VP) discussed in this article. Transductants generated with the VP acquire VP production without accompanying cell lysis, whose reproduced VP can carry out successive transduction as the serial transducing particle

$$f_{\text{Particle diameter}}(\rho) = 14027\rho^3 - 53695\rho^2 + 67125\rho - 27{,}197\,(R^2 = 0.999)$$

$$f_{\text{Particle DNA content}}\ (\rho) = -571110\rho^3 + 2394972\rho^2 - 3347855\rho + 1{,}560{,}453\ (R^2 = 0.999)$$

According to the empirical formula [6] above, AkVP could encapsulate 416 kb as the transfer cargo capacity, whereas the result of a nucleic acid type analysis for the encapsulated DNA species estimated by an in-situ lysis technique [152] and pulse field gel electrophoresis (PFGE) [153] was 198 kb. Hence, a single AkVP might have carried two segments of the 198 kb DNA cargo inside. AkEtrans produced particles without an envelope or vesicle structure (see Fig. 7.2), with a cargo capacity of 266 kb as dsDNA showed two types of DNA species estimated to be 174 and 195 kb [5]. It is unknown how and why the AKEVP was produced from *E. coli* transductant as a free single particle but not collectively packed in the envelope or vesicle. High coordinated marker transfer (i.e. linkage) was recorded to exceed 70% between every selected and unselected marker for both AKVP and AkEVP mediated transduction, even using low MOI.

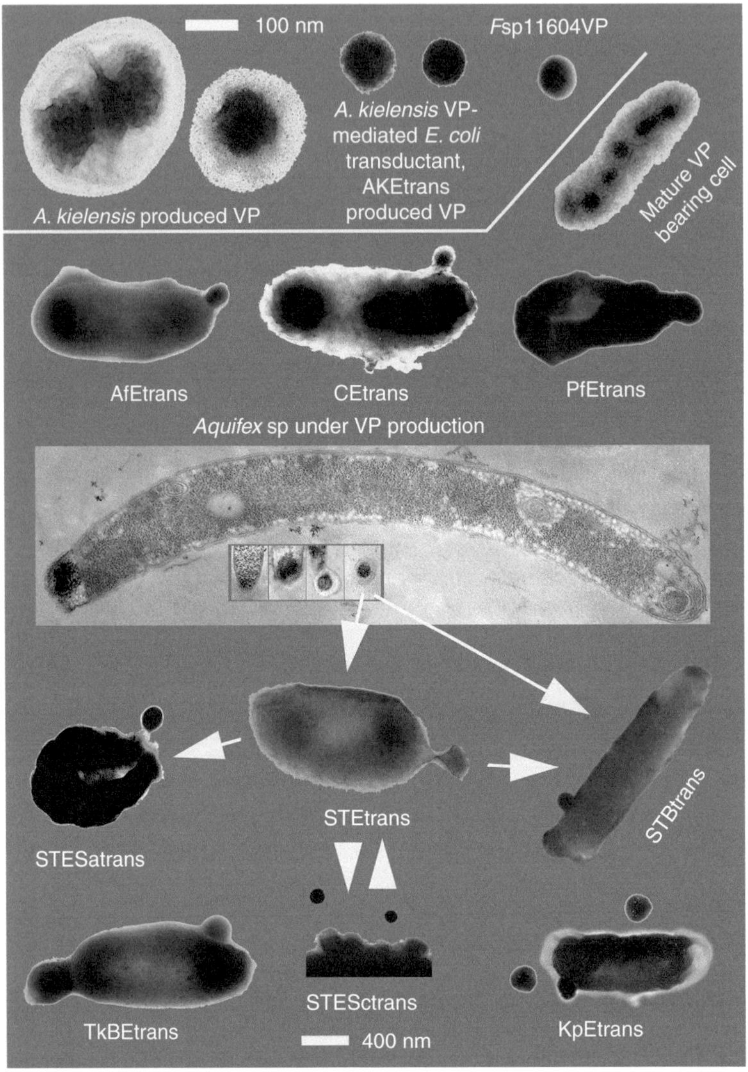

Fig. 7.2 Electron micrography (EM) images of VPs. Kiel Bay originating *Ahrensia kielensis* produced VPs are characterized for their envelope structures, which range from single to several particles. *A. kielensis* VP-mediated *Escherichia coli* transductants acquire production of VPs without the envelope, whose size is slightly smaller than that of *A. kielensis* VP. *F*spI1604 VP from *Flavobacterium* sp. I1604 originating from the Indian Ocean. Size scale, 100 nm, is placed above. Upper right: A bacterium is bearing five electron-dense bodies (EDBs) in the cell corresponding to the yields of "virus" as five particles per cell. A proportion of such cells having an intracellular mature "virus" in the population stands for the frequency of visibly infected cells (FVIC [128]). In the case of the VP lysogens, FVIC should be read as the VP induction frequency because VP production takes budding from the host without accompanying lysis. Center: A lengthy sausage-shaped *Aquifex* sp. is under VP production; EDBs are seen in the left end of the cell. Superimposed panels indicate the process of the budding production of VP, referred to as STVP, released to the milieu. STVP exhibited trans-Domain HGT via the serial transduction of VP-mediated transductants towards Archaea and Eukarya. Another example of the trans-Domain HGT is given by the VP originating from a hyperthermophilic archaeon, *Thermococcus kodakaraensis* B41, to transfer thermo-resistance towards a mesophile enterobacterium, *E. coli*, to generate TkBEtrans. For details of VP budding transductants, see Tables 7.2 and 7.3

As the mode of gene transfer, the generalized transduction was carried out because preferential marker transfer was not observed. In the case of AkVP, multiple particle infections might have occurred for its architecture; however, AkEVP did not organize a multi-particle assembly in an envelope. The distance between markers deduced from the co-transfer frequency [154] was given as *E. coli* markers: *his-arg*, 203.2 kb and *pro-arg*, 232.2 kb; however, based upon the genome sequence data of *E. coli* AB1157 (Chiura, 2016, Unpublished data), it was given as *his-arg*, 139.1 kb and *pro-arg*, 322.2 kb. Although the results of coordinated marker transfer are not always represented as the distance of respective markers in the recipient and carried DNA species molecular mass, it is plausible that significant genetic traits are transferred to the recipient beyond generations in a phenomenon termed "serial transduction" [4, 7]. Concerning the cargo content of the vehicle, Hageman and colleagues [155] reported that MVs produced by *A. kielensis* DSM No. 5890 Strain B, whose outer membrane vesicle's encapsulated genome size was 30.1 kb, encoded a prokaryotic sequence without a virus. Both strains used in these experiments are the same strain according to NCBI. Although a considerable discrepancy was found between the sizes carried in the AkVP and the MV derived from *A. kielensis* DSM, the contents of these vehicles' cargo would consistently host chromosomal genes. Furthermore, *A. kielensis* encodes GTA in the genome (NCBI Reference Sequence: NZ_ARFW00000000.1 https://www.ncbi.nlm.nih.gov/genome/15060), which might have been contributing to MV biogenesis, although the MV genome would be exceeding the GTA capacity reported thus far [156].

7.2.1 Looking for the Environmental Broad-Host-Range Serial Transducing Particles: VPs

Some marine isolates may possibly produce VPs to carry out broad-host-range serial transduction [1, 2, 6, 7, 157, 158]. Hence, "virus" fractions collected from normal and thermal [8, 9, 130] environments were examined for the broad-host-range serial transducing particles: VPs. The phylogenic position of the recipient and the original VLP host bacteria are differently classified at the family level (see Table 7.4). These findings are not explained by the general concept of lytic or lysogenic cycles of viruses infectious to bacterial cells, suggesting that some VLPs do not fit the conventional molecular view of viral genetics.

Although it would be a herculean task to set up a protocol in the laboratory to reproduce transduction consistent with the environmental situation, the procedure was done according to the empirically studied method for phage-mediated transduction. Criteria for the VP judgement are achieved for a VLP or VLP producer to accomplish the serial transduction with the acquisition of budding particle production from the generated transductant belonging to at least the family level. The budding prokaryotic virus is quite scarce, with a strictly restricted host range. Budding is suggested as the mode of exit for pleomorphic archaeal and bacterial viruses [159]: pleolipovirus, plasmavirus, and the mycoplasma phage L2 and L172 [160–163]. Practically, prokaryotic virus budding has not been established because of the current lack of knowledge concerning the exit mechanism [159]; it relies on eukaryotic virus research, which is beyond the scope of this chapter.

To ensure their purity prior to running the transduction experiment, the recovered VLPs as a fraction of 0.2 μm > specimen > 30 kDa were treated with DNase I and RNase A to avoid the possibility of transformation. This was followed by equilibrium CsCl-density-gradient ultracentrifugation [164] at 175000 × *g* overnight at 15 °C. The bands formed were separately recovered and examined for the abundance, shape, and size of VLPs using electron microscopy (EM) [2], whereas the SYBR-epifluorescence microscopy method [165] was used on board the research cruise. The purified VLP was suspended in TBT (100 mM Tris-HCl, 100 mM NaCl, and 10 mM $MgCl_2$, pH 7.4) buffer to make an appropriate dilution after CsCl removal.

For the recipient preparation, the physiological condition of the recipient should be reproducible to ensure the application of different VLP specimens from different origins. The recipient strains were grown to the mid-log phase. Then, the culture broth was substituted to 7% glycerine containing TBT buffer, dispensed to aliquots, frozen in liquid nitrogen, and kept in the deep freezer until use to endorse such requirement.

As the standardized protocol for transduction, 1 mL of thawed recipient suspension was mixed with the VLPs to obtain an appropriate MOI, and the mixture was left undisturbed at 30 °C for 15~30 min. Cells were washed with buffer and suspended in 1 mL of the same solution. Inactivation of the VLPs was attained by ultraviolet (UV) irradiation as previously described [2]. The four controls were as follows: (1) recipient cells tested with TBT buffer instead of the VLPs to determine the spontaneous revertant rate; (2) UV-inactivated VLPs and recipient cells; (3 and 4) VLPs with/without UV inactivation without the addition of recipient cells. For the transduction experiment, the recipients were incubated with or without VLPs, inoculated in LB medium, and kept at 30 °C for 2 days. In the transduction experiment, a portion of the sample was fixed with 2.5% glutaraldehyde; free particles and bacteria were enumerated by electron microscopy [166] to determine the practical MOI.

The growth profiles of generated transductants were examined, with specific reference to the particle budding production, the proportion of the particle producing population (VPIF; see below), the amount of the produced particle per cell, and the second VP-mediated transduction (see Fig. 7.1). Using this method, verification of the lethal effect on "VP infection," the period of VP production, and the yields of VPs per cell can be obtained. The lethal effect caused by the virus is the mature virion release through the host cell rupture, with infection from cell death-inducing MVs as the microbial cellular response being equivocal. In addition, if the recipient fails to make an appropriate membrane fusion with VP at the VP-mediated transduction, plausible cytosol leakage of the recipient cell would occur and cause death (see Fig. 7.2).

The results obtained to date are summarized in Tables 7.2 and 7.3. The frequency of the marker transfer with the aid of VP-mediated generalized transduction depends upon the combination of donor and recipient, varying between 3.23×10^{-2} and 3.24×10^{-9}, with a mean value of 1.14×10^{-3} and a median value of 9.22×10^{-6} CFU/VP. These values are considered to be high frequency among the values accounting

Table 7.2 Employed VP source, recipient, product, VLP size, applied DNA, kb, multiplicity of infection (MOI) and efficiency of plating (EOP) with specific reference to the particle treatment with or without ultraviolet (UV) light

Source VP	Recipient	Product	Size, nm	ρ^{25}, g/cm^3	*n*	Applied DNA, kbp	MOI	EOP, % UV+	EOP, % UV−	Reference
Ahrensia kielensis IAM12618	*E. coli* AB1157	AkEtrans_F1	123 ± 3.9	1.3131–1.3288	46	198	0.1	17.3 ± 1.6	16.25 ± 1.0	[1, 2]
AkEtrans_F1	*E. coli* AB1157	AkEtrans_F2	106 ± 4.3	1.3132 ± 0.0184	54	174, 195	0.18	32.3 ± 17.3	30.4 ± 11.6	[1, 2]
Flavobacterium sp. I1604	*E. coli* AB1157	FsI1604Etrans	52.4–58.2	1.4725–1.4539	24	61.1	1.16	62.8 ± 20.4	62.3 ± 11.6	[1, 2]
Aliivibrio fischeri NCIMB1281^T	*E. coli* AB1157	AfEtrans	18.1–159.2	1.3607–1.3980	11	48.2 ± 27.8	100	136.0 ± 14.0	116.0 ± 39.0	[6]
Vibrio sp. FK01073	*E. coli* JM109	VtcEtrans	20–220	1.1911–1.5102	3	145 ± 48	0.3	58.6 ± 6.0	55.2 ± 17.6	[129]
Vibrio sp. FK01073	*E. coli* JM109	VtcEtrans	20–220	1.1911–1.5256	3	145 ± 48	0.34	91.4 ± 9.4	69.4 ± 23.5	
Calvi seawater	*E. coli* AB1157	CEtrans_F1	38.0–132.3	1.3026–1.5757	24	370	0.1	100.0 ± 9.9	100 ± 9.9	[7]
					24	370	1	76.5 ± 9.8	64.8 ± 9.8	
					24	370	5	64.9 ± 1.7	48.2 ± 1.7	
					24	370	20	35.3 ± 1.4	20.8 ± 1.4	
					24	370	200	17.8 ± 0.5	15.2 ± 0.5	
CEtrans_F1	*E. coli* AB1157	CEtrans_F2	100–140	1.2875–1.3680	24	370	5.5	67.8 ± 1.2	66.2 ± 1.2	
Toyoha geothermal vent	*E. coli* AB1157	TYEtrans	30–320	1.3033–1.3520	12	195.7	0.28	68.6 ± 14.2	83.4 ± 4.9	[130]
Suiyo seamount	*E. coli* AB1157	WedEtrans	138.2–68.2	1.2902–1.3524	9	20–800	0.16	60.8 ± 17.3	67.8 ± 6.6	[131]
Nakanoyu '95	*E. coli* AB1157	STEtrans_95	33.8–99.1	1.3898–1.5137	9	33.0–145.0	0.26	29.0 ± 9.0	20.0 ± 7.0	[4, 8, 9]
Nakanoyu '97	*E. coli* AB1157	STEtrans_95	56.3–127.7	1.3116–1.4636	9	18.9–697.8	6.76	58.3 ± 5.3	54.2 ± 42.9	
Nakanoyu '02	*E. coli* AB1157	STEtrans_F1	43.9–146.1	1.2757–1.4923	5	406.4	0.16	19.5 ± 6.8	28.3 ± 8.7	
Nakanoyu '02	*B. subtilis* PS9	BsEtrans_F1	43.9–146.1		9	406.4	0.12	68.5 ± 6.2	29.3 ± 23.2	

(continued)

Table 7.2 (continued)

Source VP	Recipient	Product	Size, nm	ρ^{25}, g/cm^3	*n*	Applied DNA, kbp	MOI	EOP, % UV+	EOP, % UV−	Reference
STEVP_F1	*S. cerevisiae* YPH499	ScYPH499trans			12	372.1	0.06	111.0 ± 13.8	71.1 ± 30.3	[132]
	S. cerevisiae YPH500	ScYPH500trans			12		0.12	68.4 ± 63.8	75.4 ± 23.6	
	S. acidocaldarius MR87	SaEtrans			9		1	81.0 ± 29.3	63.7 ± 23.2	
ScSEY6120VP	*E. coli* AB1157	ScEtrans_Sc	128.3	1.2685–1.3945	3	372.1	0.26	170.0 ± 3.5	66.7 ± 15.7	[133]
Thermococcus kodakaraensis B41	*E. coli* AB1157	TkBEtrans	90.1–155.5	1.2265–1.3860	3	230.5–386.7	0.25	77.0 ± 24.0	46.4 ± 11.2	[131, 134]
STEVP_1	*E. coli* AB1157	STEtrans_F2	128.3		3	372.1	0.1	54.0 ± 11.0	58.0 ± 18.0	[8, 9]
STEVP_1	*B. subtilis* PS9	BsEtrans_F2	128.3	1.2056–1.3926	3	372.1	0.1	69.0 ± 6.0	29.0 ± 23.0	
STEVP_1	*E. coli* W2252: HfrC	Hfr*Met*trans	128.3		9	372.1	0.29	82.8 ± 5.4	73.5 ± 3.5	[135]
STEVP_F2	*E. coli* DH5a	DH*lac*trans	128.3	1.2056–1.3926	3	372.1	0.14	92.1 ± 14.5	73.6 ± 13.5	[136]
STEVP_1	YFPEtrans[a]	JMlactrans	110	1.2875–1.4925	3	373.33	1.04	82.2 ± 8.8	77.8 ± 5.7	[137]
JMlacVP	*E. coli* DH5a	DH*amp*^{r}*lac*trans	43–140		3	297.8	9.9	89.63 ± 9.6	79.0 ± 5.7	
STEVP	*E. coli* AB1157	STEtrans_F3	128.3	1.2056–1.3926	3	372.1	0.1	129.0 ± 40.4	68.0 ± 19.3	
P. filamentus ATCC700397^T	*E. coli* AB1157	PfEtrans	86–346	1.2870–1.3566	5	150	4.55	121.25 ± 28.1	102.5 ± 9.4	[138]
K. pneumonia subsp. rhinoscleromatis	*E. coli* AB1157	KpEtrans	42.0–294		9	175.9, 320.3	1.09	160.0 ± 17.2	128.0 ± 17.9	[139]
	*E. coli*DH5α	KpDHtrans	42.0–294	1.3190–1.3551	9		1.12	112.7 ± 3.9	111.8 ± 1.5	
	*E. coli*JE6937	KpJEtrans	42.0–294		9		1.12	187.9 ± 7.0	142.1 ± 11.8	

[a]*yfp::amp* fusion plasmid bearing *E. coli* JM109

Table 7.3 Employed particle source, recipient, and transduction frequency for specific biochemical markers

Source VP	Recipient	MOI	Tolerance, N2 fix plasmid	Leu	Pro	His	Arg	Trp	Lys/Met/Ura	All Marks
A. kielensis IAM12618	*E. coli* AB1157	0.1	$2.27 \pm 1.50 \times 10^{-4}$ 2,4D	$2.73 \pm 0.94 \times 10^{-3}$	$2.84 \pm 1.57 \times 10^{-3}$	$2.03 \pm 0.22 \times 10^{-3}$	$2.38 \pm 2.17 \times 10^{-3}$			$2.31 \pm 1.50 \times 10^{-4}$
AkEtrans	*E. coli* AB1157	0.18	$3.87 \pm 2.85 \times 10^{-4}$ 2,4D	$1.66 \pm 0.64 \times 10^{-4}$	$2.05 \pm 1.69 \times 10^{-3}$	$1.28 \pm 0.31 \times 10^{-3}$	$1.70 \pm 0.71 \times 10^{-3}$			$5.25 \pm 2.85 \times 10^{-4}$
F. spI1604	*E. coli* AB1157	1.16	$1.22 \pm 0.78 \times 10^{-4}$ 2,4D	$7.34 \pm 1.59 \times 10^{-4}$	$1.18 \pm 0.26 \times 10^{-3}$	$8.43 \pm 1.83 \times 10^{-4}$	$8.46 \pm 1.84 \times 10^{-4}$			$1.87 \pm 0.78 \times 10^{-4}$
A. fischeri NCIMB1281[T]	*E. coli* AB1157	100		$1.48 \pm 0.21 \times 10^{-6}$	$1.31 \pm 0.16 \times 10^{-6}$	$1.58 \pm 0.23 \times 10^{-6}$	$2.62 \pm 0.37 \times 10^{-6}$			1.30×10^{-4}
V. sp. FK01073	*E. coli* AB1157	0.3	1.09×10^{-5} OTC	$4.41 \pm 3.04 \times 10^{-6}$	$5.73 \pm 2.82 \times 10^{-6}$	$3.53 \pm 0.92 \times 10^{-6}$	$3.53 \pm 0.88 \times 10^{-6}$			$1.00 \pm 1.00 \times 10^{-8}$
Calvi seawater	*E. coli* AB1157	0.1		$3.33 \pm 3.22 \times 10^{-5}$	$2.22 \pm 2.38 \times 10^{-5}$	$7.41 \pm 1.76 \times 10^{-6}$	$1.48 \pm 0.52 \times 10^{-5}$			$1.00 \pm 1.00 \times 10^{-8}$
Calvi seawater	*E. coli* AB1157	1		$9.40 \pm 2.85 \times 10^{-3}$	$8.34 \pm 2.83 \times 10^{-3}$	$9.04 \pm 3.07 \times 10^{-3}$	$9.73 \pm 1.90 \times 10^{-3}$			$1.39 \pm 0.10 \times 10^{-7}$
Calvi seawater	*E. coli* AB1157	5		$1.04 \pm 1.62 \times 10^{-3}$	$7.24 \pm 2.83 \times 10^{-3}$	$9.83 \pm 1.12 \times 10^{-3}$	$6.03 \pm 0.92 \times 10^{-3}$			$1.46 \pm 0.01 \times 10^{-5}$
Calvi seawater	*E. coli* AB1157	20		$7.34 \pm 3.04 \times 10^{-4}$	$7.15 \pm 0.28 \times 10^{-4}$	$7.34 \pm 4.03 \times 10^{-4}$	$5.83 \pm 1.08 \times 10^{-4}$			$1.94 \pm 0.95 \times 10^{-5}$
Calvi seawater	*E. coli* AB1157	200		$9.14 \pm 0.001 \times 10^{-4}$	$9.53 \pm 0.001 \times 10^{-4}$	$7.24 \pm 0.001 \times 10^{-4}$	$7.18 \pm 0.001 \times 10^{-4}$			$1.02 \pm 0.01 \times 10^{-3}$
Toyoha geothermal vent	*E. coli* AB1157	5.5	5.12×10^{-8} Ter	$1.89 \pm 0.92 \times 10^{-4}$	$2.58 \pm 0.87 \times 10^{-4}$	$1.05 \pm 0.0 \times 10^{-4}$	$1.74 \pm 0.40 \times 10^{-4}$			
Suiyo Seamount	*E. coli* AB1157	0.28		$9.22 \pm 0.59 \times 10^{-6}$	$9.22 \pm 5.87 \times 10^{-6}$	$3.71 \pm 2.36 \times 10^{-6}$	$2.41 \pm 1.54 \times 10^{-5}$			
Nakanoyu ‘93	*E. coli* AB1157	0.16		$8.90 \pm 1.20 \times 10^{-7}$	$1.37 \pm 1.86 \times 10^{-6}$	$6.73 \pm 1.0 \times 10^{-7}$	$1.21 \pm 0.20 \times 10^{-6}$			
Nakanoyu ‘95	*E. coli* AB1157	0.26		$2.04 \pm 0.02 \times 10^{-4}$	$6.43 \pm 0.14 \times 10^{-4}$	$9.11 \pm 2.76 \times 10^{-5}$	$3.86 \pm 0.51 \times 10^{-4}$			

(continued)

Table 7.3 (continued)

Source VP	Recipient	MOI	Tolerance, N2 fix plasmid	Leu	Pro	His	Arg	Trp	Lys/Met/Ura	All Marks
Nakanoyu '97	*E. coli* AB1157	6.76		8.90 ± 0.89 × 10^{-7}	1.37 ± 0.87 × 10^{-6}	6.30 ± 5.13 × 10^{-7}	1.21 ± 0.20 × 10^{-6}			
Nakanoyu '97	*B. subtilis* PS9	0.16		8.20 ± 0.0 × 10^{-7}		1.17 ± 0.80 × 10^{-6}		1.21 × 10^{-6}	5.20 ± 2.00 × 10^{-7} M	
STEVP_F1	*S. cerevisiae* YPH499	0.12		4.35 ± 0.05 × 10^{-4}		3.01 ± 3.70 × 10^{-4}		2.18 ± 1.86 × 10^{-4}	3.79 ± 5.41 × 10^{-4} L	3.33 ± 2.75 × 10^{-4}
STEVP_F1	*S. cerevisiae* YPH500	0.06		8.39 ± 0.02 × 10^{-5}		1.68 ± 0.02 × 10^{-5}		1.34 ± 0.02 × 10^{-5}	2.70 ± 2.24 × 10^{-5} L	3.53 ± 0.57 × 10^{-5}
STEVP_F1	*S. cerevisiae* SEY6120	0.12		1.74 ± 0.18 × 10^{-6}		2.08 ± 0.20 × 10^{-6}		1.21 ± 0.55 × 10^{-6}	4.73 ± 1.46 × 10^{-6} L	2.44 ± 0.60 × 10^{-6}
STEVP_F1	*S. cerevisiae* W303	0.2		3.23 ± 0.006 × 10^{-2}		8.08 ± 1.14 × 10^{-3}		3.23 ± 4.57 × 10^{-3}		1.45 ± 0.19 × 10^{-2}
STEVP_F1	*S. acidocaldarius* MR87	0.19				1.64 ± 0.53 × 10^{-2}			1.28 ± 0.23 × 10^{-2} U	1.28 ± 0.81 × 10^{-2}
ScSEY6120VP	*E. coli* AB1157	1		2.03 ± 0.0 × 10^{-5}						
T. kodakaraensis B41	*E. coli* AB1157	0.26	1.27 × 10^{-7} Ter	3.30 × 10^{-4}	9.80 ± 10 × 10^{-7}	2.94 ± 1.00 × 10^{-6}	2.29 ± 1.53 × 10^{-6}			
STEVP_ F1	*E. coli* AB1157	0.25		2.10 ± 2.89 × 10^{-5}	2.05 ± 1.05 × 10^{-4}	1.21 ± 0.91 × 10^{-5}	4.44 ± 4.82 × 10^{-5}			
STEVP_F1	*B. subtilis* PS9	0.1		6.42 ± 8.58 × 10^{-8}		3.72 ± 5.25 × 10^{-8}			2.01 ± 1.79 × 10^{-8} L 2.53 ± 2.79 × 10^{-8} M	
STEVP_F1	*E. coli*W2252: HfrC	0.1							1.43 ± 0.82 × 10^{-7} M	
STEVP_F2	*E. coli* DH5a	0.29	2.83 ± 0.12 × 10^{-4} z							
STEVP_F1	YFPEtrans	0.14	1.00 × 10^{-5} z							
JMlacVP	*E. coli*DH5a	1.04	1.9 × 10^{-6} z							
			1.9 × 10^{-6} p							

STEVP_F2	*E. coli* AB1157	9.9	1.54 × 10^{-8} Ter	4.10 ± 0.01 × 10^{-6}	4.10 ± 0.16 × 10^{-6}	3.03 ± 0.03 × 10^{-5}	4.20 ± 0.376 × 10^{-6}			
P. filamentus ATCC700397^T	*E. coli* AB1157	0.1	1.80 ± 0.40 × 10^{-6} Ps	1.77 ± 0.61 × 10^{-5}	4.14 ± 0.56 × 10^{-6}	1.27 ± 0.45 × 10^{-5}	3.03 ± 1.09 × 10^{-5}			4.64 ± 1.40 × 10^{-8}
K. pneumonia subsp. rhinoscleromatis	*E. coli* AB1157	4.55	4.37 ± 2.52 × 10^{-7} N2	5.04 ± 0.58 × 10^{-7}	2.52 ± 0.58 × 10^{-7}	2.02 ± 0.58 × 10^{-7}	1.73 ± 0.15 × 10^{-7}			
	E. coli DH5a	1.09	2.38 ± 0.03 × 10^{-7} N2	2.87 ± 0.50 × 10^{-7}	8.62 ± 8.62 × 10^{-8}	2.10 ± 0.13 × 10^{-7}	2.20 ± 0.13 × 10^{-7}			
	E. coli JE6937	1.12	1.84 ± 0.80 × 10^{-8} N2	2.87 ± 0.50 × 10^{-8}	8.62 ± 8.62 × 10^{-9}	2.30 ± 1.00 × 10^{-8}	2.87 ± 0.50 × 10^{-8}			

Tolerance, N2 fix plasmid column keys: *2,4D* 2,4-D utilization, *OTC* OTC resistance acquisition, *Ter* thermo-resistance acquisition, *z lac* Z acquisition, *p* plasmid acquisition, *Ps* psychro tolerance acquisition

Table 7.4 Occurrence of VP-mediated HGT between phylogenetically different organisms

Domain:	Bacteria	Bacteria	Bacteria	Bacteria	Bacteria	Bacteria	Bacteria	Archaea	Archaea	Eukarya
Kingdom:	Gram-negative bacteria	Gram-negative bacteria	Gram-positive bacteria	Gram-negative bacteria	Gram-negative bacteria	Gram-negative bacteria	Gram-negative bacteria	Crenachaeota	Euryachaeota	Fungi
Phylum:	Proteobacteria	*Aquificae*	*Firmicutes*	*Proteobacteria*	*Proteobacteria*	Baceroidetes	Baceroidetes	*Crenachaeota*	*Euryachaeota*	*Ascomycota*
Class:	a-proteobacteria	*Aquificae*	*Bacilli*	*g-proteobacteria*	*g-proteobacteria*	Flavobacteriia	Flavobacteriia	*Thermoprotei*	*Thermococci*	*Saccharomycotia*
Order:	Rhodobacterales	*Aquificales*	*Bacillales*	*Vibrionales*	*Enterobacteriales*	Flavobacteriales	Flavobacteriales	*Sulforobales*	*Thermococ-cales*	*Saccharomycetales*
Family:	Rhodobactera-ceae	*Aquificaceae*	*Bacillacea-sea*	*Vibrionaceae*	*Enterobactera-ceae*	*Flavobacteria-ceae*	*Flavobacteria-ceae*	*Sulfolobase-aes*	*Termococca-ceae*	*Saccharomyceta-ceae*
Genus:	*Ahrensia*	*Aquifex*	*Bacillus*	*Aliivibrio*	*Escherichia*	*Flavobacterium*	*Polaribacter*	*Sulfolobus*	*Thermococcus*	*Saccharomyces*
Species:	*A. kielensis*	*A.* sp.	*B. subtilis* PS9	*A. fischeri*	*E. coli* AB1157	*Flavobacerium* sp. I1604	*P. filamentus* ATCC700397	*S. acidocaldarius* MR87	*T. kodakaraensis* B41	*S. cerevisiae* SEY499, SEY6120, YPH500, W303
Sampled location	Kiel Bay, DEU	Nakanoyu Hot Spring, JPN	IAM, JPN	Yellow Sea, KOR	NIG, JPN	Bay of Bengal Centre, Indian Ocean	Arctic Ocean, ATCC, USA	Yellow Stone National Park, USA	Suiyo Sea Mount, Izu-Ogasawara Arc, Pacific Ocean	NAIST, JPN

for HGT [42, 44, 167]. A summary of the VP-mediated HGT occurrence between phylogenetically different organisms is given in Table 7.4, which shows that VPs are susceptible to a broad range of recipients. This fact implies that VPs should be quite prevalent on the planet (see Fig. 7.1 and 7.2).

7.2.2 VPs in the Environment

Cells undergoing viral induction can be recognized with transmission electron microscopy (TEM) [3, 168] in whole-cell TEM images of the induction; the proportion of the population with such intracellular particles is given as the frequency of visibly infected cells (FVIC) [128]. The method was first developed using thin-sectioning of picoplankton cells, which makes it possible to enumerate the fraction of cells infected within aquatic microbial consortia. This approach has demonstrated infection of several planktonic microorganisms in an environmental situation (in situ) and has been used to estimate the virioplankton production rate. Proctor and Fuhrman [169] demonstrated that intracellular viruses could be detected in cyanobacteria and heterotrophic bacteria collected from several marine environments, ranging from mesotrophic coastal sites to the oligotrophic ocean. The incidence of cells containing viruses [101] ranged from 0.9 to 4.3% of heterotrophic bacteria and 0.8 to 2.8% of cyanobacteria. In another study examining bacterial populations associated with sinking particles, a similar FVIC (0.7–3.7%) was recorded. Bacterial populations within this unique marine niche may also be controlled by viral lysis [170].

The FVIC within bacterioplankton was also determined by merely examining whole bacterial cells at a high accelerating voltage (80 kV) using TEM [171]. This method requires much less effort than the thin sectioning of bacterioplankton, but concern has been expressed over its accuracy [76]. Weinbauer [128] and colleagues found that whole-cell estimates of FVIC were on average 79% of the value of estimates based on thin sections [101]. Hence, whole-cell estimation consistently underestimates the level of virus-mediated mortality. FVIC values are inevitably converted to a proportion of the entire bacterioplankton community, estimating the overall level of virus-mediated lysis in situ; however, cells bearing mature viral particles would be ruptured according to the current concept. Proctor [172] and colleagues hypothesized that infected heterotrophic marine bacteria exhibit mature phages after 73% to 86% of the latent period has passed. Therefore, the total abundance of infected cells is expected to exceed the number of visibly infected cells by between 3.7- and 7.14-folds [101]. Based on FVIC data, a phages infecting average of 17% of the bacteria comprising bacterioplankton communities from a variety of aquatic environments [101].

Although VPs are distinct from common lytic viruses, this method can be applied for determining the extent of VP induction because the time of the particle production does not accompany the host cell disruption. Hence, the timing of the start of VP production can be judged from the time of the intracellular formation of electron-dense bodies (EDBs). Furthermore, GTA behaves the same as a virus, whereas the majority of MVs are considered to be "outer membrane vesicles," with the relationship between EDBs and MVs still being uncertain. In the author's experience, FVIC accounted for 6.02 ± 3.45% (*n* = 3591) in the marine environment (Pacific Ocean, Mediterranean Sea) and 6.92 ± 3.93% (*n* = 3925) in the thermal environment (geothermal vent, hydrothermal vent). Indeed, FVIC represents the total cell proportion engaging in virus and VP production. Estimates of VLP and cell abundance are 2.3×10^7 VLP/mL and 1.8×10^6 cells/mL in marine environments and 1.9×10^7 VLP/mL and 3.6×10^6 cells/mL in thermal environments, respectively. The somatic gene transfer frequency of VPs has been roughly estimated as ~4×10^{-4} CFU/VP for the marine environment and ~3.6×10^{-3} CFU/VP for the thermal environment (Chiura, Unpublished data). Based upon this frequency and VP yields per cell of 3, the number of cells being committed to VP production would be 3.0×10^3 cells/mL in the marine environment and 2.3×10^4 cells/mL in the thermal environment. Consequently, VP lysogens in the environmental microbial assemblage would be estimated as 0.2% and 0.6% of the population.

The term FVIC is not appropriate to express the extent of VP induction in VP lysogens, so the term "VP-induction frequency" (VPIF)" is used hereafter. Both natural and VP transductant strains generally start particle production by host cells becoming stationary. Furthermore, the yields of VPs per cell are in the range of three and five particles, whose period and productivity per cell appear to be strictly controlled to fit the cell phase. In most cases, the abundance of free (extracellular) VPs do not exceed the host cell abundance, although this was not the case for VP production from *Vibrionaceae* (*Aliivibrio fischeri* NCIMB1281^T = ORI No.194: basonym, *Vibrio fischeri*, and *Vibrio* sp. FK01073). The number of free particles appeared to be associated with the number of VP-producing cells in the population—that is, the extent of the VPIF would determine the number of extracellular particles. Therefore, *Vibrionaceae* exhibited a higher frequency of VPIF. *A. fischeri* entered a logarithmic phase of growth for up to 12 h immediately after the initiation of culture, showing a generation time of approximately 1.84 h, and then entered a stationary phase to attain a stabilized cell abundance at approximately 1.2×10^9 cells/mL until the end of the culture. Initiation of the production of particles, referred to as AfVPs, by budding was determined from the particle per cell ratio and particle induction frequency to occur at approximately 15 h after initiation. The number of free particles continued to increase up to approximately 50 h after initiation, then stabilized at 1.20×10^{10} AfVP/mL. Three sharp peaks of induction (VPIF, period, %: 24 h = 15.8; 100 h = 19.1; 200 h = 23.8) were recorded, whereas no marked change in the cell population was observed to the end of culture. The average number of mature AfVPs per cell was approximately 3 (median ± SD, 3.0 ± 0.9, *n* = 475). Mucoidal substances heavily surrounded cells with budding particles and other particles. Budding particles from the cells showed a clear boundary of electron-dense bodies [6].

The oligotrophic West Mediterranean seawater (Calvi, Corsica, France; VFIC = 4.2 ± 2.5, *n* = 1800) originating VLP fraction exhibited $\rho^{25} = 1.2684 - 1.4055$ g/cm^3, with a diameter ranging between 38 and 132 nm (diameter median ± SD = 67.6 ± 35.4 nm, *n* = 317). A portion ($\rho^{25} = 1.3135 - 1.3304$ g/cm^3)

of 118.3–126.7 nm VLPs were infected with *E. coli* AB1157 at MOI from 0.1 to 200 to examine the efficiency of plating (EOP) and the transduction frequency; this resulted in an EOP from 100 to 17.2% regardless of the VLP UV-irradiation. Transduction frequencies from 3.5×10^{-6} to 9.8×10^{-3} CFU/VP were observed. The generated transductants (CEtrans-F1) acquired budding particle production (produced particles per cell of approximately 3). A correlation between MOI and transduction frequency was observed, and the highest performance was recorded between MOI values of 5 and 20. Produced particles from CEtrans-F1 showed varied diameters ranging from 37.4 to 185.4 nm. Subsequently, the purified portion (ρ^{25}=1.3285 – 1.3057 g/cm^3) of 119.2–130.7 nm in diameter was infected at an MOI of 5.5 to *E. coli* AB1157. Consequently, a comparable EOP to the first experiment was observed, while the transduction frequency was reduced by approximately one order of magnitude from the first transduction. The generated second transductants (CEtrans-F2) acquired consistent particle production with CEtrans-F1. For the particle production from transductants, VPIF was between 15% and 20% during the stationary phase.

P. filamentus ATCC700397^T, which was isolated from Arctic ice and cultured in 1/2 ZoBell at 10 °C, showed budding particle production. The produced particle size was distributed between 86 and 346 nm in diameter and contained approximately 150 kb as dsDNA (*P. filamentus*: 5×10^8 cells/mL, 2×10^9 VLP/mL, VP yields: 1.1 VP/cell). However, during the stationary phase, VPIF maintained an approximately 14% higher yield of free particles when approximately 120% of the cell population was given. The particles were purified in two bands by equilibrium CsCl density gradient ultracentrifugation to give 1.2893 ± 0.0102 g/cm^3 as the buoyant density. The upper particles were infected with *E. coli*. Consequently, the recipient lethality was not observed. As for gene transfer, three transductants with complete marker reversion exhibiting a frequency at $5.0 \pm 1.6 \times 10^{-5}$ CFU/VP, referred to as PfEtrans, were obtained [138]. A generated *E. coli* transductant, PfEtrans, exhibited enhanced growth in 1/2 ZoBell at 10 °C, exceeding parental *P. filamentus* by attaining ~2×10^9 cells/mL. VP yields per cell were consistent with the parental strain at 1.1 and VPIF during the stationary phase kept recording as high as approximately 30%, while yields of free particles were suppressed to approximately 20% of the cell population. Therefore, the continual reincorporation of produced particles to the surrounding cells must have occurred.

Fluctuating particle abundance, like AfVP, was as well observed for a VP transductant, TYEtrans, which was generated with the aid of VPs collected at the Toyoha [173] mine (42°48'N, 141°2'E, Hokkaido, Japan). The parental recipient, *E. coli* AB1157, reached a stationary phase at approximately 9 h after initiation (1.7×10^9 cells/mL) at 30 °C, whereas TYEtrans entered the stationary phase at approximately 48 h (6.0×10^8 cells/mL, approximately 35% of the parental recipient population); the same level of population was observed until the end of culture (195 h). The minimum number of free particles (TYEVP) was recorded just after starting the culture (particle number: 1.13×10^4 particles/mL). The highest particle population was observed at approximately 48 h, amounting to 5.68×10^9 TYEVP/mL (VP/Cell = 9.50) when the transductant cells reached a stationary phase. Approaching 70 h, the free particle number decreased by three orders of magnitude and remained at the similar level, between 1 × 10^6 and 1 × 10^7 TYEVP/mL, up to 144 h. A second substantial increase of TYVP was observed at 170 h, which amounted to approximately 9.4 × 10^8 TYEVP/mL, followed by a decrease in the free particle number again to 1 × 10^6 TYEVP/mL at 195 h, exhibiting a similar trend as observed between 70 h and 144 h (VP/Cell = 0.01 and 0.05) [130].

A decrease of the free particle population without a change in cell numbers might have caused the reincorporation of free particles into the cells. The oscillating behavior of free particle abundance might reflect occasional repetition of the release from and reincorporation of VPs to the cell. Such reincorporated free particles must have been providing a template strand eligible for chromosomal mismatch repair [174], having the highest performance of the repair system.

An extraordinarily broad range of susceptible recipients was shown by a thermophile-originating particle collected from a hot spring called "Nakanoyu" (36°12'N, 137°36'E) in Nagano prefecture, Japan [8]. Sampling was occasionally done from 1993 to 1997, with 1997 having the highest water temperature during the period. The characteristic large sausage-shaped bacteria (see Fig. 7.2 [8]) of the sulfur-turf mats have been uncultivable to date. 16S rDNA cloning and sequencing studies have shown them to be members of the order *Aquificales,* which is the deepest-branching lineage of the domain *Bacteria* [175]. The cell population (2~6 × 10^5 cells/mL) was kept at the same order of magnitude, while VLP abundance was shown to be two orders of magnitude higher in the 1997 sampling (1.4 × 10^7 cells/mL, *Aquificales* cell No. ± SD, 95.7 ± 15.6%, n = 7) compared with other times (1~7 × 10^5 cells/mL, *Aquificales* ca. 6.4%). The thermophiles might have provided HGT with a relaxed broad-host-range. However, VLPs originating from the marine isolates of the most recent lineage of the domain *Bacteria* were incapable of transferring genes across the border of phyla [8], which could not have been accomplished by AKVP.

Both intra- and extra-cellular particles collected from the sulfur-turf cells exhibited comparable physicochemical parameters [8]. The second band of equilibrium CsCl density ultracentrifugation with the highest proportion (STVLP: ρ^{25}, 1.2892 ± 0.0188; amount, 2.23 × 10^{11}; particle size, 104.4 ± 9.2 nm in diameter; proportion, 40.1%; protein/nucleic acid ratio, 23.93) was applied to gene transfer experiments toward recipients of *E. coli* AB1157 and *Bacillus subtilis* PS9 (*hisA metB5 thr-5 leuA8 trpCll lys21 purA non A-1/B-1 rfmR* *sp*β$^-$). Results indicated that STVLPs had a lethal effect on both recipients regardless of UV treatment, as well as a gene transfer capability [9] that resembled the case of those of marine origin [1, 2, 157, 158]. Consequently, VLPs originating from the "oldest" thermophile were capable of transferring genes across the borders of phyla. Hence, the existence of broad-host-range gene transporters among VLPs originating from thermophiles was confirmed.

It is notable that *B. subtilis* and *E. coli* transductants generated with the aid of STVLPs acquired the ability to produce particles [9]. The production of particles started from the host attaining the stationary phase, like that in the original host

Aquifex strain, whose yield also was 3–5 particles per cell [8, 9]. Nucleic acid cargo in STVLP is susceptible to DNase, whose size as linear dsDNA was approximately 406.4 kb and RNA was not detected [8]. When the particles (STEVP) reproduced from *E. coli* transductants (STEtrans) were again infected with a *B. subtilis* auxotrophic mutant, they showed a lethal effect regardless of UV treatment of the particles, together with a gene transfer frequency between 2×10^{-8} and $\times 10^{-7}$ transductants/particle. The transduction frequency was slightly low and incomparable to that of the original particles from *Aquificales* used directly towards *B. subtilis* and *E. coli*.

Gene transfer from gram-positive to gram-negative bacteria does not take place because of the differences in applicable systems in the respective bacteria [94], with a few exceptions among artificial plasmid vectors with the natural transposon Tn916 [176]. Therefore, the finding is the first real evidence that VPs have the potential to enhance biodiversity among the bacterial community.

Furthermore, applications of STEVP to recipients of Archaeon—*Sulfolobus acidocaldarius* MR87 (*his ura*) [177, 178], Eukarya: *Saccharomyces cerevisiae* SEY499, SEY6120, YPH500, W303 (*lys trp his leu*) (provided by Takeuchi M, NAIST, 1999)—and the transductants were obtained with a frequency of 1×10^{-8}~1×10^{-3} CFU/particle together with particle production from these transductants. *S. acidocaldarius* MR87 transductant produced particles ranging between 40 and 145 nm, whose productivity was reduced by approximately 22% of parental STEtrans. The *S. cerevisiae* SEY 6210 *lys*$^+$ transductant, ScEtrans_Sc, produced two types of particles (ScEtrans VP, ρ^{25} = 1.2415 and 1.2968) that were again infected with *E. coli* AB1157 to produce transductants with particle production. The yield of particles per cell was estimated as 3 ± 2 (Fig. 7.2 [132]). The particles are likely to "infect" recipients belonging to different domains, which implies that at least some environmental VLPs share similar characteristics—a phenomenon that is not explained by the general features of viruses that infect microorganisms. A greater extent of recipient chromosomal marker transfer was recorded by several orders of magnitude compared with that reported for the marine environment [41, 73, 140].

Another example of trans-domain gene transfer was demonstrated by an archaeon-originating particle and the recipient *E. coli*. The survey area was extended to hydrothermal vents at Suiyo seamount located in Izu-Ogasawara arc. It showed the existence of a microbial population, even in the hyper-thermal fluid [3], with ~5×10^4 cells and ~3×10^4 VLPs/mL. A hyperthermophilic archaeon, *Thermococcus kodakaraensis*, was successfully obtained from an APSK06 (28°34.313'N, 140°38.617'E, 1386 m deep) boring core. The permissive growth temperature ranged between 70 °C and 90 °C and was recorded for an 80 °C culture with elemental sulfur, with 80% of particle abundance to the cell. A 70 °C culture with elemental sulfur gave the highest cell yields and 132% of particle abundance to the cell. Therefore, the strain was cultivated in the sulfur-supplemented medium at 70 °C. VLP production started when the stationary phase attained *T. kodakaraensis*, whose cell and particle abundances at 480 h of culture were 1.8×10^8 and 4.2×10^9, respectively [131, 179]. The median particle size was 125.19 ± 36.45 nm (n = 351) and yield per cell was approximately 1.5, with VPFI of approximately 19% during the stationary phase.

Gene transfer frequencies of particles from APSK07 fluid and *T. kodakaraensis* were between 2.6×10^{-4} CFU/particle and 3.0×10^{-6} CFU/particle. The production of particles from generated transductants was observed under TEM, with the particle yield per cell resembling that of the previously described marine strains and *Aquifex* sp. at 3–5 [131]. The size of the particles produced from the transductant, TkBEtrans, became larger (median ± SD = 132.55 ± 23.6, $n = 272$). Hence, a hyperthermophilic archaeon produced VP-transferred genetic traits towards mesophilic *E. coli* across the Domain border.

7.3 How Do VPs Produced in the Cell Release from the Host and Infect a New Host?

Electron microscopy observations suggest that membrane fusion occurs between the recipient cell and VPs (Fig. 7.3) at the initial stage of infection [6] because membrane fusion between the VP "capsid" and recipient cell was observed upon "infection." Before membrane fusion, the VP pierces the recipient cell wall to create a place for membrane fusion [182]. Therefore, a membrane-fusion error upon the VP's "infection" would cause cytoplasmic content efflux [183, 184]. VPs have shown an extraordinarily relaxed susceptible recipient range (Table 7.4). An interaction between the particle and recipient cell might determine the orientation for the recipient's survival. As shown in Table 7.2, some VPs had a lethal effect on the recipients regardless of the strength of the UV treatment, reducing the EOP by up to one order of magnitude. VP infection of the recipient could be accomplished by the membrane fusion [6]. The lethal effect would be a consequence of a failure to form complete membrane fusion between the recipient cell and the particle, which must have lethally leaked the cytoplasmic content.

The contents of nucleic acid, carbohydrate, proteinase/peptidase, and glycanase activities were examined for particles and vesicles derived from STEtrans and DH*lac*trans (see Tables 7.2 and 7.3). The VP was predicted to have lipids as its surface component (Fig. 7.3), although an insufficient amount of samples prevented the lipid component from being analyzed. VPs are supposed to carry cellular cytoplasmic material as proteinaceous material and DNA; however, an RNA component has not been detected to date. A specific and strictly controlled VP cargo sorting and intracellular trafficking mechanism is expected to exist. However, it is currently difficult to infer an appropriate expression feasible for the strict control mechanism that produces 3 VPs/cell per generation. The results are given in Table 7.5. The surface structure of VPs must be composed of the multifunctional domain to accomplish infection of a variety of organisms.

The VP assemblage seems to carry all host genomic information almost exclusively, whereas the yields per cell are controlled strictly to be small unless a host control disorder is introduced, such as a *rel* mutation. Prior to the budding release of VPs, condensation of DNA cargo, which is likely a part of the host

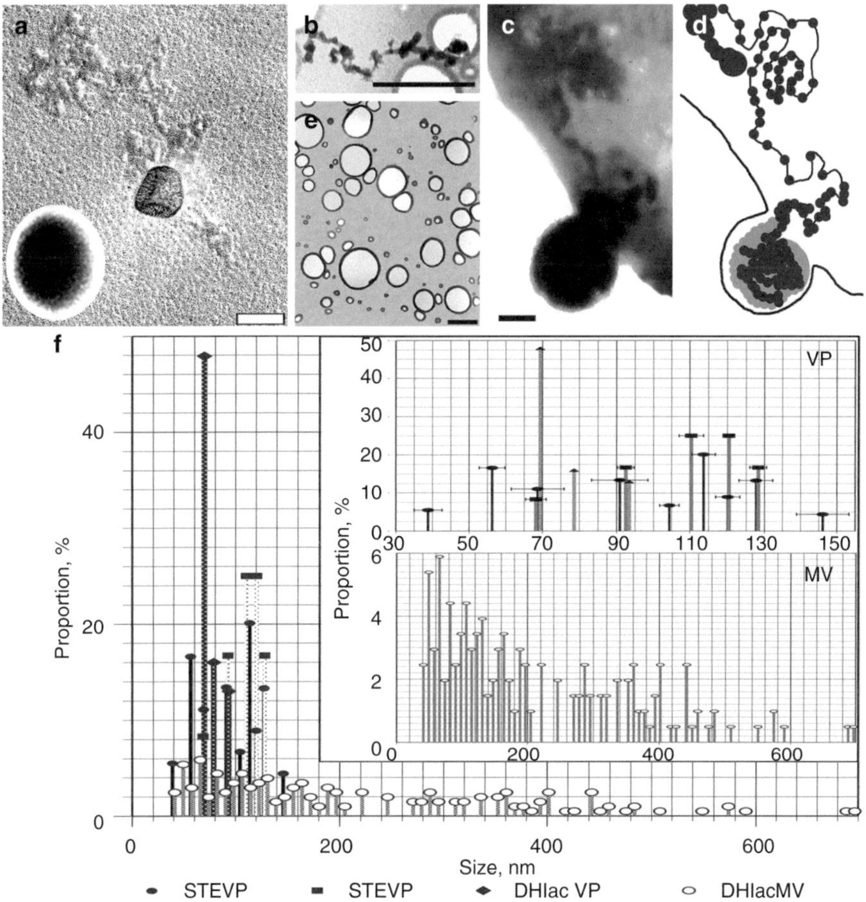

Fig. 7.3 VP treatment of organic solvent, membrane vesicle (MV) production from DH*lac*trans, membrane fusion and translocation of DNA cargo to the recipient at VP infection, and pleomorphic size distribution of VPs after being experienced serial transduction. Electron micrography (EM) image of an organic solvent treated VP shows in (**a**). Purified VPs were dialysed against an up series of ethanol and chloroform and a descending series of ethanol and then treated VPs were placed on the EM grid [166] to observe. VP must have been contained in the lipid component for its coat because the content was extruded from the particle. In (**b**), extruding DNA is shown together with bead-like structure from a membrane vesicle from *relA1* bearing transductant, DH*lac*trans, whose shape was not a string like Coliphage T4 but a tangled structure like a nucleosome observed for an archaeon [180]. (**c, d**) Membrane fusion and translocation of DNA cargo to the recipient at VP infection, DNA strand looks like an archaeon nucleosome structure [180, 181]. In (**d**), a schematic representation of (**c**) is shown. In (**e**), a *relA1*-bearing transductant produced MV together with VP, whose proportion was about 1/10 of the real VPs. In (**f**), the pleomorphic size distribution of VPs is shown after serial transduction. VPs showed production of several discretely different particle sizes, even a selected size VP employed for the transduction. Although the host strictly controls the VP yields per cell, an introduction of *relA1* resulted in the turbulence of cellular control to produce many different sized empty MVs. Generally, the proportion of DNA bearing MVs is quite small; the DNA cargo size is also as small as 10 kb [22, 61, 63]. MVs produced by the *relA1*-bearing transductant must have been produced to encapsulate DNA cargo as intracellularly produced EDB

Table 7.5 Chemical constituents, proteinase/peptidase, and glycanase activities

Amount per particle	STEVP	DH*lac*VP	DH*lac*MV
Protein/nucleic acid ratio	16.7	21.4	3862.9
Nucleic acid, $\times 10^{-18}$ g	8.12	13.9	0.102
Proteinaceous material, $\times 10^{-15}$ g	13.6	29.8	39.4
Total carbohydrate, $\times 10^{-15}$ g	42.6	1.89	4.22
Proteinase/peptidase, $\times 10^{-15}$ unit	1100	27,500	13.7
Glycanase as mannase, $\times 10^{-15}$ unit	51.1	207.2	23.7
Leu equivalent free NH-, $\times 10^{-15}$ g	0.254	35.1	2.58
Glc equivalent reducing power, $\times 10^{-15}$ g	23.8	1.89	5.01
Uronic acid, $\times 10^{-16}$ g	0.38	–	–

Amount of nucleic acid and the proteinaceous materials was determined using nomograph according to Adams [185] by reading the absorbance at 260 nm and 280 nm. The protein content was determined using the Bio-Rad Protein Microassay kit according to the manufacturer's specification with BSA as the standard by reading the absorbance at 595 nm. The total sugar content of the subsamples was determined by the phenol-sulphuric acid method [186] by reading the absorbance at 480 nm for hexose and 490 nm for uronic acid, with glucose as the standard. Free amino acid was determined by the ninhydrin method [187] with leucine as the standard by reading the absorbance at 570 nm. Glucose equivalent reducing power was measured using the Somogyi-Nelson method according to Somogyi [188] with a modification by Chiura and Kita-Tsukamoto [189] with glucose standard read at 500 nm absorbance. A Shimadzu Type UV260 spectrophotometer (Shimadzu Co., Kyoto, Japan) was employed for photometric analysis. The substrate for the enzyme activity, *E. coli* AB1157, was grown to the mid-log phase, then diethylpyrocarbonate was added to make 0.1% v/v to stop the cell growth, washed with 100 mM phosphate buffer saline (PBS), and centrifuged. Fixed cells were dialyzed against PBS until no nucleic acid, carbohydrate and/or protein was detected in the dialysate, and finally lyophilized. The dried cell was suspended in an appropriate liquid at the time of enzyme assay [190]

chromosome, makes EDS followed by intracellular trafficking in proximity of the cell membrane (Fig. 7.5). In general, transducing particles are formed when the phage packaging mechanism seizes upon host DNA, instead of concatemeric phage DNA, as a packaging substrate [191, 192]. The association of transducing host chromosomal DNA with a protease-sensitive component is estimated as 500 kDa of protein [46], with a substantial fraction (>75%) of the transduced DNA appearing to adopt a circular conformation maintained by an attached protein. Unfortunately, this protein has not been studied further, although the protein has been suggested to be a packaging enzyme that remains associated with the transducing DNA after completing cleavage [193]; this, however, remains to be established. There are two plausible hypotheses on how the DNA packaging mechanism makes transducing particles: one is a sequence-specific manner from DNA ends, whereas the other is a consequence of nonspecific cutting by packaging enzymes. The accumulated evidence clearly favors a pac site mechanism for P22 transducing particle formation, but a non-sequence-specific mechanism for the formation of P1 transducing phages [194, 195]. The findings would also apply to the case of VPs, as VP-encapsulated DNA seems to be heavily associated with a proteinaceous material (Figs. 7.3, 7.4, and 7.5).

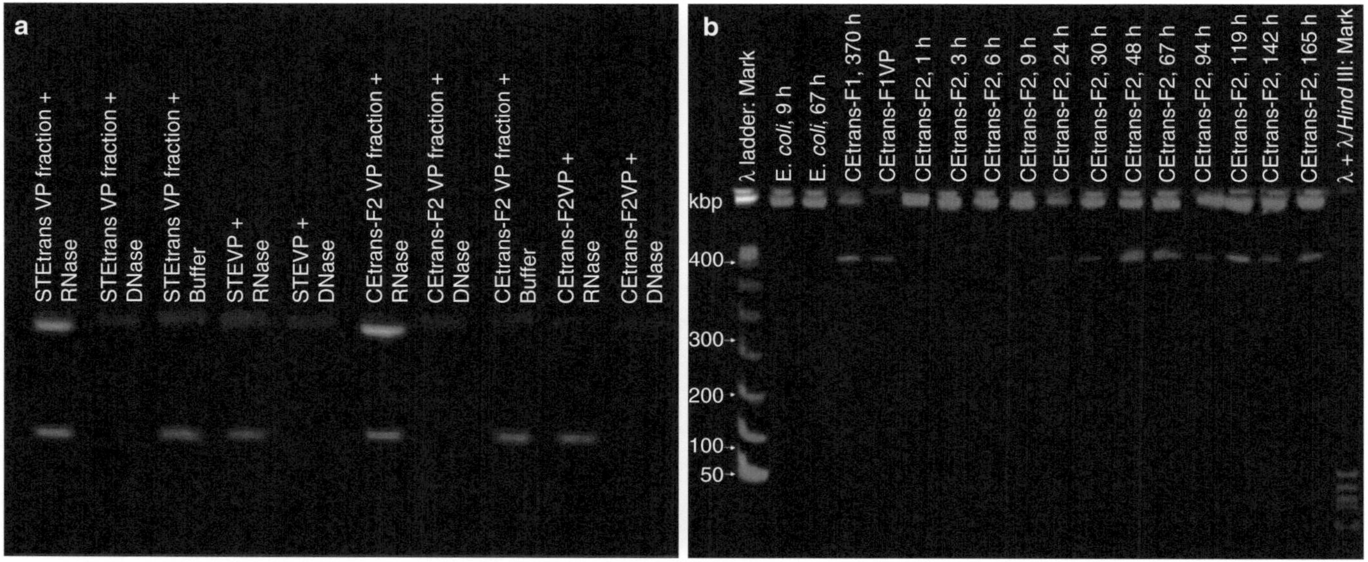

Fig. 7.4 VP cargo nucleic acid type, cargo production profile along with the cell growth, the difference between the recipient and transductants in the logarithmic phase, and transductant-specific membrane localized proteinaceous species. (**a**) CEtrans-F2 released from the first CEtrans-F1VP were again subjected to in situ lysis and obtained nucleic acids treated with DNase I and RNase A. DNase treatment showed no bands on the electrophoretogram; however, RNase treatment showed no change. STEtrans and STEVP were also treated in the same manner, of which VPs consisted of DNA [4]. (**b**) Intermittently withdrawn samples from the culture were subjected to in situ lysis (4 μg as nucleic acid per piece) and nucleic acids were examined. VP cargo DNA appeared in the cell after the stationary phase was attained. Although particle size showed pleomorphism, cargo DNA size was observed to be somewhat uniform throughout the producing period. Dalton markers: λ ladder and λ genome+λ/*Hin*d III. VP encapsulated nucleic acid size for both CEtrans-F1 and CEtrans-F2 was estimated to be 368.5 ± 8.7 kbp (n = 49), which showed a resemblance to that of STEVP 372.09 ± 10.12 (n = 46). (**c**) The logarithmic phase cell provided a difference between transductants (CEtrans-F2, STEtrans) and parental strain AB1157. Broad-faint bands, whose size ± SD (n = 3) was 11.78 ± 7.87 kbp, were detected only for the transductants on the electrophoretogram. (**d**, **e**) VP production periods exhibited specific molecular species of membrane fraction only for the transductant, and *relA1*-bearing transductants showed different molecular species as above. Small DNA species found in the preceding period of VP production might have been related to proteinaceous species found explicitly for the transductants' membrane fraction

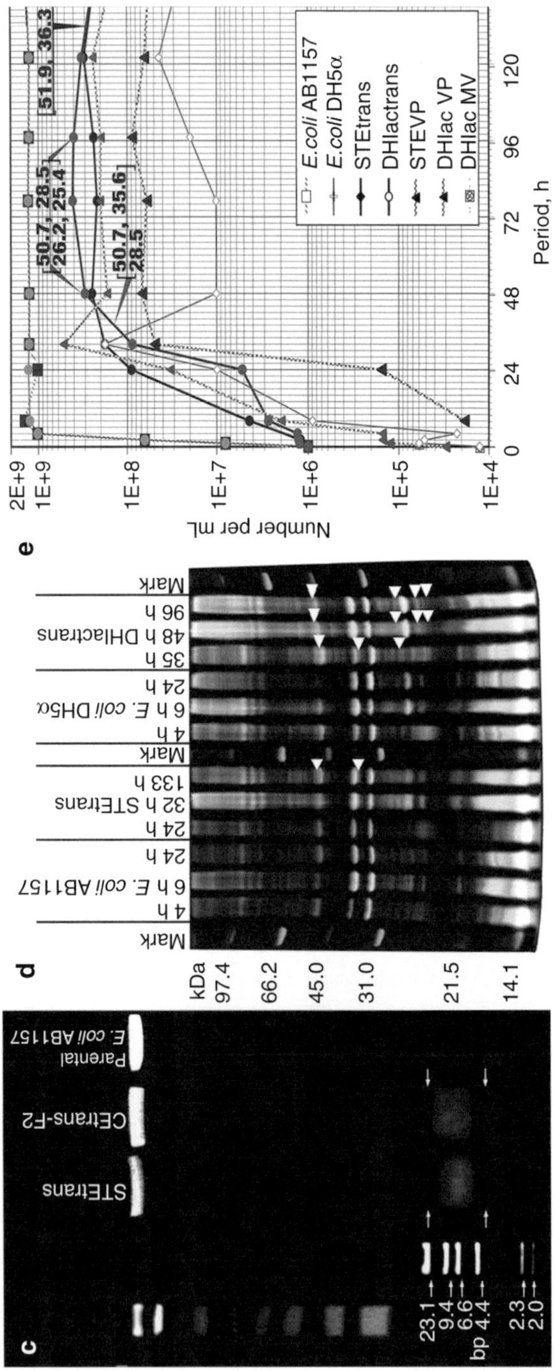

Fig. 4 (continued)

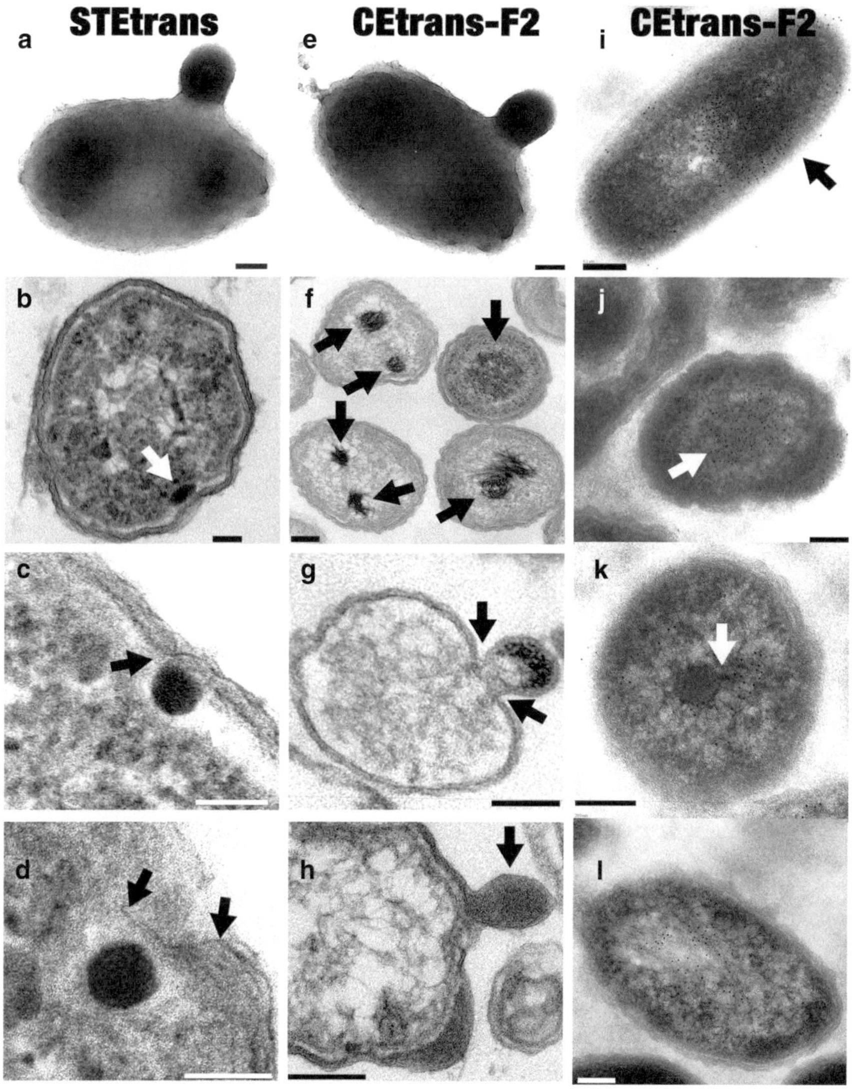

Fig. 7.5 Transductant cells in the process of budding. In (**a**) to (**d**), STE-trans was used as a reference; in (**e**) to (**l**), CEtrans-F1 has used a reference. (**a**) STEtrans under particle budding. (**b**) Before the budding, the electron-dense body (EDB) shifted to the vicinity of the cell wall, and then the EDB attached to the cell wall (**c**). Following this step, the phase change of membrane from bi- to monolayer was observed (**d**). CEtrans-F1 is shown around the budding particle (**e**). Another three EBDs can be seen in the cell. Thin sections of CEtrans-F1 demonstrated EDBs. (**f**) Different stages of EDB formation in the cells. (**g**) Particle under budding. A change of membrane phase is seen at the budding position. (**h**) The final stage of budding. Note "capsid" structure is different from the cell membrane. Epon-etched thin sections of the CEtrans-F1: (**i**–**k**) sections with central EDB and a dense cluster of gold-labelled DNA-antibodies. (**l**) Control: thin-section without EDB with sparse gold-labelled DNA antibodies. All sections belong to the same assay. Scale bars: **a**, **d**, 100 nm; **b**–**d**, **f**–**l**, 200 nm

Some MVs are known to inhabit a variety of enzymes [196]. The hydrolytic enzyme activities of VPs reside in an association of phage endolysins, which recalls the production process of some MVs [197]. Endolysin is a well-conserved enzyme that is typically involved in the release of dsDNA phages [198]. Endolysin opens holes in the cell wall, through which MVs are formed presumably due to turgor pressure and eventually released [28, 29]. Toyofuku et al. indicated that the simultaneous release of abundant MVs and the matrix in *P. aeruginosa* biofilm would occur by endolysin-induced explosive cell lysis in a subpopulation of the cells to supply public goods for the remaining cells [62]. Furthermore, endolysin was recently shown to induce MVs in Gram-positive bacteria through a distinct mechanism of explosive cell lysis [28, 29]. Hence, endolysin is a universal trigger for MV formation. Cell death is involved in MV formation, which renews the concept of cell death in microbial communities [199]; however, MVs were produced only from growing cells but not from lysing cells. From this standpoint, MV formation is considered to be a consequence of cell lysis that is distinct from the case of VPs.

MVs from a variety of organisms form the whole Domain of life and are called by several names, including membrane vesicles, outer membrane vesicles [OMVs], exosomes, and shedding microvesicles [57]. Bitto et al. demonstrated that chromosomal DNA is packaged into OMVs shed by bacteria during the exponential phase. Most of this DNA was present on the external surfaces of OMVs [35], with smaller amounts located internally. The DNA within the internal compartments of *P. aeruginosa* OMVs were consistently enriched in specific regions of the bacterial chromosome, encoding proteins involved in virulence, stress response, antibiotic resistance, and metabolism [200]. Roier et al. proposed a novel and highly conserved bacterial OMV biogenesis mechanism for Gram-negative bacteria based on phospholipid accumulation in the outer leaflet of the OM [201]. A proposed phospholipid transporter system of the VacJ/Yrb ABC (ATP-binding cassette) transport was shown to be involved in OMV formation [202]. However, the mechanism for intracellular trafficking of the cargo DNA in accounting for HGT is still insubstantial.

The conserved functions and mechanistic strategies of MV release are similar, including the use of eukaryotic endosomal sorting complexes required for transport (ESCRT) proteins and ESCRT protein homologues to facilitate these processes in archaea and eukaryotic microbes [57]. The ESCRT machinery is made up of cytosolic protein complexes: ESCRT-0, ESCRT-I, ESCRT-II, and ESCRT-III. Together with several accessory proteins, these ESCRT complexes enable a unique mode of membrane remodeling that results in membranes bending/budding away from the cytoplasm [203, 204]. These ESCRT components have been respectively studied in numerous eukaryotes, including yeast and humans [205].

The change in the intracellular structure before budding is illustrated in Figs. 7.4 and 7.5. DNA cargo must have been prepared around the center of the cell; however, how and what mechanism would proceed to select the DNA cargo content accompanied by budding production are unclear. OMV was reported to have continually budding production with a morphologically similar shape to VPs from γ-proteobacteria, *Acinetobacter baylyi* [63]. OMVs could be grouped into three size populations, with mean diameters ranging from 13 to 304 nm. OMVs were extracted from bacterial cultures and tested for their ability to vector gene transfer into populations of *E. coli* and *A. baylyi*, including naturally transformation-deficient mutants of *A. baylyi*, and determine the movement of DNA into OMVs using FITC-labelled and anti-dsDNA antibodies to track the movement of OMVs. Exposure to OMVs isolated from plasmid-containing donor cells resulted in HGT to *A. baylyi* and *E. coli* at transfer frequencies ranging from 1×10^{-6} to 1×10^{-8}, with transfer efficiencies of approximately 1×10^{3} and 1×10^{2}/μg of vesicular DNA, respectively. The intracellular production looked at several points in the cell, which was different from that observed for VP, as described below. It is uncertain whether this finding could be comparable to the content described previously because no DNA content per OMV was provided. Nonetheless, this report provides the only data on quantitatively estimated MV-mediated gene transfer. To the author's knowledge, there have been no reports on MV-mediated "serial transduction" attempts to date.

DNA containing extracellular vesicles (see Table 7.1), referred to as ToEV, produced by budding was observed for *T. onnurineus*" NA1^{T}. The DNA content in ToEV amounted 40 to 20 kb [22]. In comparing the encapsulated DNA in VP with such MVs, VP carried substantial DNA content, as shown in Table 7.2.

The VP production process was studied using ultra-thin sections to analyze features of the membrane fraction at the time of particle production. The materials CEtrans_F2 and STEVP (Table 7.1) were used because their morphological characteristics, production profiles of VP, and cargo size DNA (CEtrans_F2: 368.5 ± 8.7 kb, $n = 49$; STEVP: 372.09 ± 10.12 kb, $n = 46$) coincided with each other, even though the environmental situation of sampling sites differed [7, 8]. Consequently, the particle production initiated around the center of the cell by an adhering electron-dense material around the chromosome, following a hurricane-like structure formation. The shape was changed by condensation to create an EDB, which migrated to the vicinity of the cell membrane. Finally, budding took place. VP production in STEtrans was also consistently observed for intracellular trafficking of EDB from the cellular center towards the cell membrane for budding. The mode of VP production was found to be consistent between STEtrans and CEtrans. The acquisition of membrane vesicle production of VP-mediated transductant was observed when the *rel*-mutant was used for the recipient, whose population size was approximately 1/10 of that of "bona fide" VP particles.

To collect information on the cell membrane during VP production, 8 M urea dissolved the cell lysate membrane fraction of STEtrans, DH*lac*trans, and the parental *E. coli* and was applied to 5.2 M urea denatured-polyacrylamide gel electrophoresis (PAGE) (Fig. 7.4d). The denatured-PAGE showed that specific proteins in the membrane protein fraction corresponded to the period of particle production only for the transductant. The cytoplasmic extrachromosomal nucleic acid fraction before the VP budding release period was also examined for STEtrans, CEtrans_F2, and the recipient *E. coli* AB1157. Small DNA species specific for this period were observed, whereas the RNA fraction showed no difference (Fig. 7.4c). Such a VP transductant-specific molecular species must have been committed to intracellular EDB production. Successful intracellular trafficking from the central part towards budding proximity site might have been imitated in ESCRT of Archaea and Eukarya. Further biochemical studies are needed to elucidate the molecular mechanisms of VP production.

A standard feature of all VPs is that the particles reproduced from the transductant do not have a uniform size and are composed of varying types of particle species, even though particles of a specific size are selected for transduction. In the marine strain, approximately 20 species of VP were produced; a much smaller number of VP species was produced by the thermophile strain, whose trend for particle production of the transductant appeared to be inherited from the parental strain. Such pleomorphic particle production can be found in MV production; however, the molecular mechanism of MV depends upon the host cell belonging to a variety of phylogenic positions, and a commonly applicable pathway is still uncertain. Some similarity of pleomorphic particle production can be found among budding viruses. Usually, the capacity of a viral capsid is well regulated, and the packaging of a nucleic acid is performed by a head-full mechanism [94]. A eukaryotic paramyxovirus with an ssRNA genome (*Myxoviridae*) produces a varied particle size and comprises several pleomorphic viruses that may occur as irregular spheres of size, ranging from approximately 150 to nearly 300 nm in diameter [206, 207], or even as long filaments [207]. Although a prokaryotic budding virus is rare, the mode of exit is shared by pleomorphic archaeal and bacterial viruses (pleomorphiclipovirus, plasmavirus) [208] and mycoplasma phages [160] out of 70 lipid-containing viruses known to date (Archaea: 41; Bacteria, Gram-positive: 21, Gram-negative: 8; lipid enveloped: 9; Archaea 7; Gram-positive: 2) [159].

Genomic sequencing is currently in progress for recipient *E. coli* AB 1157, PfEtrans, VP source *P. filamentus*, STEtrans, and DH*lac*trans followed by comparative genomic analyses combining data with the generalized transducing phage *Myoviridae* P1, budding prokaryotic virus *Plasmavirus Acholeplasma phage* L2 [160], its *host Acholeplasma laidlawii* PG-8A, and the polymorphic virus *Pleolipoviridae* HGPV- 1, HHPV-1, HIS 2, HRPV-1, 2, 3, 6 [209]. Preliminary findings revealed that none of the complete gene set of viruses was discovered in the VP transductants' genome; however, *dnaB*-like Phage P1 replicative helicase, Phage P1 DNA invertase, Phage P1 IS5 transposase family mobile element, Phage P1 ssDNA binding protein, and Phage P1 methyl-directed repair DNA adenine transferase were shared with all VP donor microorganisms and generated transductants.

7.4 Function-Related Gene Transfer

The VP-mediated HGT described previously is mainly for host somatic gene rescue; however, the simultaneous transfer of multiple genes might have impacts that are much more profound on the evolutionary processes. Some examples of VP-mediated function transfer are described in the following sections.

7.4.1 Nitrogen Fixation Acquisition of *Escherichia coli* by VPs Originating from *Klebsiella pneumoniae* subsp. *rhinoscleromatis* [139]

From the rhizosphere of the Sago palm tree, the Sago2 strain was isolated, which has a 16S DNA sequence with 99.93% similarity to *Klebsiella pneumoniae* subsp. *rhinoscleromatis*. Sago2 cultured in LB at 30 °C showed budding spherical particle (SGVP) production; the particles were purified by equilibrium CsCl density gradient ultracentrifugation. The particles had densities between 1.3183 and 1.2722 and diameters between 103.8 and 164.1 nm. They were used for transduction at a MOI of 2 by infecting the following recipients at 30 °C for 15 min: *E. coli* AB1157, DH5α [F^-; *ø80d, lacZ M15 endA1 recA1 hsdR17* (rk^-, mk^-) *supE44 thi-1* λ^- *gyrA96 relA1 (lacZYA-argF)U169*]; and JE6937 [F^-; *strR*]. Irrespective of UV irradiation, SGVPs showed no lethal effect on the recipients.

The selection of transductants was made on N^- agar plates supplemented with and without 20-μg/mL arginine and incubated anaerobically at 30 °C. The strains, which could form distinctive colonies within 5 days of incubation, were regarded as the transductants. SGVPs without UV inactivation were successfully transferred with N_2 fixation to generate on N^- agar plates. The results for DH5α were: four colonies (SGDHtrans), N_2 fixation gene transfer frequency = $2.83 \pm 3.24 \times 10^{-7}$ CFU/SG-VP ($n = 3$). The results for the JE6937 recipient were: four colonies (SGEtrans), N_2 fixation gene transfer frequency = $1.84 \pm 0.80 \times 10^{-8}$ CFU/SGVP ($n = 3$). To examine the N_2 fixation ability of SGVP-mediated transductants, seven clones of SGDHtrans and four clones of SGEtrans were subjected to an acetylene reduction assay, with the parental recipients as the negative control and Sago2 as the positive control.

Consequently, anaerobically incubated Sago2 in LB and N^- medium showed nitrogenase activity. All the SGDHtrans clones exhibited nitrogenase activity when anaerobically cultured in LB; however, two of them did not show nitrogenase activity in N^- medium. As for SGEtrans, no clones exhibited nitrogenase activity in LB and N^- medium. Hence, different genotypic variations of transductants could be generated by VPs; still, more acceptance of incorporating gene sets must have been affected by recipient genetic constitution.

7.4.2 Thermo- and Psychro-Tolerance Acquisition

P. filamentus ATCC700397T, which was isolated from Arctic ice, cultured in 1/2 ZoBell at 10°C showed budding particle production. Produced particle size in diameter distributed between 86 and 346 nm that contained ca. 150 kb as dsDNA (*P. filamentus*: 5×10^8 cells/mL, 2×10^9 VLP/mL, VP yields: 1.1 VP/cell). Although VPIF during the stationary phase kept ca. 14% high yields of free particles by ca. 120% of the cell population was given. The particles were purified in two bands by equilibrium CsCl density gradient ultracentrifugation to give 1.2893 ± 0.0102 g/cm^3 as the buoyant density. The upper particles were infected with *E. coli* AB1157. Consequently, the recipient lethality was not observed. As for gene transfer, three transductants with the complete marker reversion exhibiting the frequency at 5.0 ± 1.6×10^{-5} CFU/VP, referred to as PfEtrans, were obtained [138]. PfEtrans exhibited enhanced growth in 1/2 ZoBell at 10 °C, exceeding parental *P. filamentus* to attain ~2×10^9 cells/mL. A hypothesis was examined that the gene transfer due to VPs originating from *P. filamentus* lowers the range and optimum temperature of growth of the recipient *E. coli*. The optimum growth temperature of *P. filamentus* is 10 °C, and PfEtrans exhibited a cell population in the stationary phase that enhanced growth to the maximum of *P. filamentus* cell population at 10 °C. PfEtrans was examined in the still culture in 1/2 ZoBell in 3 days, consequently showing that the optimal temperature was reduced to 30 °C and it was possible to grow at 0 °C. Furthermore, culture in 1/2 ZoBell at 10 °C showed a maximum cell density of ~2×10^9 cells/mL with VP production, by which serial transduction was also demonstrated [138]. Genomic sequence analyses for *P. filamentus* with *E. coli* AB1157 clarified how PfEtrans acquire cold and high osmotic tolerance that resulted in the enhancement of genes relating to stress response and osmotic regulators (Chiura, Yoshizawa, Kogure, Unpublished). (Fig. 7.6).

Numerous factors are assumed to control the temperature range and the optimum growth. Hence, the growth–temperature relationship has been examined. To determine whether thermo-resistance gene(s) could be transferred towards mesophilic recipients, *Aquifex* sp. and *T. kodakaraensis* B4 originating VP-mediated *E. coli* transductants, STEtrans and TkBEtrans, were grown in conditions exceeding the permissive temperature for *E. coli* (50, 56, and 70 °C) for 15 days. Consequently, the parental recipient, *E. coli* AB1157, decreased its population to less than a few percent of the inoculum by 48 h; then, the living cells became undetectable by LIVE/DEAD in all conditions [131, 134] (Fig. 7.6). It was known that heat-shock response would be activated when *E. coli* cells were transferred from a permissive state to a high temperature; however, the situation exceeding 50 °C could not be rescued even with such a response. Hence, prolonged treatment above 42 °C would cause the disorganization of cell structure, resulting in cell death. On the contrary, STEtrans and TkBEtrans treated under the same conditions exhibited different profiles. At the initiation of the treatments under the restricted condition for *E. coli*, the population decreased by 5–25% in 2–5 days because transductants were generated at 30 °C. After that, the population increased by 2- to 140-fold or at least maintained the initial inoculum size.

TkBEtrans was inoculated at an initial population size of 1.23×10^7 cells/mL. The cells decreased to 6.31×10^5 cells/mL (approximately 5% of the inoculant) at day 5, and then the cells grew to 7.48×10^7 cells/mL (approximately six-fold of the inoculant) at day 13 at 50 °C. The generation time of the transductant in the period of propagation was estimated to be approximately 28 h. At 56 °C, the cells decreased to approximately 3×10^6 cells/mL (~25%) at day 5, and then the cells grew to 1.95×10^7 cells/mL (~1.6-fold of the inoculant); the generation time in the growing period was estimated to be approximately 73 h. At 70 °C, the cells decreased to 1.53×10^6 cells/mL (~12% of the inoculant) at day 3, and then grew to 2.43×10^7 cells/mL (~2-fold of the inoculant) at day 13, of which generation time during the propagation was estimated to be approximately 266 h.

Change in the population under high-temperature incubation for STEtrans is also given in Fig. 7.6. For STEtrans (initial population size: 4.20×10^7 cells/mL), the

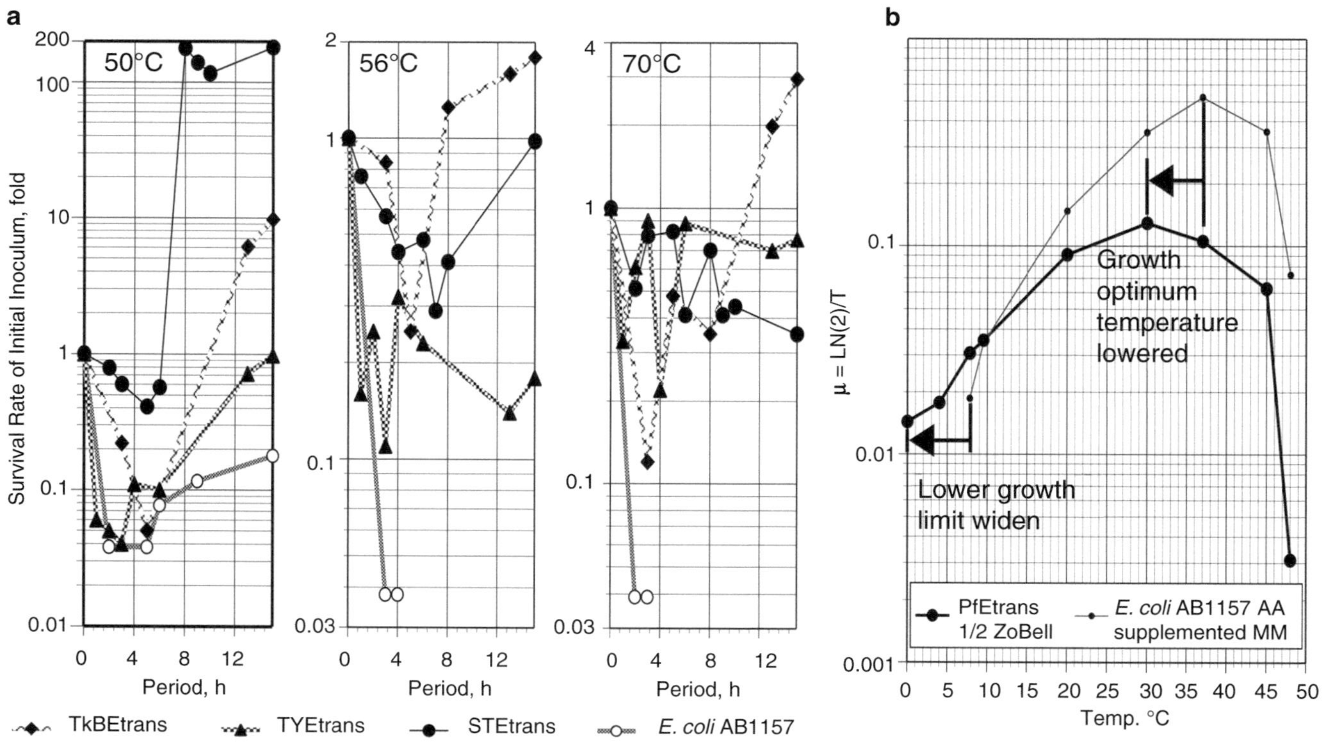

Fig. 7.6 **Acquisition of heat resistance and change in the optimum and permissive growth range.** *E. coli* AB1157 transductants generated by the VP originating from thermophiles provided thermo-resistance to the mesophilic enterobacterium *E. coli* recipient (**a**) [134]. VP originating from the psychrophile lowered the optimum and permissive growth range of *E. coli* (**b**) [138]

population was observed to decrease to approximately 2×10^7 cells/mL (~50%) toward the second day. It maintained this level for another 4 days, after which the cells grew to approximately 6×10^9 cells/mL (~140-fold of the inoculant) in the following 2 days; finally, the population became stable. The generation time of the transductant during the growth was estimated to be approximately 6 h. At 56 °C, the cell number at day 2 was observed to be approximately 2×10^7 cells/mL (~50%), and its population became constant afterwards. At 70 °C, approximately 2×10^7 cells/mL (~50%) of cells were observed on day 2, and its population became constant afterwards.

This trans-phyla thermo-tolerance transfer is a consequence of broad-host-range transduction from a thermophile to a non-thermophilic enterobacterium. The transductants used above were first selected by the prototrophic reversion of the auxotrophic recipient at 30 °C and resided at the mesophilic temperature for several generations. Even after such a period when the transductants remained at a permissive temperature, the thermo-tolerance trait of the transductant suggesting VPs contributes to enhanced host adaptation.

Although it is often assumed that phages do not intrude on the genus barrier [92], this concept has been questioned. The host ranges of marine phages have been shown to vary by Wichels et al. [210]. Jensen et al. [211] disputed that narrow host-ranges are an isolation artefact. Substantial variability in host ranges was also reported by Suttle [212] for cyanophages. Viruses might exhibit a broader host range in deep waters than in surface waters [97]. However, viruses also can influence the stable stockpile of genes in prokaryotic species on ecological scales [213].

7.4.3 Relationship between Cell Population Density and Induction Frequency of VP Lysogen: A Marine-Originating VP-Mediated Transductant, CEtrans [214]

In a natural water column, FVIC was observed to be an equilibrium because of cell and virus interactions; however, there is no substantial evidence to interpret such a situation. VP lysogens started particle production when the host attained the stationary phase, whose VPIF is commonly quite low (the Pacific Ocean and Mediterranean Sea average, 6.92 ± 3.93%, n = 3925). Namely, less than 10% of the population would commit to particle production. Some VP lysogens, such as *Vibrionaceae,* produce VP exceeding the host population, which has been ascribed to their high VPIF [6]. One of the well-known characteristics of *Vibrionaceae* is its quorum-sensing (QS) behavior. However, the current understanding of QS is that the signal is not only solubilized in an aqueous environment but is also packaged in MVs and delivered in varying propensities to different bacteria [28, 29, 39, 40, 197]. As described in Sect. 7.6, VP transfers plasmid and cytoplasmic material; VP can transfer cytoplasmic material, so VP production must be intimately connected with the community situation. In this situation, a change in VPIF under different VP lysogen abundances around the host secreted diffusible substance. CEtrans_F1 was grown in minimal medium (MM) after Davis for approximately 2000 h and then the cells were removed to recover the supernatant, followed by a subsequent filter with 0.2 μm and 30 kDa cut-off membranes to collect the permeate. This permeate was used as the base material. The conditioned medium (CM) was prepared by adding the nutrient supplement as MM (Fig. 7.7).

CEtrans_F1 was cultured as described above for 48 h, corresponding to the period to enter the stationary phase; then, centrifugally collected cells were washed with Davis salt. Two sets of 1×10^6 and 1×10^8 cells/mL as the initial cell abundance were prepared. For one set of 1×10^7 and 1×10^8 cells/mL, the specimen was cultured in the MM as the fresh medium (FM); another set of cells was cultured in the CM. The same conditions were used for both sets of specimens. Growth profiles were monitored. Subsamples were intermittently withdrawn to examine VPIF and the particle yield per cell by EM. As a result, VP abundance for the CM series was 10-fold higher than that of FM at the beginning; after 161 h of incubation, it increased to 100-fold higher than that of FM (Fig. 7.7a).

The initially high VPIF for every specimen was ascribed to the inoculant being started with VP production collected from the 48 h culture. The high initial cell abundance seemed to enhance VP induction; furthermore, the CM showed excessive VP induction. In the usual situation of small cell abundance, VPIF was usually suppressed less than 20%, while more than 80% of the population committed to VP production with a high cell abundance in the stationary phase (Fig. 7.7b). The particle yields per cell (2.7 ± 0.7. $n = 96$) were somewhat consistent with each other; however, the extended production period and higher VPIF resulted in high yields of produced VPs for the CM series (Fig. 7.7c). The result suggests that a diffusible substance was profoundly affected by the induction of VPs. A QS system is considered to be a bi-directional function towards lysogenized entities and their host. QS is a system for bacteria to communicate with each other with a diffusible substance [215].

7.5 Molecular Evidence for VP-Mediated Gene Transfer

Attempts have been made to perform polymerase chain reaction (PCR) for molecular evidence of VP-mediated gene transfer and construct a clone library from VP-derived DNA for gene determination [136]. STEVP was infected with *E. coli* DH5α (Δ*lacZYA-argF relA1 recA*) and selected by lactose utilization acquisition; arginine prototrophism resulted in the generation of *E. coli* DH*lac*trans. DH*lac*trans produced DH*lac*VP (size ± SD: 69.4 ± 2.1, 78.6 ± 1.1, and 93.4 ± 4.0 nm; $n = 480$) with smaller diameter and more compact encapsulated DNA (size ± SD: 74.9 ± 6.7, 108.5, ± 4.5, and 181.9 ± 23.4 kb; $n = 4$) than parental STEVP (size ± SD: 128.3 ± 2.3 nm, $n = 1225$; DNA size ± SD: 389.7 ± 6.5 kb, $n = 49$). Particle production commenced immediately after inoculation, accompanied by transparent membrane vesicles (DH*lac*MV), whose proportion to the VP with electron-dense

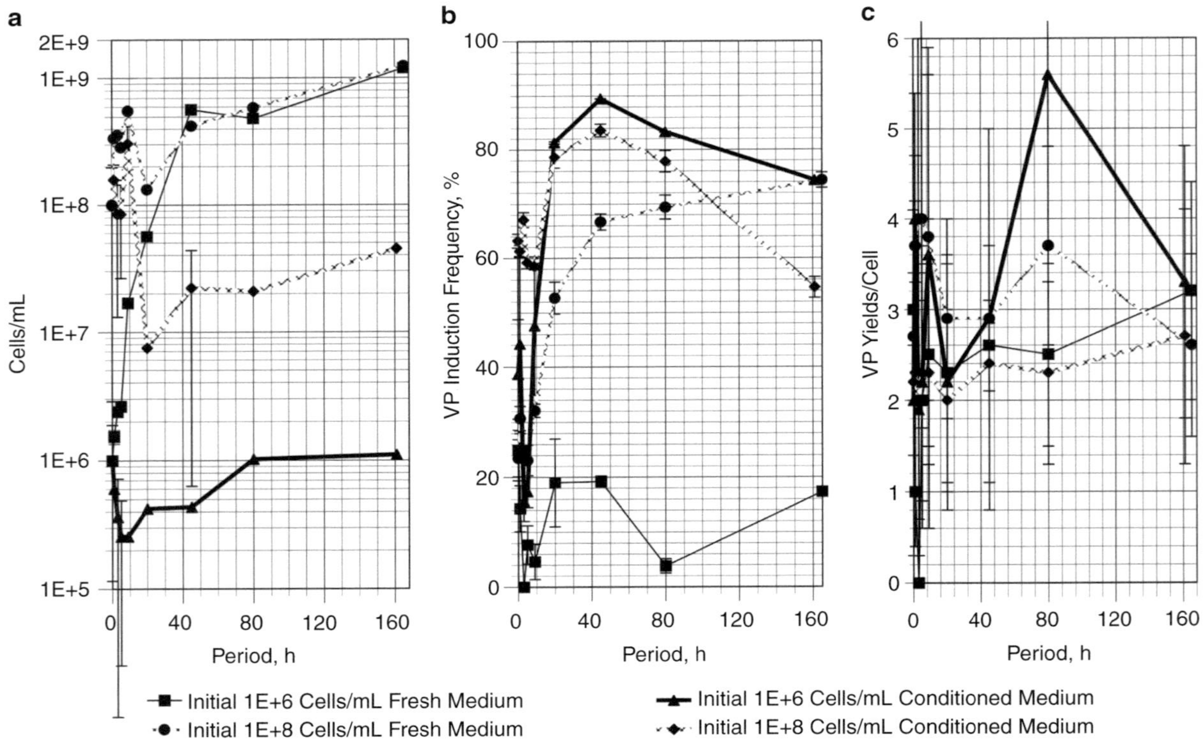

Fig. 7.7 **Quorum sensing-like response of VP production with a specific reference to VPIF and VP yields per cell.** The prolonged culture medium used as the material of the "conditioned"-medium and the low initiating population density induced growth suppression (**a**). The sever effect of the conditioned-medium to VPIF was observed for the low initiating population density (**b**). Furthermore, VP yields per cell were also enhanced by the low initiating population density (**c**). The findings suggest that the QS mechanism might have been involved in VP production

content was approximately 10%; this can be ascribed to *relA1* because a comparable phenomenon has been observed for STEVP-infected *E. coli* JM109 transductants (Chiura, Unpublished). Production of the membrane vesicle concomitantly changed with a change in particle production to maintain the MV/VP ratio (Fig. 7.4f).

The *recA* mutation is a point mutation caused by a substitution of glycine to aspartic acid ($G_{160} \rightarrow D_{160}$), which prevents the homologous recombination of exogenous fragments from the host chromosome ("Strain-DH5a"; https://cogsc2.biology.yeal.edu/Strain.php?=43590). Hence, the cargo DNA strand stays in the extrachromosomal state after STEVP infection in a DH5α recipient. Lactose utilization made the selection of the transductant; the cargo content must have been only for the lactose metabolism with additional VP production-responsible genes. Generated DH*lac*trans reproduced DH*lac*VP to accomplish the serial transduction. Therefore, the cargo DNA size is postulated to decrease as a consequence of packaging the Δ*lacZYA-argF* frame of 78 kbp (Fig. 7.8b) to complement the host, with an additional 16 kbp of DNA encoding the VP production-responsible genes (PPRG) originating from STEVP. VP cargo size decreased, which supports the above hypothesis, as shown in Fig. 7.8a.

Three-step PCR was used for verification of *lacZ* with *lacZ*-specific primers on VP-derived extrachromosomal DNA and total DNA extracted from DH5α and DH*lac*trans. Positive DNA amplification was observed in the VP transductant DH*lac*trans and VP-derived extrachromosomal DNA (Fig. 7.8c, d). Nonetheless, DH*lac*VP cargo content was found to be composed of the whole range of *E. coli* chromosomal genes (Kawarabayashi, personal communication). Hence, the above hypothesis is disproven. DH*lac*trans was subjected to a genome sequence together with STEtrans and *E. coli* AB1157. The sequence data of DH5α was obtained from the NCBI database (*Escherichia coli* strain DH5alpha chromosome, complete genome https://www.ncbi.nlm.nih.gov/nuccore/CP025520.1?report=genbank). The comparative genomic analysis is currently in progress (Table 7.6).

Preliminary findings revealed that the *recA1* mutation in DH5α remains unchanged in DH*lac*trans and no introduction of wild *recA* was found from STEtrans. The total amino acid sequence identity between DH5α and DH*lac*trans is 99.39%. In all, 26 unique genes, of which 25 are hypothetical proteins, are exclusively found in DH*lac*trans, in addition to 79 non-synonymous substitutions. Genomic analysis revealed the genetic robustness of the recipient as a biological entity, together with the flexibility to adapt to environmental change. There must be an unknown scheme for genetic recombination upon intracellular VP production. Including the material from the experiment described in Sect. 7.4.2 on thermo- and psychro-tolerance acquisition, all host and recipient strains were found to share a generalized transducing phage P1 recombinase gene; therefore, such a recombinase might have been committed in recombination with host chromosomal fragment replication to produce the VP cargo. Thus, the gene set for the VP production apparatus may have been widely distributed among cellular organisms in advance throughout evolutionary history.

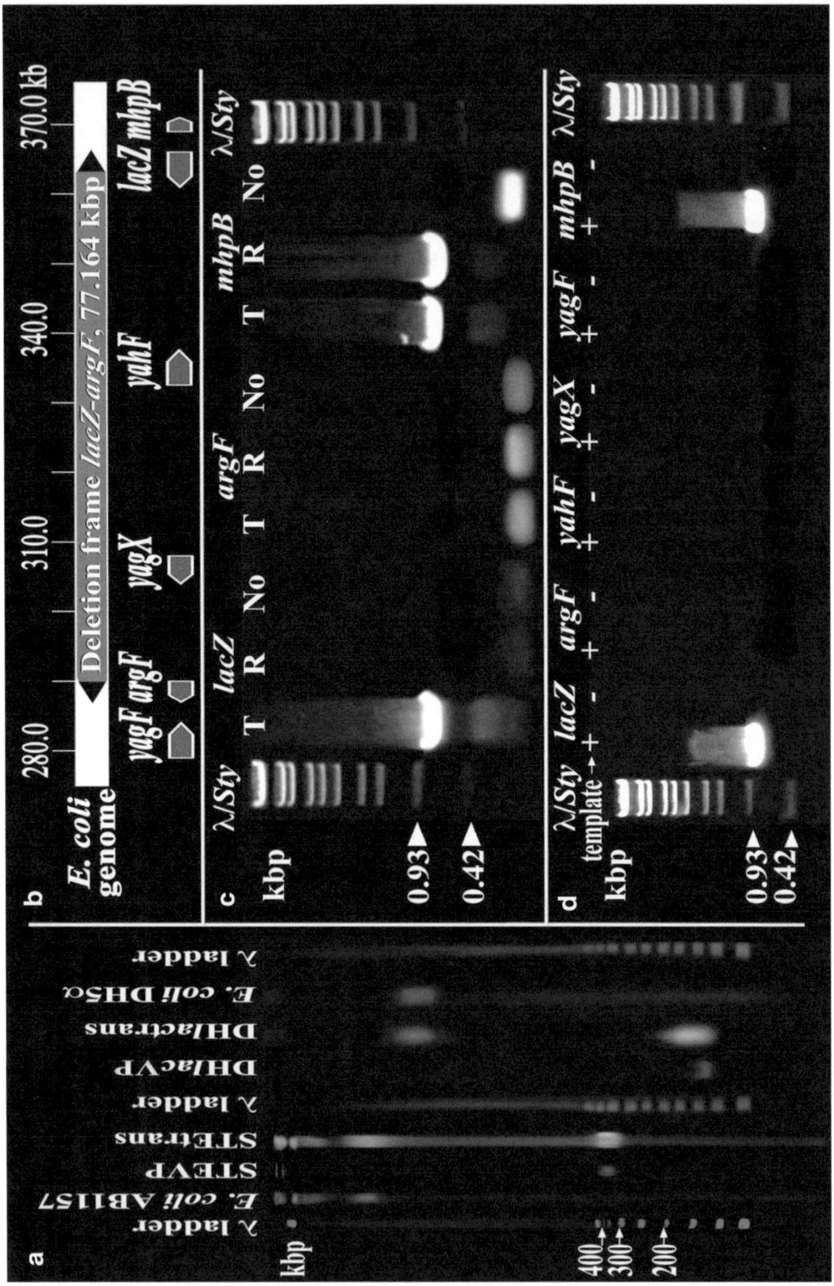
a
kbp
400
300
200
λ ladder
E. coli AB1157
STEVP
STEtrans
λ ladder
DHlacVP
DHlactrans
E. coli DH5α
λ ladder
b
E. coli genome
280.0
310.0
340.0
370.0 kb
Deletion frame lacZ-argF, 77.164 kbp
yagF argF
yagX
yahF
lacZ mhpB
c
kbp
0.93
0.42
λ/Sty
lacZ
T R No
argF
T R No
mhpB
T R No
λ/Sty
d
kbp
template
0.93
0.42
λ/Sty
lacZ
argF
yahF
yagX
yagF
mhpB
+ -
λ/Sty

Fig. 7.8 A change in VP cargo size depended on the transduction recipient, loci of target genes in *E. coli* chromosome, and PCR amplification of target genes in DH5α and DH*lac* trans. (**a**) After in situ lysis, the treated gel plugs were loaded on 1.0% agarose ME gel and run in a gel with 0.5× TBE buffer at 14 °C; switching time, 50–90 s; runtime, 22 h; angle, 120°, and voltage gradient, 6 V/cm. A higher yield of ~120 kb DNA fragments was gained in samples washed with salt. Lanes: λ ladder, size marker (48.5 kb – 1.2 Mb, FMC, USA); *E. coli* AB1157, the transduction recipient of STVP; STEVP, VP produced by STEtrans; STEtrans: a transductant of *E. coli* AB1157 generated with the aid of STVP-mediated transduction originated from *Aquifex* sp.; DH*lac*VP, VP produced by DH*lac*trans; DH*lac*trans, a transductant of *E. coli* DH5α generated with the aid of STEVP-mediated transduction originated from STEtrans. Note that the cargo size of DH*lac*VP decreased by 1/3 from STEVP. The recipient *E. coli* AB1157 is an auxotrophic mutant but has wild *rel*, whereas *E. coli* DH5α has a deletion of *lacZ-argF* as depicted above and mutations of *relA1* and *recA*. The recipient with the *relA* mutation was found to decrease the cargo size of reproduced VPs from VP-mediated transduction. (**b**) PCR primers were designed to amplify genes shown as arrows. Sequence information was obtained from the database "Profiling of *E. coli* Chromosome" (PEC: http://www.shigen.nig.ac.jp/ecoli/pec/index.jsp). DH5α has a deletion frame over the *lacZ-argF* region (broken line). Genes *yagX* and *yahF* were located inside, whereas *yagF* and *mhpB* were located outside the deletion frame. The sharp end of the individual gene indicates the transcription direction. All genes described here are non-essential. (**c**) Amplification of total DNA prepared from the recipient *E. coli* DH5α (R) and the transductant DH*lac*trans (T) by using specific primers for marker genes, *lacZ*, *argF*, and *mhpB*. The gene *mhpB* locates ~7 kb downstream of *lacZ* in *E. coli* genome and therefore should be contained in both R and T DNA. The reaction mixture was directly loaded on 1.0% agarose gel. Template-primer pairs are shown as follows: Lane, T + *lacZ* template; R + *lacZ* template; −w/o addition of *lacZ* template; T + *argF*; R + *argF*; −w/o addition of *argF* template; T + *mhpB* template; R + *mhpB* template; −w/o addition of *mhpB* template. λ/*Sty* size marker: 19.33, 7.74, 6.22, 4.26, 3.47, 2.69, 1.88, 1.49, 0.93, 0.42, and 0.07 kbp. (**d**) DNA extracted from DH*lac*trans was amplified with primers (+) for six positional marker genes regarding non-template reactions (−). Lane, (+ and −) underneath respective maker gene stands for +, template added and −, without addition of the template. λ/*Sty*: size marker

Table 7.6 Genomic feature of the recipients and the transductants

Genome	*E. coli* DH5α	DH*lac*trans	STEtrans	*E. coli* AB1157
Size, bp	46,05,446	45,86,117	46,07,192	45,96,195
GC content	50.8	50.8	50.8	50.8
No. of contigs (with PEGs)	1	33	26	2
No. of subsystems	585	583	591	591
No. of coding sequences	4433	4397	4368	4477
No. of RNA	108	120	112	107

7.6 VPs Also Transfer Plasmid and Cytoplasmic Material

VPs from the marine strains exhibited 2,4D utilization and drug resistance transfer. Hence, cellular proteinaceous material and plasmids for drug resistance transfer to the recipient, mediated by VPs, in addition to the host chromosomal fragments examined. To survey the cytoplasmic substance transfer capability of VPs, conjugative plasmid pYFP-Bluescript (*yfp*::*amp*^{r}) was constructed from *Bam*HI-*Not*I double-digested fragments pEYFP-Mito (YFP) and Blue-script II kS (*amp*^{r}) to generate fusion plasmid pYFP-Bluescript (*yfp*::*amp*^{r}), applied to *E. coli* JM109 (*recA1, endA1, gyrA96, thi, hsdR17* ($r_k^- m_k^+$), *e14*$^-$ (*mcrA*$^-$), *supE44, relA1, Δ(lac-proAB)/F'* [*traD36, proAB*$^+$, *lac I*^{q}, *lacZΔM15*]) and selected white colonies. The obtained transformants, *E. coli* JM109 (*yfp*::*amp*^{r}), exhibited ampicillin resistance and yellow fluorescence under B excitation. Then, STEVP was infected with *E. coli* JM109 (*yfp*::*amp*^{r}) at a MOI of 3.0 to generate the lactose utilization transductants (JM*lac*trans) with a transduction frequency of 1.0×10^{-5} CFU/particle [137].

JM*lac*trans acquired budding particle (JM*lac*VP) production together with empty vesicles starting immediately after the culture; like DH*lac*trans, this was ascribed to the *relA1* mutation of the recipient. The abundance of of JM*lac*VPs in the 96-h culture was ~1×10^{11} particle/mL, with 1/10 abundance of the vesicles. Collected and purified JM*lac*VPs exhibited yellow fluorescence under B excitation. Consequently, VPs originating from the thermophile were demonstrated to encapsulate cellular proteinaceous material (YFP) in the particle. Applying JM*lac*VPs to an *E. coli* DH5α recipient at a MOI of 9.9 resulted in the generation of ampicillin-resistant *E. coli* DH5α with a transduction frequency of 1.9×10^{-6} CFU/particle.

Some novel properties of VP-mediated broad-host-range gene transfer originating from *Aquifex* sp. after several serial transductions showed transferability of cellular proteinaceous material and plasmids for drug resistance. Thus, the JM*lac*VP characterized above might be a novel pathway for gene and material flux in the environment, contributing to the dispersal of drug resistance among environmental microbes [4, 5, 9].

7.7 VP Production-Responsible Gene(s) Are Recombined with the Host Chromosome

The author defined VPs as being capable of infecting, transferring, and budding reproduction in a recipient cell and exceeding the phylogenetic family level from the original host, as previously described. Plentiful VLPs are found in the oceans that exceed the cell population by tenfold, but their original host is uncertain because a VLP is defined as a particle that is morphologically like a virus under electron microscopy. Other than the bona fide virus, VPs and MVs compose VLP assemblages, whose cargo is considered to be information substances. In other words, the VLP assemblage is an enormous gene pool composed of total organismal genes pursuing HGT in situ. The organismal range—of which individual VLPs can infect, propagate, and maintain—corresponds to the host range. The author has isolated

VLPs from the natural environment originating from various hosts to verify as VPs by applying them to various recipients, whose hosts are phylogenetically distant from the family level. Although the first VP phenomenon was demonstrated by adopting VPs originating from a marine proteobacteria (AkVP), the VP is not capable of transferring genes toward exceeding phylum. An extreme example of a VP capable of transferring genes towards the entire Domain is produced by an *Aquifex* sp. (STVP). Thus, there must be a level of accessibility for the VP to the recipient cell that is still unknown. If VPs have a VP production gene (PPRG), then elucidation of the PPRG redefines the new HGT scheme establishment. Would PPRG be transferred exclusively by VP infection of the recipient? If PPRG is integrated with the chromosome, then conjugative transfer would be possible. If PPRG is like a plasmid, there must be an incompatibility group. If PPRG is like a bona fide virus, then there must be new category group. Thus, the possibility of the conjugative transfer of VP production must be considered.

A particular genetic subsystem for particle production must be inherited in the progeny transductant. Thus, the locus of the VP production-related gene, namely PPRG, was investigated by employing a classical microbial genetics method. First, STEVP was infected at a MOI of 0.29 in an *E. coli* W2252 (HfrC; *metB* λ^- Sm^s) recipient and selected by *met* prototrophic reversion; *met*$^+$ *E. coli* transductants were successfully obtained with a transduction frequency of $1.43 \pm 0.82 \times 10^{-7}$ CFU/particle ($n = 3$), referred to as Hfr*met*trans. Then, Hfr*met*trans (donor) and *E. coli* AB1157 (F^-; *leuB proA hisG argE* Sm^r, recipient) were applied to conjugation at donor/recipient ratios of 1:30, 1:50, and 1:90. Combination HfrC without VP infection and the recipient F^- was used as the control [135].

For *arg* and *his* markers of Hfr*met*trans, conjugative transfer frequencies toward AB1157 were comparable to that of the *E. coli* W2252 control, whereas *leu* marker frequency decreased by four orders of magnitude; furthermore, *pro* marker frequency was undetectable. Moreover, about half the proportion of the generated Hfr*met*trans-*E. coli* AB1157 transconjugants were incapable of particle production. Gene transfer of *E. coli* W2252 at conjugation begins at 13 min on the linkage map (*thrA*: 0 min) and proceeds in the counterclockwise direction [216]. Hence, *pro* and *leu* markers should be translocated to the recipient before *arg* and *his* markers from the donor. Namely, DNA originating from STEVP is likely to integrate near *proA* (5.6 min) and *leuB* (1.7 min) loci. Likewise, *proA* and *leuB* were recombined between *hisG* (45.0 min) and the terminal of conjugation (13 min) because the generated transductant showed prototrophic reversion for *pro* and *leu* markers. STEVP can transfer a dsDNA fragment of approximately 370 kb [130] and the distance between *proA* and *leuB* markers is 180.9 kb [217]. Therefore, STEVP would have the capacity for multiple recombinations at the transduction of *E. coli* W2252. PPRG would probably be transferred at the beginning of conjugation, being likely to locate near the conjugative terminal point because the coordinated marker transfer with *leu*, *arg*, and *his* markers was observed without the *pro* marker. VP-mediated generalized transduction has been observed [4]. Generalized recombination involves large regions of homologous DNA sequences; however, site-specific recombination involves considerably smaller segments of DNA, in which the recombination event occurs at a specific sequence—that is, the recognition sequence [218]. Based on this result, *proA* and *leuB* were integrated between 13 and 45 min on the linkage map of *E. coli*. This result indicates that the recombination in the VP-mediated transduction should not be ascribed to homologous recombination.

Furthermore, the existence of the recognition sequence or insertion sequence was postulated. From the above findings, the bacteriophage P22 (*attP22*: 5.6 min) attachment site is located within the region of 1.7–5.6 min on the *E. coli* linkage map [217, 219, 220], which would be associated with the PPRG integration region. VP is distinct from the P22 phage regarding shape, DNA molecular mass, host range, and so on. However, generalized transduction occurred by a wrapping choice mechanism [191, 192, 221], which might indicate that a comparable mechanism is involved in VP-mediated transduction. Indicated site-specific recombination may provide a reference to explain the genetic transfer processes in VP-mediated transductions.

Consequently, PPRG is likely to locate between 1.7 and 5.6 min on the transductant chromosome. A sulfur-reducing, hyperthermophilic archaeon, *Thermococcus onnurineus* NA1^T, produces extracellular vesicles referred to as ToEV by budding [22], with a TEM image resembling that described for VP above. The nucleic acid retained in the ToEVs amounted 40 to 20 kb. Interestingly, the host *T. onnurineus* chromosomal region corresponding to 9.4 kb is not encapsulated in ToEV. The region encodes hydrogenase, formate- and alcohol-dehydrogenase, oxidoreductase, Fe and S binding protein; however, why these enzyme genes are not encapsulated in the particles is unknown. As described previously, the conjugation of *E. coli* AB1157 with Hfr*met*Etrans was not able to generate *proAB* transconjugant. The region of *proAB* encodes dehydrogenase, protein-kinase, and DNA binding protein. The lack of transconjugant generation for that region might imply that the sequence region in the chromosome somehow prevented transfer from the encapsulation to the particle. A comparative genome analysis for such a region might provide some insight.

7.8 Summary and Concluding Remarks

7.8.1 Discovery

Plasmid-like elements were extracted from marine isolates by 2,4 D utilization and prototrophic reversion of auxotrophic recipient by transformation. Electron microscopy observations of PLE revealed the existence of particles instead of DNA strings being postulated intracellularly as maturing phage-like particles. Thus, the term "plasmid-like elements" was changed to "phage-like particles."

The spontaneous release of PLPs to the culture broth started when the host cell entered the stationary phase. Collected and purified PLPs accomplished transductant generation of the auxotrophic recipient beyond the family border. The spontaneous release without host cell disruption by PLPs from the transductant was observed, which was inconsistent with the concept of virus-mediated transduction. This curious and unusual manner of gene transfer was denoted as serial transduction, and the mediator for this gene transfer phenomenon is referred to as the broad-host-range VP.

7.8.2 Definition

VP-mediated transductants can produce new progeny particles by budding, which are again infectious to carry on the same transfer gene toward the next recipients. The resulting second-generation VP-mediated transductants again produce particles like the original donor, which again infect and induce successive transductants; this phenomenon is termed "serial transduction." Under current technical restraints, confirmation of this phenomenon is exclusively provided by the budding particle production from the transductant.

7.8.3 Distribution

VPs and VP producers must be ubiquitous on this planet because they have been found in a variety of environments, including marine, thermal, and terrestrial.

7.8.4 Physico-Chemical Characteristics

Based on studies to date, the morphology of VPs takes a spherical shape with a diameter between 20 nm and 500 nm, resembling a spherical virus. The collection and purification of VPs may be achieved by a method used for environmental viruses because the VPs reside in a virus-like particle fraction (20 nm < VLPs <100 kDa) in the environment. The concentration of VPs in environmental samples and the culture supernatant of producers was found by the tangential flow technique followed by equilibrium CsCl density gradient ultracentrifugation. VPs are considered to have lipid components in their surface structure; the buoyant density of VPs is lighter than that of a bona fide virus.

VPs are composed of cargo DNA combined with proteinaceous material and cytoplasmic substances without RNA, whose surface must have embedded enzyme domains (proteinase/peptidase and glycanase) [222] to achieve membrane fusion the infection of the recipient. The production of a VP appears as an electron-dense body under electron microscopy, starting around the center of the producing cell by a spooling DNA strand covered with proteinaceous material to make a reel structure. Then, the reel structure is translocated to the vicinity of the cell membrane, from where the budding-like extrusion of the VP occurs. Finally, pinching releases the VP to the external environment. The yield of VPs per cell was found to be an average of three particles, so the host strictly controls the production of VPs. The produced VPs exhibit a pleomorphic discrete size distribution that is even used to select a distinct range of VP sizes. A VP's cargo DNA could range from 5 kb to 2 Mb in size, whereas the volume of a VP does not always correspond to the molecular type of DNA estimated by PFGE. Large VPs may encapsulate multiple segments of somewhat similar sizes.

7.8.5 VPs as a Biological Function

No lethal effects (or a slight lethal effect) were observed for the VP on the recipient, whereas enhanced growth was found in some cases. A trend of lethal effects may be found in a smaller subset of close phylogenic positions between the producer and the recipient. Gene transfer was observed for a single gene locus as well as gene cluster, contributing to adaptations such as temperature and osmolality tolerance.

7.8.6 Resemblances and Differences of VPs with the Budding Virus and the MV

The host range of an HGT mediator is defined as the breadth of organisms that are capable of infecting and expressing their function, with limits on the host cell ascribed to the mediator, host, or environmental characteristics. However, no standard definition of a broad-host-range mediator currently exists because of the viewpoints of different disciplines.

The author empirically defines a broad-host-range vector particle as a VP that is capable of infecting, transferring, and budding reproduction in recipient cell and exceeds the phylogenetic family level from the original host.

VPs are characterized by a massive chromosomal gene transfer ability together with cytosolic substances, including plasmids and proteins. VPs are distinct from budding viruses and MVs. Although pleomorphic viruses have been recently conferred, pleomorphic viruses have a strict host range. None of the pleomorphic or budding virus genes are found in the VP-mediated transductants' genomes (Chiura, 2018, Unpublished data). The production scheme of MVs is considered to be entirely different based on the DNA cargo size and yields per cell. The proportion of DNA containing MVs is reportedly quite small.

The molecular mechanism for VP production must be partly shared with budding viruses and MVs, whereas the scheme for VP production is considered to be unique. A VP production scheme would be an intermingled function taken from the virus and variety of the cells during the evolution of life to be maintained throughout the cell.

Considering that VP budding release is specific to the stationary phase, a particular purpose for the VP lysogenic host must have existed for the VP production. Focusing on *Aquifex* sp., the originating STVP encapsulated approximately 400 kb DNA species; the host strain's genome size of 1.5 Mb corresponds to approximately 3.2 VPs yields per cell. STVP-mediated transduction toward *E. coli* generated VP lysogens of STEVP. Encapsulated DNA species amounted to the same as STVP, approximately 400 kb, whereas VP yields per cell were consistent with the original host. Three times more VP production would be necessary to cover the whole *E. coli* genome size of 4.6 Mb. To compensate for the shortage of the amount, STEVP should encapsulate three sets of 400-kb DNA cargo in one particle. As seen above, AkVP originating from *A. kielensis* is considered to carry plural segments of DNA cargo, which might have induced pleomorphism of the particle.

VP assemblage is considered to be the gene pool of VP lysogens because the DNA cargo in VPs consists of the host genome chromosome. Microorganisms in situ are continually suffering from environmental stresses, with the most severe impairment being chromosomal damage. Among the DNA repair systems, mismatch repair is the most effective system [174] to maintain the sequence fidelity; however, the system could not be achieved without the template strand. The most prominent role of VPs in the environment is likely to be the provider for the template strand to cure cells in need, in addition to carrying genetic information to be shared in the microbial assemblage to achieve environmental adaptation enhancements.

Acknowledgments The author is grateful to A Ellinger, M Fukui, A Hiraishi, H Hoaki, S Hagemann, J Ishibashi, K Kogure, S-J Kim, Y Kumagai, Y Kawarabayashi, A Maruyama, M Maruyama, T Nakagawa, RW Ridge, U Simidu, S Suzuki, T Urabe, B Velimirov, S Yoshizawa for their collaboration, discussion, and helpful comments. Sincere thanks to M Arata, K Hirota, E Ito, K Ito, K Inoue, M Imada, S Imazu, T Ichinose, Y Inokuchi, D Koketsu, D Kato, F Maruyama, S Morimoto, Y Morikawa, D Nakata, H Naito, K Nakamura, Late M Segawa, Y Suzuki, M Takeuchi, T Takemura, N Uchiyama-Arai and other Lab members at ICU and AORI, the University of Tokyo for important collaboration and assistance with the experiments, and R/V Tanseimaru, Shinseimaru, Shinryumaru, Toyoha Mines Co. Ltd. and Nikko Tankai Co. Ltd. for assistance with sampling. The research presented here was supported in part by Grant-in-Aid for Scientific Research Nos. 10490012, 12490009, 14208063, 14405016, 15310007, 16310031 Archaean Park Project, Int'l Study Programme 259 (1995-97), 15H0549 from the Japan Society for the Promotion of Science (JSPS), the Donations to Encourage Research by Kyowa Hakko Kogyo Ltd., Life Cycle Association Ltd., and Amano Enzyme Inc. Last and certainly not least, thanks are given to the author's family.

References

1. Chiura HX, Takagi J. Phage-like particles production and gene transfer by marine bacteria. Bull Jap Soc Microb Ecol. 1994;9:74–90.
2. Chiura HX. Generalized gene transfer by virus-like particles from marine bacteria. Aquat Microb Ecol. 1997;13:75–83.
3. Chiura HX. Marine lysogenic virus. Kaiyo. 2003;33:137–45.
4. Chiura HX. Novel broad-host range gene transfer particles in nature. Microbes Environ. 2004;19:249–64.
5. Chiura HX. Contribution of serial transduction for horizontal gene transfer in the natural environment. Actinomycetotologica. 2008;20:s13–4.
6. Chiura HX, Uchiyama N, Kogure K. Broad-host range gene transfer particle produced by *Aliivibrio fischeri*. Microbes Environ. 2009;24:322–9.
7. Chiura HX, Kogure K, Hagemann S, Ellinger A, Velimirov B. Evidence for particle-induced horizontal gene transfer and serial transduction between bacteria. FEMS Microbiol Ecol. 2011;76:576–91. https://doi.org/10.1111/j.1574-6941.2011.01077.x.
8. Chiura HX, Yamamoto H, Koketsu D, Naito H, Kato K. Virus-like particle derived from a bacterium belonging to the oldest lineage of the domain bacteria. Microbes Environ. 2002;17:48–52.
9. Chiura HX. Broad host range xenotrophic gene transfer by virus-like particles from a hot spring. Microbes Environ. 2002;17:53–8.
10. Cury J, Touchon M, Rocha EPC. Integrative and conjugative elements and their hosts: composition, distribution and organization. Nucleic Acids Res. 2017;45(15):8943–56. https://doi.org/10.1093/nar/gkx607.

11. Derbyshire KM, Gray TA. Distributive conjugal transfer: new insights into horizontal gene transfer and genetic exchange in *Mycobacteria*. Microbiol Spectr. 2014;2(1):61–79. https://doi.org/10.1128/microbiolspec.MGM2-0022-2013.
12. Waters VL. Conjugation between bacterial and mammalian cells. Nat Genet. 2001;29:375–6. https://doi.org/10.1038/ng779.
13. Nonaka L, Isshiki T, Suzuki S. Distribution of the oxytetracycline resistance determinant Tet 34 among bacterial isolated from diseased fish. Microbes Environ. 2002;17:26–31.
14. Nonaka L, Suzuki S. New Mg^{2+} dependent oxytetracycline resistance determinant Tet 34 in *Vibrio* isolates from marine fish intestinal contents. Antimicrob Agents Chemother. 2002;46:1550–2.
15. Nonaka L, Ikeno K, Suzuki S. Distribution of tetracycline resistance gene, *tet*(M), in gram-positive and gram-negative bacteria isolated from sediment and seawater at a coastal aquaculture site in Japan. Microbes Environ. 2007;22:355–64.
16. Lang AS, Beatty J. Importance of widespread gene transfer agent genes in α-proteobacteria. Trends Microbiol. 2007;15(2):54–62. https://doi.org/10.1016/j.tim.2006.12.001.
17. McDaniel L, Young E, Delaney J, Ruhnau F, Ritchie K, Paul J. High frequency of horizontal gene transfer in the oceans. Science. 2010;330(6000):50. https://doi.org/10.1126/science.1192243.
18. Stanton T. Prophage-like gene transfer agents—novel mechanisms of gene exchange for *Methanococcus*, *Desulfovibrio*, *Brachyspira*, and *Rhodobacter* species. Anaerobe. 2007;13(2):43–9. https://doi.org/10.1016/j.anaerobe.2007.03.004.
19. Dubey G, Ben-Yehuda S. Intercellular nanotubes mediate bacterial communication. Cell. 2011;144(4):590–600. https://doi.org/10.1016/j.cell.2011.01.015.
20. Yang F, Moss LG, Phillips GN Jr. The molecular structure of green fluorescent protein. Nat Biotechnol. 1996;14:1246–51.
21. Chatterjee SN, Chaudhuri K. Outer membrane vesicles of bacteria. Heidelberg: Springer; 2012.
22. Choi DH, Kwon YM, Chiura HX, Yang EC, Bae SS, Kang SG, Lee J-H, Yoon HS, Kim S-J. Extracellular vesicles of the hyperthermophilic archaeon *Thermococcus onnurineus* $NA1^T$. Appl Environ Microbiol. 2015;81:4591–9. https://doi.org/10.1128/AEM.00428-15.
23. Kulp A, Kuehn MJ. Biological functions and biogenesis of secreted bacterial outer membrane vesicles. Annu Rev Microbiol. 2010;64:163–84. https://doi.org/10.1146/annurev.micro.091208.073413.
24. Lee EY, Choi DY, Kim DK, Kim JW, Park JO, Kim S, Kim SH, Desiderio DM, Kim YK, Kim KP, Gho YS. Gram-positive bacteria produce membrane vesicles: proteomics-based characterization of *Staphylococcus aureus*-derived membrane vesicles. Proteomics. 2009;9:5425–36.
25. McBroom AJ, Kuehn MJ. Chapter 2.2.4, Outer membrane vesicles. In: Finlay BB, editor. EcoSal - *Escherichia coli* and *Salmonella*: cellular and molecular biology. Washington: ASM Press; 2005.
26. Rivera J, Cordero RJB, Nakouzi AS, Frases S, Nicola A, Casadevall A. *Bacillus anthracis* produces membrane-derived vesicles containing biologically active toxins. Proc Natl Acad Sci U S A. 2010;107:19002–7.
27. Soler N, Marguet E, Verbavatz J-M, Forterre P. Virus-like vesicles and extracellular DNA produced by hyperthermophilic archaea of the order *Thermococcales*. Res Microbiol. 2008;159:390–9. https://doi.org/10.1016/j.resmic.2008.04.015.
28. Toyofuku M, Cárcamo-Oyarce G, Yamamoto T, Eisenstein F, Hsiao M, Kurosawa C-C, Gademann K, Pilhofer M, Nomura N, Eberl L. Prophage-triggered membrane vesicle formation through peptidoglycan damage in Bacillus subtilis. Nat Commun. 2017;8:481. https://doi.org/10.1038/s41467-017-00492-w.
29. Toyofuku M, Morinaga K, Hashimoto Y, Uhl J, Shimamura H, Inaba H, Schmitt-Kopplin P, Eberl L, Nomura N. Membrane vesicle-mediated bacterial communication. ISME J. 2017;11:1504–9.

30. Yaron S, Kolling G, Simon L, Matthews K. Vesicle-mediated transfer of virulence genes from *Escherichia coli* O157:H7 to other enteric bacteria. Appl Environ Microbiol. 2000;66:4414–20. https://doi.org/10.1128/AEM.66.10.4414-4420.2000.
31. Beveridge T. Structures of gram-negative cell walls and their derived membrane vesicles. J Bacteriol. 1999;181:4725–33.
32. Brown L, Kessler A, Cabezas-Sanchez P, Luque- Garcia JL, Casadevall A. Extracellular vesicles produced by the gram-positive bacterium *Bacillus subtilis* are disrupted by the lipopeptide surfactin. Mol Microbiol. 2014;93:183–98.
33. Brown L, Wolf JM, Prados-Rosales R, Casadevall A. Through the wall: extracellular vesicles in gram-positive bacteria, mycobacteria and fungi. Nat Rev Microbiol. 2015;13(10):620–30. https://doi.org/10.1038/nrmicro3480.
34. Haurat MF, Elhenawy W, Feldman Mario F. Prokaryotic membrane vesicles: new insights on biogenesis and biological roles. Biol Chem. 2015;396:95. https://doi.org/10.1515/hsz-2014-0183.
35. Kadurugamuwa JL, Beveridge TJ. Membrane vesicles derived from *Pseudomonas aeruginosa* and *Shigella flexneri* can be integrated into the surfaces of other gram-negative bacteria. Microbiology. 1999;145:2051–60.
36. Kolling GL, Matthews KR. Export of virulence genes and Shiga toxin by membrane vesicles of *Escherichia coli* O157:H7. Appl Environ Microbiol. 1999;65:1843–8.
37. Li Z, Clarke AJ, Beveridge TJ. Gram-negative bacteria produce membrane vesicles which are capable of killing other bacteria. J Bacteriol. 1998;180:5478–83.
38. Renelli M, Matias V, Lo RY, Beveridge TJ. DNA-containing membrane vesicles of *Pseudomonas aeruginosa* PAO1 and their genetic transformation potential. Microbiology. 2004;150:2161–9. https://doi.org/10.1099/mic.0.26841-0.
39. Tashiro Y, Ichikawa S, Nakajima-Kambe T, Uchiyama H, Nomura N. *Pseudomonas* quinolone signal affects membrane vesicle production in not only gram-negative but also gram-positive bacteria. Microbes Environ. 2010;25:120–5. https://doi.org/10.1264/jsme2.ME09182.
40. Tashiro Y, Uchiyama H, Nomura N. Multifunctional membrane vesicles in *Pseudomonas aeruginosa*. Environ Microbiol. 2012;14:1349–62. https://doi.org/10.1111/j.1462-2920.2011.02632.x.
41. Jiang SC, Kellogg CA, Paul JH. Characterization of marine temperate phage-host systems isolated from Mamala Bay, Oahu, Hawaii. Appl Environ Microbiol. 1998;64:535–42.
42. Jiang SC, Paul JH. Gene transfer by transduction in the marine environment. Appl Environ Microbiol. 1998;64:2780–7.
43. Lang AS, Beatty JT. The gene transfer agent of *Rhodobacter capsulatus* and "constitutive transduction" in prokaryotes. Arch Microbiol. 2001;175(4):241–9.
44. Morrison WD, Miller RV, Sayler GS. Frequency of F116-mediated transduction of *Pseudomonas aeruginosa* in a freshwater environment. Appl Environ Microbiol. 1978;36:724–30.
45. Ripp S, Ogunseitan OA, Miller RV. Transduction of a freshwater microbial community by a new *Pseudomonas aeruginosa* generalized transducing phage, UT1. Mol Ecol. 1994;3:121–6.
46. Sandri RM, Berger H. Bacteriophage P1-mediated generalized transduction in *Escherichia coli*: structure of abortively transduced DNA. Virology. 1980;106:30–40.
47. Saye DJ, Ogunseitan OA, Sayler GS, Miller RV. Transduction of linked chromosomal genes between *Pseudomonas aeruginosa* during incubation in situ in a freshwater habitat. Appl Environ Microbiol. 1990;56:140–5.
48. Watson BNJ, Staals RHJ, Fineran PC. CRISPR-Cas-mediated phage resistance enhances horizontal gene transfer by transduction. MBio. 2018;9:e02406-17. https://doi.org/10.1128/mBio.02406-17.
49. Arber W. Horizontal gene transfer among bacteria and its role in biological evolution. Life. 2014;4:217–24. https://doi.org/10.3390/life4020217.
50. Bertani G. Transduction-like gene transfer in the methanogen *Methanococcus voltae*. J Bacteriol. 1999;181:2992–3002.

51. Eiserling F, Pushkin A, Gingery M, Bertani G. Bacteriophage-like particle associated with the gene transfer agent of *Methanococcus voltae* PS. J Gen Virol. 1999;80:3305–8.
52. Humphrey SB, Stanton TB, Jensen NS, Zuerner RL. Purification and characterization of VSH-1, a generalized transducing bacteriophage of *Serpulina hyodysenteriae*. J Bacteriol. 1997;179:323.
53. Solioz M, Marrs B. The gene transfer agent of *Rhodopseudomonas capsulata*. Arch Biochem Biophys. 1977;181:300–7.
54. Wood AG, Whiteman WB, Konisky J. Isolation and characterization of an archaebacterial virus-like particle from *Methanococcus voltae* A3. J Bacteriol. 1989;171:93–8.
55. Lang AS, Beatty JT. Genetic analysis of a bacterial genetic exchange element: the gene transfer agent of *Rhodobacter capsulatus*. Proc Natl Acad Sci U S A. 2000;97:859–64.
56. Westbye AB, Leung MM, Florizone S, Taylor TA, Johnson JA, Fogg PC, Beatty JT. Phosphate concentration and the putative sensor-kinase protein CckA modulate cell lysis and release of the *Rhodobacter capsulatus* gene transfer agent (RcGTA). J Bacteriol. 2013;195(22):5025–40.
57. Deatherage BL, Cookson BT. Membrane vesicle release in bacteria, eukaryotes, and archaea: a conserved yet underappreciated aspect of microbial life. Infect Immun. 2012;80:1948–57. https://doi.org/10.1128/IAI.06014-11.
58. Forterre P, Soler N, Krupovic M, Marguet E, Ackermann HW. Fake virus particles generated by fluorescence microscopy. Trends Microbiol. 2013;21:1–5. https://doi.org/10.1016/j.tim.2012.10.005.
59. Biller SJ, Schubotz F, Roggensack SE, Thompson AW, Summons RE, Chisholm SW. Bacterial vesicles in marine ecosystems. Science. 2014;343:183–6.
60. Soler N, Krupovic M, Marguet E, Forterre P. Membrane vesicles in natural environments: a major challenge in viral ecology. ISME J. 2015;9:793–6. https://doi.org/10.1038/ismej.2014.184.
61. Biller SJ, McDaniel LD, Breitbart M, Rogers E, Paul JH, Chisholm SW. Membrane vesicles in seawater: heterogeneous DNA content and implications for viral abundance estimates. ISME J. 2017;11:394–404. https://doi.org/10.1038/ismej.2016.134.
62. Toyofuku M, Roschitzki B, Riedel K, Eberl L. Identification of proteins associated with the *Pseudomonas aeruginosa* biofilm extracellular matrix. J Proteome Res. 2012;11:4906–15.
63. Fulsundar S, Harms K, Flaten GE, Johnsen PJ, Chopade BA, Nielsen KM. Gene transfer potential of outer membrane vesicles of *Acinetobacter baylyi* and effects of stress on vesiculation. Appl Environ Microbiol. 2014;80:3469–83. https://doi.org/10.1128/AEM.04248-13.
64. Mashburn LM, Whiteley M. Membrane vesicles traffic signals and facilitate group activities in a prokaryote. Nature. 2005;437:422–5. https://doi.org/10.1038/nature03925.
65. Davis DM, Sowinski S. Membrane nanotubes: dynamic long- distance connections between animal cells. Nat Rev Mol Cell Biol. 2008;9:431–43.
66. Hayashi T, Makino K, Ohnishi M, Kurokawa K, Ishii K, Yokoyama K, et al. Complete genome sequence of enterohemorrhagic *Escherichia coli* O157:H7 and genomic comparison with a laboratory strain K-12. DNA Res. 2001;8:11–22.
67. Shaefer R, Hinnen R, Franklin RM. Bacteriophage of *Halobacterium salinarium*. Nature. 1974;243:681–3.
68. Stackebrandt E, Goebel BM. Taxonomic note: a place for DNA-DNA reassociation and 16S rRNA sequence analysis in the present species definition in bacteriology. Appl Environ Microbiol. 1994;44:846–9.
69. Zambrynski P, Tempe J, Schell J. Transfer and function of T-DNA genes from *Agrobacterium* Ti and Ri plasmids in plants. Cell. 1989;56:193–201.
70. Schell J, Van Montagu M. The Ti-plasmid of *Agrobacterium tumefaciens*, a natural vector for the introduction of nif genes in plants? Basic Life Sci. 1977;9:159–79.
71. Stachel SE, Timmerman B, Zambryski P. Generation of single-stranded T-DNA molecules during the initial stages of T-DNA transfer from *Agrobacterium tumefaciens* to plant cells. Nature. 1986;322:706–12.

72. Straube WL, Deming WJ, Somerville CC, Colwell RR, Baross JA. Particulate DNA in smoker fluids: evidence for existence of microbial populations in hot hydrothermal system. Appl Environ Microbiol. 1990;56:1440–7.
73. Hennes KP, Simon M. Significance of bacteriophages for controlling bacterioplankton growth in a mesotrophic lake. Appl Environ Microbiol. 1995;61:333–40.
74. Hennes KP, Suttle CA. Direct counts of viruses in natural waters and laboratory cultures by epifluorescence microscopy. Limnol Oceanogr. 1995;40:1050–5.
75. Levy SB, Miller RV. Gene transfer in the environment. New York: McGraw-Hill; 1989.
76. Steward GF, Smith DC, Azam F. Abundance and production of bacteria and viruses in the Bering and Chukchi seas. Mar Ecol Prog Ser. 1996;131:287–300.
77. Mathias CB, Kirschner AKT, Velimirov B. Seasonal variations of virus abundance and viral control of the bacterial production in a backwater system of Danube river. Appl Environ Microbiol. 1995;61:3734–40.
78. Levin BR, Lenski RE. Coevolution in bacteria and their viruses and plasmids. In: Futuima DJ, Slatkin M, editors. Coevolution. Sunderland: Sinauer; 1983. p. 99–127.
79. Ohnishi M, Kurokawa K, Hayashi T. Diversification of *Escherichia coli* genomes: are bacteriophages the main contributors? Trends Microbiol. 2001;9:481–5.
80. Yeats S, McWilliam P, Zillig W. A plasmid in the archaebacterium *Sulfolobus acidocaldarius*. EMBO J. 1982;9:1035–8.
81. Young JPW, Wexler M. Sym plasmid and chromosomal genotypes are correlated in field population of *Rhizobium leguminosarum*. J Gen Microbiol. 1988;134:2731–9.
82. Paul JH. Microbial gene transfer. J Mol Microbiol Biotechnol. 1999;1:45–50.
83. Grissa I, Vergnaud G, Pourcel C. The CRISPRdb database and tools to display CRISPRs and to generate dictionaries of spacers and repeats. BMC Bioinformatics. 2007;8:172. https://doi.org/10.1186/1471-2105-8-172.
84. Barrangou R. The roles of CRISPR-Cas systems in adaptive immunity and beyond. Curr Opin Immunol. 2015;32:36–41. https://doi.org/10.1016/j.coi.2014.12.008.
85. Barrangou R, Fremaux C, Deveau H, Richards M, Boyaval P, Moineau S, et al. CRISPR provides acquired resistance against viruses in prokaryotes. Science. 2007;315(5819):1709–12. https://doi.org/10.1126/science.1138140.
86. Horvath P, Barrangou R. CRISPR/Cas, the immune system of bacteria and archaea. Science. 2010;327(5962):167–70. https://doi.org/10.1126/Science.1179555.
87. Marraffini LA, Sontheimer EJ. CRISPR interference limits horizontal gene transfer in staphylococci by targeting DNA. Science. 2008;322(5909):1843–5. https://doi.org/10.1126/science.1165771.
88. Quemin ERJ, Chlanda P, Sachse M, Forterre P, Prangishvili D, Krupovic M. Eukaryotic-like virus budding in Archaea. mBio. 2016;7(5):e01439–16. https://doi.org/10.1128/mBio.01439-16.
89. Jouvenet N, Zhadina M, Bieniasz PD, Simon SM. Dynamics of ESCRT protein recruitment during retroviral assembly. Nat Cell Biol. 2011;13(4):394–401. https://doi.org/10.1038/ncb2207.
90. Hibbing ME, Fuqua C, Parsek MR, Peterson SB. Bacterial competition: surviving and thriving in the microbial jungle. Nat Rev Microbiol. 2010;8:15–25. https://doi.org/10.1038/nrmicro2259.
91. Suttle CA. Marine viruses-major players in the global ecosystem. Nat Rev Microbiol. 2007;5:801–12. https://doi.org/10.1038/nrmicro1750.
92. Ackerman H-W, DuBow MS. Viruses of prokaryotes. In: General properties of bacteriophages, vol. 1. Boca Raton: CRC Press; 1987. p. 49–85.
93. Ambile-Cuevas CF, Chicurel ME. Bacterial plasmid and gene flux. Cell. 1992;70:189–99.
94. Birge EA. Bacterial and bacteriophage genetics. 3rd ed. New York: Springer; 1994.
95. Streips UN, Yasbin RE. Modern microbial genetics. New York: Wiley-Liss; 1991.
96. Fry JC, Day MJ. Overview of gene transfer in aquatic habitats. In: Guerrero R, Pedros-Alio C, editors. Trends in microbial ecology. Barcelona: Spanish Society for Microbiology; 1993. p. 315–8.

97. De Corte D, Sintes E, Winter C, Yokokawa T, Reinthaler T, Herndl GJ. Links between viral and prokaryotic communities throughout the water column in the (sub)tropical Atlantic Ocean. ISME J. 2010;4(11):1431–42.
98. Fuhrman JA, Ouverney CC. Marine microbial diversity studied via 16S rRNA sequences: cloning results from coastal waters and counting of native archaea with fluorescent single cell probes. Aquat Ecol. 1998;32:3–15.
99. Fuhrman JA. Marine viruses and their biogeochemical and ecological effects. Nature. 1999;399:541–8.
100. Miller RV, Ripp S, Relicon J, Ogunseitan OA, Kokjohn TA. Virus-mediated gene transfer in freshwater environment. In: Gauthier MJ, editor. Gene transfers and environment. Berlin: Springer; 1992. p. 51–62.
101. Wommack KE, Colwell RR. Virioplankton: viruses in aquatic ecosystems. Microbiol Mol Biol Rev. 2000;64:69–114.
102. Koonin EV. Horizontal gene transfer: essentiality and evolvability in prokaryotes, and roles in evolutionary transitions. F1000Res. 2016;5. https://doi.org/10.12688/f1000research.8737.1.
103. Lawrence JG, Ochman H. Molecular archaeology of the *Escherichia coli* genome. Proc Natl Acad Sci U S A. 1998;95:9413–7.
104. Martin W. Mosaic bacterial chromosomes: a challenge *en route* to a tree of genomes. Bioassays. 1999;21:99–104.
105. McInerney J, Cummins C, Haggerty L. Goods-thinking vs. tree-thinking. Mob Genet Elem. 2011;1(4):304–43. https://doi.org/10.4161/mge.19153.
106. Doolittle WF. Phylogenetic classification and the universal tree. Science. 1999;284:2124–9.
107. Eisen JA. Horizontal gene transfer among microbial genomes: new insights from complete genome analysis. Curr Opin Genet Dev. 2000;10(6):606–11.
108. Chiura H, Noro Y, Kanayama S, Ueda Y, Simidu U, Takagi J. Site-specific deoxyribonuclease produced by a marine bacterium, *Flavobacterium* I 16-04. Agric Biol Chem. 1988;52:2107–9.
109. Chiura HX, Kamiyama T, Hirano H, Futagami M, Watahiki M, Kobayashi K, Simidu U, Takagi J. Purification and characterization of *AspMD*1, an isoschizomer of *Sau3*AI, from a marine bacterium, *Alcaligenes* sp. MD1. Nucleic Acids Res. 1992;20:1996.
110. Chiura HX, Kato K, Takagi J. PLPs released by a marine bacterium. In: 4th international symposium on contamination of the environment by viruses and method of control, Die Universität für Bodenkultur Wien, Wien, Austria, August 1993.
111. Östling J, Goodman A, Kjelleberg S. Behaviour of IncP-1 plasmids and a miniMu transposon in a marine *Vibrio* sp.: isolation of starvation inducible *lac* operon fusions. FEMS Microbiol Ecol. 1991;86:83–94.
112. Chaudhry GR, Chapalamadugu S. Biodegradation of halogenated organic compounds. Microbiol Rev. 1991;55:59–79.
113. Reineke W, Knackmus H-J. Microbial degradation of haroaromatics. Annu Rev Microbiol. 1988;42:263–87.
114. van der Meer JR, de Vos WM, Harayama S, Zehnder AJB. Molecular mechanisms of genetic adaptation to xenobiotic compounds. Microbiol Rev. 1992;56:677–94.
115. Amy PS, Schulke W, Fraizer LM, Seidler RJ. Characterization of aquatic bacteria and cloning of genes specifying partial degradation of 2,4-dichlorophenoxyacetic acid. Appl Environ Microbiol. 1985;49:1237–45.
116. Chaudhry GR, Huang GH. Isolation and characterization of a new plasmid from a *Flavobacterium* sp. which carries the genes for degradation of 2,4-dichlorophenoxyacetic acid. J Bacteriol. 1988;170:3897–902.
117. Don RH, Pemberton JM. Properties of six pesticide degradation plasmids isolated from *Alcaligenes paradoxus* and *Alcaligenes eutrophus*. J Bacteriol. 1981;45:681–6.
118. Don RH, Weightman AJ, Timmis KN. Transposon mutagenesis and cloning analysis of the pathways for degradation of 2,4-dichlorophenoxyacetic acid and 3-chlorobenzonate in *Alcaligenes eutrophus* JMP134(pJP4). J Bacteriol. 1985;161:85–90.
119. Friello OA, Chakrabarty AM. Transposable mercury resistance. In: Stuttard C, Rozee KR, editors. Plasmids and transposons. New York: Academic Press; 1980. p. 249–60.

120. Mazodier P, Davies J. Gene transfer between distantly related bacteria. Annu Rev Genet. 1991;25:147–71.
121. Silver S, Misra TK. Plasmid-mediated heavy metal resistances. Annu Rev Microbiol. 1988;42:717–43.
122. Brown MJ, Lund PA, Nibhriain N. Mercury resistance in bacteria. In: Hapwwod DA, Chater KE, editors. Genetics of bacterial diversity. New York: Academic; 1998. p. 121–39.
123. DiGiovanni DG, Neilson JW, Pepper IL, Sinclair NA. Gene transfer of *Alcaligenes eutrophus* JMP134 plasmid pJP4 to indigenous soil recipients. Appl Environ Microbiol. 1996;62:2521–6.
124. Yabuuchi E, Kosako Y, Yano I, Hotta H, Nishiuchi Y. Transfer of two *Burkholderia* and an *Alcaligenes* species to *Ralstonia* gen. Nov.: proposal of *Ralstonia pickettii* (Ralston, Palleroni and Doudoroff 1973) comb. Nov., *Ralstonia solanacearum* (Smith 1896) comb. Nov. and *Ralstonia eutropha* (Davis 1969) comb. Nov. Microbiol Immunol. 1995;39(11):897–904. https://doi.org/10.1111/j.1348-0421.1995.tb03275.x.
125. Ahrens R. Taxonomische Untersuchungen an sternbildenden *Agrobacterium* - Arten aus der westlichen. Ostsee Kiel Meeresforsch. 1968;24:147–73.
126. Uchino Y, Hirata A, Yokota A, Sugiyama J. Reclassification of marine *Agrobacterium* species: proposals of *Stappia stellulata* gen. nov., comb. nov., *Stappia aggregata* sp. nov., nom. rev., *Ruegeria atlantica* gen. nov., comb. nov., *Ruegeria gelatinovora* comb. nov., *Ruegeria algicola* comb. nov., and *Ahrensia kieliense* gen. nov., sp. nov., nom. rev. J Gen Appl Microbiol. 1998;44(3):201–10.
127. Chiura HX, Rao Bhamidimarri SM, Yu PL. Plasmid for the degradation of 2,4-dichlorophenoxyacetic acid (2,4-D). In: Yu PL, editor. Fermentation technology: industrial application. London: Elsevier; 1990. p. 59–62.
128. Weinbauer MG, Höfle MG. Significance of viral lysis and flagellate grazing as factors controlling bacterioplankton production in a eutrophic lake. Appl Environ Microbiol. 1998;64:431–8.
129. Izumi A, et al. Broad-host-range vector particle-mediated horizontal drug resistance gene transfer. In: ISME 2006, Vienna, Austria, 20–25 August 2006.
130. Chiura HX, Umitsu M. Isolation and characterisation of broad-host range gene transporter particles from geo-thermal vent of the Toyoha mine. Microbes Environ. 2004;19:20–30.
131. Chiura HX, Umitsu M, Fukazawa Y, Nakata D, Nakamura K, Tomaru A, Okita N, Kawarabayashi Y, Hoaki T. Broad-host range vector-particle from thermal vents. In: The 6th international marine biotechnology conference (IMBC2003), Makuhari, Chiba, 21–27 September 2003.
132. Chiura HX, Imada M, Maruyama F, Yasuoka N. Transkingdom gene transfer using virus-like particles originating from thermophile. In: The 6th international marine biotechnology conference (IMBC2003), Makuhari, Chiba, 21–27 September 2003.
133. Chiura HX, Sakai A, Takasaki Y, Hoaki T. Are broad-host-range gene transfer particles ubiquitous existences in the environment? In: The annual meeting of the Japanese Society of Microbial Ecology, Tsu, Mie, 15–17 November 2002.
134. Sugitate T, Chiura HX. Functional gene transfer towards a broad range of recipients with the aid of vector particle originating from a thermophile. In: The proceedings of international symposium on extremophiles and their applications, vol. 2005. Yokosuka: Extremobiosphere Research Center, JAMSTEC; 2007. p. 141–7. http://www.jstage.jst.go.jp/article/isea/2005/0/141/.
135. Inoue K, Sugitate T, Chiura HX. Study of particle production responsible gene locus by bacterial conjugation. ISME 2006, Vienna, Austria, 20–25 August 2006.
136. Inokuchi Y, Chiura HX. Search for a broad-host-range gene transfer particle (VP) production responsible gene of VP-transductant: *Escherichia coli* DH-*lac*-trans. In: MBSJ 2006 Forum, Nagoya, Aichi, Nagoya Congress Centre 6–8 December 2006.
137. Sugitate T, Ito K, Chiura HX. Novel Characteristics of Broad-host range gene transfer particles from thermophiles. In: 8th international conference on thermophiles research, Gold Coast, Australia, 18–22 September 2005.

138. Chiura HX, Yoshizawa S, Kogure K. Vector particles originating from *Polaribacter* changed the minimum and optimum growth temperature of *Escherichia coli*. In: ISME 2012, Copenhagen, Denmark, 19–24 August 2012.
139. Chiura HX, Kogure K. Nitrogen fixation acquisition of *Escherichia coli* by vector particles originating from *Klebsiella pneumonia subsp. rhinoscleromatisi*. In: XV ISME, Seoul, Korea, 24–29 August 2014.
140. Gauthier MJ, Breittmayer VA. Gene transfer in marine environments. In: Fry JC, Day MJ, editors. Bacterial genetics in natural environments. London: Chapman and Hall; 1990. p. 100–10.
141. Stewart GJ, Carlson CA. The biology of natural transformation. Annu Rev Microbiol. 1986;40:211–35.
142. Stewart GJ, Carson CA, Ingraham JL. Evidence for an active role of donor cells in natural transformation of *Pseudomonas stutzeri*. J Bacteriol. 1983;156:30–5.
143. Hirsch PR. Factors limiting gene transfer in bacteria. In: Fry JC, Day MJ, editors. Bacterial genetics in natural environments. London: Chapman and Hall; 1990. p. 31–40.
144. Bathe S, Lebuhn M, Ellwart JW, Wuertz S, Hausner M. High phylogenetic diversity of transconjugants carrying plasmid pJP4 in an activated sludge-derived microbial community. FEMS Microbiol Lett. 2004;235:215–21. https://doi.org/10.1016/j.femsle.2004.04.038.
145. Sen D, Van der Auwera GA, Rogers LM, Thomas CM, Brown CJ, Top EM. Broad-host-range plasmids from agricultural soils have IncP-1 backbones with diverse accessory genes. Appl Environ Microbiol. 2011;77:7975–83. https://doi.org/10.1128/AEM.05439-11.
146. Szpirer C, Top EM, Couturier M, Mergeay M. Retrotransfer or gene capture: a feature of conjugative plasmids, with ecological and evolutionary significance. Microbiology. 1999;145:3321–9.
147. Thomas CM. Paradigms of plasmid organization. Mol Microbiol. 2000;37:485–91. https://doi.org/10.1046/j.1365-2958.2000.02006.x.
148. Dahlberg C, Bergström M, Andreasen M, Christensen BB, Molin S, Hermansson M. Interspecies bacterial conjugation by plasmids from marine environments visualized by *gfp* expression. Mol Biol Evol. 1998;15:385–90.
149. Dahlberg C, Bergström M, Hermansson M. In situ detection of high levels of horizontal plasmid transfer in marine bacterial communities. Appl Environ Microbiol. 1998;64:2670–5.
150. Lejeune P, Mergeay M, Van Gijsegem F, Faelen M, Gerits J, Toussaint A. Chromosome transfer and R-prime plasmid formation mediated by plasmid pULB113 (RP4::Mini-Mu) in *Alcaligenes eutrophus* CH34 and *Pseudomonas fluorescens* 6.2. J Bacteriol. 1983;155:1015–26.
151. Black LW. DNA packaging in dsDNA bacteriophages. In: Calendar R, editor. The bacteriophages, vol. 2. New York: Plenum Publishing Corp; 1988. p. 321–73.
152. Charvin M, Rastogi N, Lévy-Frébault VV. An easy and rapid method for isolation of entire mycobacterial genome for application in pulsed-field gel electrophoresis. Current Microbiol. 1991;22(5):327–31. https://doi.org/10.1007/BF02091963.
153. McEllistrem MC, Stout JE, Harrison LH. Simplified protocol for pulsed-field gel electrophoresis analysis of *Streptococcus pneumoniae*. J Clin Microbiol. 2000;38(1):351–3.
154. Yen H-C, Marrs B. Map of genes for carotenoid and bacteriochlorophyll biosynthesis in *Rhodopseudomonas capsulata*. J Bacteriol. 1976;126:619–29.
155. Hagemann S, Stoeger L, Kappelmann M, Hassl I, Ellinger A, Velimirov B. DNA-bearing membrane vesicles produced by Ahrensia kielensis and *Pseudoalteromonas marina*. J Basic Microbiol. 2013;54(10):1062–72.
156. Lang AS, Zhaxybayeva O, Beatty JT. Gene transfer agents: phage-like elements of genetic exchange. Nat Rev Microbiol. 2012;10(7):472–82.
157. Chiura HX, Kato K, Takagi J. Phage-like particles released by a marine bacterium. Wien Mitteil. 1995;128:149–57.
158. Chiura HX, Velimirov B, Kogure K. Virus-like particles in microbial population control and horizontal gene transfer in aquatic environments. In: Bell CR, Brylinsky M, Johnson-Green P, editors. Microbial biosystems: new frontiers. Halifax: Atlantic Canada Society for Microbial Ecology; 2000. p. 167–73.

159. Atanasova NS, Senčilo A, Pietilä MK, Roine E, Oksanen HM, Bamford DH. Chapter one. Comparison of lipid-containing bacterial and archaeal viruses. In: Maramorosch K, Mettenleiter T, editors. Advances in virus research, vol. 92. 1st ed. Amsterdam: Elsevier; 2015. p. 1–61. Table 1: 4–16. https://doi.org/10.1016/bs.aivir.2014.11.005.
160. Dybvig K, Nowak JA, Sladeck TL, Maniloff J. Identification of an enveloped phage, mycoplasma virus L172, that contains a 14-kilobase single-stranded DNA genome. J Virol. 1985;53:384–90.
161. Pietilä MK, Roine E, Paulin L, Kalkkinen N, Bamford DH. An ssDNA virus infecting archaea: a new lineage of viruses with a membrane envelope. Mol Microbiol. 2009;72:307–19. https://doi.org/10.1111/j.1365-2958.2009.06642.x.
162. Pietilä MK, Roine E, Sencilo A, Bamford DH, Oksanen HM. Pleolipoviridae, a newly proposed family comprising archaeal pleomorphic viruses with single-stranded or double-stranded DNA genomes. Arch Virol. 2016;161(1):249–56.
163. Putzrath RM, Maniloff J. Growth of an enveloped mycoplasma virus and establish-ment of a carrier state. J Virol. 1977;22:308–14.
164. Davis WR, Bostein D, Roth JR. A manual for advanced bacterial genetics. Cold Spring Harbor: Cold Spring Harbor Laboratory; 1980. p. 148–52.
165. Chen F, Lu JR, Binder B, Hodson RE. Enumeration of viruses in aquatic environments using SYBR gold stain: application of digital image analysis and flow cytometer. Appl Environ Microbiol. 2001;67:539–45.
166. Børsheim KY, Bratbak G, Heldal M. Enumeration and biomass estimation of planktonic bacteria and viruses by transmission electron microscopy. Appl Environ Microbiol. 1990;56:352–6.
167. Kenzaka T, Tani K, Nasu M. High-frequency phage-mediated gene transfer in freshwater environments determined at single-cell level. ISME J. 2010;4:648–59.
168. Weinbauer MG, Peduzzi P. Frequency, size and distribution of bacteriophages in different marine bacterial morphotypes. Mar Ecol Prog Ser. 1994;108:11–20.
169. Proctor LM, Fuhrman JA. Viral mortality of marine bacteria and cyanobacteria. Nature (London). 1990;343:60–2.
170. Proctor LM, Fuhrman JA. Roles of viral infection in organic particle flux. Mar Ecol Prog Ser. 1991;69:133–42.
171. Weinbauer MG, Fuks D, Peuzzi P. Distribution of viruses and dissolved DNA along a coastal trophic gradient in the northern Adriatic Sea. Appl Environ Microbiol. 1993;59:4074–82.
172. Proctor LM, Okubo A, Fuhrman JA. Calibrating estimates of phage-induced mortality in marine bacteria: ultrastructural studies of marine bacteriophage development from one-step growth experiments. Microb Ecol. 1993;25:161–82.
173. Nakagawa T, Hanada S, Maruyama A, Marumo K, Urabe T, Fukui M. Distribution and diversity of thermophilic sulfate-reducing bacteria within a Cu-Pb-Zn mine (Toyoha, Japan). FEMS Microbiol Ecol. 2002;41(3):199–209. https://doi.org/10.1111/j.1574-6941.2002.tb00981.x.
174. Iyer R, Pluciennik A, Burdett V, Modrich P. DNA mismatch repair: functions and mechanisms. Chem Rev. 2006;106(2):302–23. https://doi.org/10.1021/cr0404794.
175. Yamamoto H, Hiraishi A, Kato K, Chiura HX, Maki Y, Shimizu A. Phylogenetic evidence for the existence of novel thermophilic bacteria in hot spring sulphur-turf microbial mats in Japan. Appl Environ Microbiol. 1998;64:1680–7.
176. Lorenz MG, Wackernagel W. Bacterial gene transfer by natural genetic transformation in the environment. Microbiol Rev. 1994;58:293–316.
177. Grogan DW. Selectable mutant phenotypes of the extremely thermophilic archaebacterium *Sulfolobus acidocaldarius*. J Bacteriol. 1991;173:7725–7.
178. Grogan DW. Exchange of genetic markers at extremely high temperatures in the archaeon *Sulfolobus acidocaldarius*. J Bacteriol. 1996;178:3207–11.
179. Sugitate T, Inaba N, Kurusu Y, Hoaki T, Chiura HX. Thermo-resistance acquisition of a mesophilic bacterium with the aid of vector particles originating from thermophiles. In: American Geophysical Union Autumn Meeting, San Francisco, USA, 13–17 December 2004. https://adsubs.harvard.edu/?#abs/2004AGUFM.V41B1371S/abstract.

180. Shioda M, Sugimori K, Shiroya T, Takayanagi S. Nucleosomelike structures associated with chromosomes of the archaebacterium Halobacterium salinarium. J Bacteriol. 1989;171:4514–7.
181. Takayanagi S, Morimura S, Kusaoke H, Yokoyama Y, Kano K, Shioda M. Chromosomal structure of the halophilic archaebacterium *Halobacterium salinarium*. J Bacteriol. 1992;174:7207–16.
182. Moak M, Molineux IJ. Peptidoglycan hydrolytic activities associated with bacteriophage virions. Mol Microbiol. 2004;51:1169–83.
183. Dreiseikelmann B. Translocation of DNA across bacterial membranes. Microbiol Rev. 1994;58(3):293–316. https://www.ncbi.nlm.nih.gov/pmc/articles/PMC372970/pdf/microrev00022-0011.pdf
184. Kivelä HM, Daugelavicius R, Hankkio RH, Bamford JK, Bamford DH. Penetration of membrane-containing double-stranded-DNA bacteriophage PM2 into *Pseudoalteromonas* hosts. J Bacteriol. 2004;186:5342–54.
185. Adams E. Biochem Z. 1942;310:384.
186. Kates M. Techniques of lipidology. Isolation, analysis and identification of lipids. Amsterdam: Elsevier; 1975. p. 83. Translated under the title *Tekhnika lipidologii*. Moscow: Mir; 1972.
187. Troll W, Cannan RK. A modified photometric ninhydrin method for the analysis of amino and imino acids. J Biol Chem. 1953;200:803–11.
188. Somogyi M. A new reagent for the determination of sugars. J Biol Chem. 1945;60:61–8.
189. Chiura HX, Kita-Tsukamoto K. Purification and characterisation of a novel agarase secreted by a marine bacterium, *Pseudoalteromonas* sp. strain CKT1. Microbes Environ. 2000;15:11–22.
190. Saito M, Sugitate T, Chiura HX. Study of broad-host range vector particle-host cell surface interaction upon initial stage of infection. In: The annual meeting of the Japanese Society of Microbial Ecology, Fukuoka, Fukuoka, 30 October–1 November 2005.
191. Ikeda H, Tomizawa J. Transducing fragments in generalized transduction by phage P1. I. Molecular origin of the fragments. J Mol Biol. 1965;14:85–109.
192. Ikeda H, Tomizawa J. Transducing fragments in generalized transduction by phage P1. II. Association of DNA and protein in the fragments. J Mol Biol. 1965;14:110–9.
193. Yarmolinsky MB, Sternberg N. Bacteriophage P1. In: Calendar R, editor. The bacteriophages, vol. 1. New York: Plenum Press; 1988. p. 291–438.
194. Schmieger H. The fate of the bacterial chromosome in P22-infected cells of *Salmonella typhimurium*. Mol Gen Genet. 1971;110:238–44.
195. Schmieger H, Buch U. Appearance of transducing particles and the fate of host DNA after infection of *Salmonella typhimurium* with P22-mutants with increased transducing ability (HT mutants). Mol Gen Genet. 1975;140:111–22.
196. Hasegawa Y, Futamata H, Tashiro Y. Complexities of cell-to-cell communication through membrane vesicles: implications for selective interaction of membrane vesicles with microbial cells. Front Microbiol. 2015;6:633. https://doi.org/10.3389/fmicb.2015.00633.
197. Toyofuku M, Nomura N. What will membrane vesicles (MVs) bring to bacterial communication? Microbes Environ. 2017;32(3):185–7. https://doi.org/10.1264/jsme2.ME3203rh.
198. Oliveira H, Melo DL, Santos SB, Nóbrega FL, Ferreira EC, Cerca N, Azeredo J, Kluskens LD. Molecular aspects and comparative genomics of bacteriophage endolysins. J Virol. 2013;87:4558–70.
199. Attar N. Biofilms: have exploding cells blown up MV dogma? Nat Rev Microbiol. 2016;14:334–5.
200. Bitto NJ, Chapman R, Pidot S, Costin A, Lo C, Choi J, D'Cruze T, Reynolds EC, Dashper SG, Turnbull L, Whitchurch CB, Stinear TP, Stacey KJ, Ferrero RL. Bacterial membrane vesicles transport their DNA cargo into host cells. Sci Rep. 2017;7(1):7072. https://doi.org/10.1038/s41598-017-07288-4.
201. Roier S, Zingl FG, Cakar F, Schild S. Bacterial outer membrane vesicle biogenesis- a new mechanism and its implications. Microb Cell. 2016;3(6):257–9. https://doi.org/10.15698/mic2016.06.508.

202. Roier S, Zingl FG, Cakar F, Durakovic S, Kohl P, Eichmann TO, Klug L, Gadermaier B, Weinzerl K, Prassl R, Lass A, Daum G, Reidl J, Feldman MF, Schild S. A novel mechanism for the biogenesis of outer membrane vesicles in gram-negative bacteria. Nat Commun. 2016;7:10515. https://doi.org/10.1038/ncomms10515.
203. Babst M. MVB vesicle formation: ESCRT-dependent, ESCRT-independent and everything in between. Curr Opin Cell Biol. 2011;23(4):452–7. https://doi.org/10.1016/j.ceb.2011.04.008.
204. Schmidt O, Teis D. The ESCRT machinery. Curr Biol. 2012;22(4):R116–20. https://doi.org/10.1016/j.cub.2012.01.028.
205. Hurley JH, Hanson PI. Membrane budding and scission by the ESCRT machinery: it's all in the neck. Nat Rev Mol Cell Biol. 2010;11(8):556–66. https://doi.org/10.1038/nrm2937.
206. Fenner F, McAuslan BR, Mims CA, Sambrook J, White DA. The biology of animal viruses. 2nd ed. New York: Academic; 1974.
207. Flint SJ, Enquist LW, Krug RM, Racaniello VR, Sakalka AM. Principle of virology. Washington: American Society Microbiology; 2000.
208. Senčilo A, Paulin L, Kellner S, Helm M, Roine E. Related haloarchaeal pleomorphic viruses contain different genome types. Nucleic Acids Res. 2012;40:5523–34. https://doi.org/10.1093/nar/gks215.
209. Luk AWS, Williams TJ, Erdmann S, Papke RT, Cavicchioli R. Viruses of *Haloarchaea*. Life. 2014;4:681–715.
210. Wichels A, Biel SS, Gelderblom HR, Brinkhoff T, Muyzer G, Schuett C. Bacteriophage diversity in the North Sea. Appl Environ Microbiol. 1998;64:4128–33.
211. Jensen E, Schrader H, Rieland B, Thompson T, Lee K, Nickerson K, Kokjohn T. Prevalence of broad-host-range lytic bacteriophages of *Spherotilus natans*, *Escherichia coli*, and *Pseudomonas aeruginosa*. Appl Environ Microbiol. 1998;64:575–80.
212. Suttle C. Cyanophages and their role in the ecology of cyanobacteria. In: Whitton B, Potts M, editors. The ecology of cyanobacteria. Dordrecht: Kluwer Academic Publishers; 2000. p. 563–89.
213. Weinbauer MG, Rassoulzadegan F. Are viruses driving microbial diversification and diversity? Environ Microbiol. 2003;6:1–11.
214. Chiura HX, Velimirov B, Kogure K. Relationship between cell population density and induction frequency of lysogen: studied for marine originating broad-host range gene transfer particle (VP)-mediated transductant. In: IUMS 2011 Sapporo. International Congress of Applied Microbiology, Sapporo, Hokkaido, October 2011.
215. Fuqua WC, Winans SC, Greenberg EP. Quorum sensing in bacteria: the LuxR-LuxI family of cell density-responsive transcriptional regulators. J Bacteriol. 1994;176:269–75.
216. Nishimura A, Iyama S. Catalogue of *Escherichia coli* genetic stocks in National Institute of Genetics. Shizuoka: National Institute of Genetics, Genetic Stocks Research Center; 1986.
217. Berlyn MB. Linkage map of *Escherichia coli* K-12, edition 10: the traditional map. Microbiol Mol Biol Rev. 1998;62:814–984.
218. Birge EA. Bacterial and bacteriophage genetics. 4th ed. New York: Springer; 2000.
219. Canchaya C, Proux C, Fournous G, Bruttin A, Brüssow H. Prophage genomics. Microbiol Mol Biol Rev. 2003;67:238–76.
220. Poteete RA. Bacteriophage P22. In: Calendar R, editor. The bacteriophages, vol. 2. New York: Plenum Press; 1988. p. 647–82.
221. Susskind MM, Botstein D. Molecular genetics of bacteriophage P22. Microbiol Rev. 1978;42:385–413.
222. Nishima W, Kanamaru S, Arisaka F, Kitao A. Screw motion regulates multiple functions of T4 phage protein gene product 5 during cell puncturing. J Am Chem Soc. 2011;133:13571–6. https://doi.org/10.1021/ja204451g.

Bacterial Transposable Elements and IS-Excision Enhancer (IEE)

8

Masahiro Kusumoto and Tetsuya Hayashi

Abstract

Insertion sequence (IS) elements are the simplest transposable elements (or mobile genetic elements) and are widely distributed in bacteria. The transposition and proliferation of IS elements induce not only insertional gene inactivation and modification of gene expression but also a wide range of genomic rearrangements, such as deletions, inversions, and duplications. IS-mediated bacterial genome diversification has been extensively studied in enterohemorrhagic *Escherichia coli* (EHEC). Excision of IS elements occurs frequently in O157, the major serogroup of EHEC isolated from humans, and IS-excision enhancer (IEE) promotes IS excision from the O157 genome in a transposase-dependent manner. IEE promotes the excision of IS elements belonging to several IS families, and various types of genomic deletions are also generated via IEE-promoted IS excision in O157. In addition, IEE has been found in specific lineages of enterotoxigenic *E. coli* (ETEC) strains isolated from swine, in which the *iee* genes are located on integrative elements that are similar to SpLE1 of EHEC O157. *iee*-positive ETEC lineages also contain multiple copies of IS elements at genomic locations that exhibit significant variations between strains, as observed in O157. These data and the phylogeny of IEE homologs found in a broad range of bacteria suggest that IEE has coevolved with IS elements and plays pivotal roles in bacterial genome evolution by inducing IS removal and genomic deletion.

M. Kusumoto (✉)
National Institute of Animal Health, National Agriculture and Food Research Organization, Tsukuba, Ibaraki, Japan
e-mail: kusu555@affrc.go.jp

T. Hayashi
Department of Bacteriology, Faculty of Medical Sciences, Kyushu University, Fukuoka, Japan

H. Nishida, T. Oshima (eds.), *DNA Traffic in the Environment*,
https://doi.org/10.1007/978-981-13-3411-5_8

Keywords
Insertion sequence (IS) · Transposase · IS-excision enhancer (IEE) · IS-mediated bacterial genome diversification · IEE-promoted IS excision · Enterohemorrhagic *Escherichia coli* · O157 · IS*629* · SpLE1

8.1 Transposition of Insertion Sequence (IS) Elements in Bacteria

8.1.1 IS Elements

Transposable elements (TEs) are widely distributed in eukaryotic and prokaryotic genomes. They are considered selfish (or parasitic) genetic elements, but they also play important roles in genome evolution [1]. IS elements, which are the simplest TEs, are generally 700–2500 bp in size and encode only the transposase (TPase) that catalyzes their own transposition [2]. The transposition and proliferation of IS elements induce not only insertional gene inactivation and modification of gene expression [1] but also a wide range of genomic rearrangements, such as deletions, inversions, and duplications [3, 4].

In bacteria, several thousand types of IS elements have thus far been identified from various species and strains [5] and classified into approximately 20 families based on the sequences of their TPases and terminal inverted repeats (TIRs) as well as several other features [2]. The mechanism of transposition differs between IS families, but IS elements generally transpose either by leaving the original copy in the donor DNA (termed copy-and-paste, or replicative, transposition) or by eradicating it from the donor DNA (cut-and-paste, or non-replicative, transposition) [6], as shown in Fig. 8.1. Most IS elements create

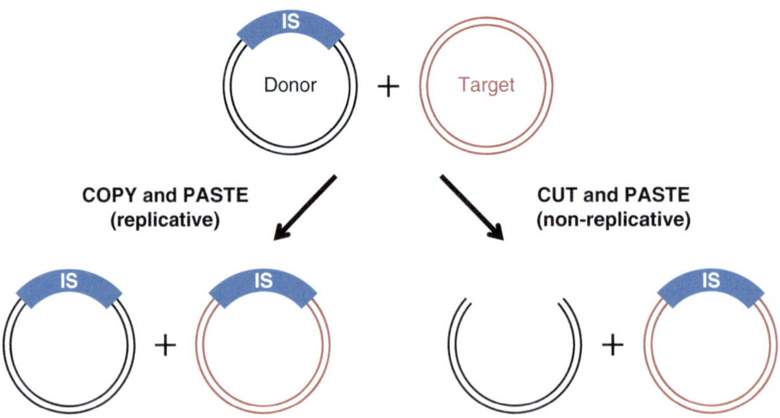

Fig. 8.1 Two types of transposition mechanisms of IS elements. Different transposition reactions generate different products. In copy-and-paste (or replicative) transposition, an IS element transposes to the target DNA and leaves the original copy in the donor DNA. In cut-and-paste (or non-replicative) transposition, the IS element is excised from the donor DNA and inserted into the target DNA

duplicated sequences 2–14 bp in length at the site of insertion, called target site duplications (TSDs), and the length of the TSD differs depending on the type of IS element is inserted.

8.1.2 Transposition Mechanism of the IS*3* Family

The mechanisms of IS transposition have been intensively studied for representative IS elements from several families; among these are the members of the IS*3* family. The process of transposition in this large IS family is initiated by the formation of a "figure-eight" intermediate, which is followed by the generation of a circular DNA molecule (called an IS circle) that is randomly inserted into the target DNA [2]. Although the entire process is not yet fully understood at the molecular level, it has been demonstrated that a replicative pathway is used to generate the IS circle from the figure-eight intermediate for IS*911*. Thus, IS*911* transposes by the copy-and-paste mechanism [7].

If the copy-and-paste mechanism is the major pathway for IS*3* family transposition, the excision of IS*3* family members rarely occurs during the normal transposition process. Indeed, it has long been believed that the excision of IS elements is a rare genetic event in bacteria [8] because end-joining systems, which are required to reseal the donor DNA for its survival after IS excision, have been identified only in a limited number of bacterial species [9]. Thus, very little attention has been paid to the genetic events that occur with the donor DNA upon IS excision. TE excision in bacteria has been described in several reports, but these excision events are considered TPase-independent. Instead, they depend on host factors required for replication slippage and repair functions [10, 11]. In practice, these excision events have been observed at low frequencies.

However, several lines of evidence show that the excision of IS*629*, a member of the IS*3* family, occurs frequently in the enterohemorrhagic *Escherichia coli* (EHEC) O157 strains that have been recently reported (see Sects. 8.2 and 8.3).

8.2 IS-Mediated Bacterial Genome Diversification in EHEC O157

8.2.1 IS*629*: The Most Abundant IS Element in EHEC O157

EHEC O157 produces highly potent cytotoxins (Shiga toxins Stx1 and/or Stx2) and causes diarrhea, hemorrhagic colitis, and hemolytic uremic syndrome; thus, it is regarded as one of the most serious food-borne infections worldwide [12]. O157 strains contain many IS elements, and these elements play important roles in O157 genome diversification. For example, the O157 strain RIMD0509952 (referred to as O157 Sakai) contains 25 types of IS elements (116 copies in total), the most abundant of which is an IS*3* family member, IS*629* (also called IS*1203*v). Twenty-three copies are present on the chromosome and a large virulence plasmid, pO157 [13, 14]. Of these IS*629* copies, 17 (74%) are apparently intact, whereas only 22 copies

(24%) are intact among all other types of IS elements, suggesting that active copies of IS*629* are maintained in O157.

The intact IS*629* consists of two open reading frames (ORFs), *orfA* and *orfB*, which overlap in the −1 frame, and TIRs located on the ends of the IS element. IS*629* TPase is produced as a fused protein, OrfAB, by translational frameshifting between the two ORFs [15], which is a common feature of the IS*3* family [16]. The −35 and −10 promoter regions and the Shine-Dalgarno sequence for the TPase are involved with the TIR located upstream of *orfA* [17]. Both TIRs are recognized by the TPase, which contains a DNA-binding motif in the N-terminal domain encoded by *orfA*, and this process is required for the transposition of IS*3* family members [2].

8.2.2 Structural Polymorphisms in O157 Genomes

The chromosome of O157 Sakai contains 1.4 Mb of sequences that are not present in the *E. coli* laboratory strain K-12. These O157 Sakai-specific sequences are scattered throughout a 4.1-Mb chromosome backbone that is shared with K-12 [13]. Most of the large O157 Sakai-specific sequences are prophages and prophage-like integrative elements (Sp1–Sp18 and SpLE1–SpLE6, respectively), and many of the virulence-related genes in O157 have been introduced into the O157 genome by these mobile genetic elements (MGEs) or the pO157 plasmid [13, 14]. K-12 also contains 11 prophages and prophage-like integrative elements, implying that bacteriophages are major contributors to the genomic diversification of *E. coli* [18]. High degrees of genomic diversity have been identified in the O157 lineage by whole genome PCR scanning (WGPS) analysis, a long-range PCR-based method to compare genome structures [19], and comparative genomic hybridization (CGH) analysis using an O157 oligo DNA microarray [20]. Among the numerous structural polymorphisms identified, large-size structural polymorphisms (LSSPs) are present in regions corresponding to prophages, indicating that prophages are a major contributor to genomic diversity within the O157 lineage [19]. This finding is supported by the results of genomic comparisons of O157 substrains obtained from rounds of repeated subculturing in vitro, i.e., spontaneous recombination between homologous prophage regions caused large-scale inversions within the O157 chromosome [21].

Numerous small-size structural polymorphisms (SSSPs), ranging from a few hundred base pairs to several kilobase pairs, have also been identified in the O157 lineage. In a systematic genome-wide analysis of SSSP-containing genomic regions in nine O157 strains, it was found that a large portion of the SSSPs (130 of 165) were generated by genetic events associated with only two types of IS elements, IS*629* and IS*Ec8* [22]. IS*Ec8* is the second-most common IS element following IS*629* (11 copies) in the O157 Sakai genome, and most IS*Ec8* insertions identified were simple insertions. In contrast, IS*629* induced a wide variety of genomic rearrangements, including simple insertions, simple excisions, and various deletions of IS-flanking regions ranging in size from 1 bp to several kilobase pairs. These findings indicate that IS*629* (and also IS*Ec8*), along with bacteriophages, serves as an important driving force to generate the genomic diversity of O157 strains.

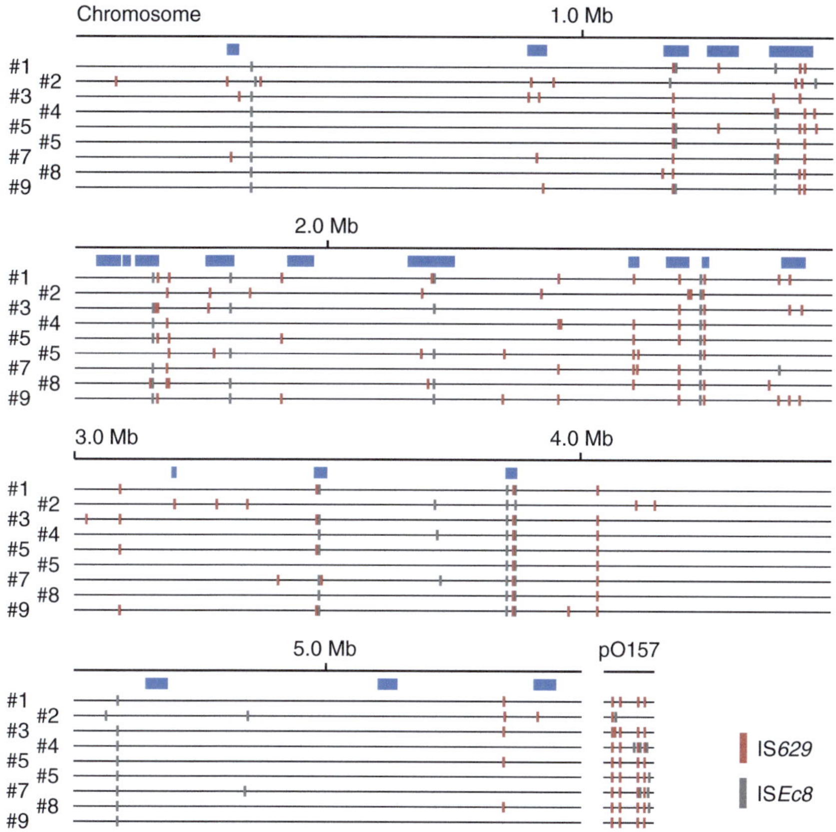

Fig. 8.2 Distribution of IS*629* and IS*Ec8* in the nine O157 genomes. The locations of IS*629* and IS*Ec8* on each chromosome and plasmid are shown. Red vertical lines, insertion sites for IS*629*; gray vertical lines, insertion sites for IS*Ec8*; blue rectangles, regions corresponding to prophages and prophage-like integrative elements

Insertion sites for IS*629* and IS*Ec8* exhibit a highly biased distribution in O157 genomes (Fig. 8.2): these IS elements are much more frequently located in O157-specific regions than in the chromosome backbone [22]. The genomic locations of other types of IS elements in the O157 Sakai genome are also highly biased toward O157-specific regions, and thus this trend is not specific to IS*629* and IS*Ec8* [22]. As shown in Fig. 8.2, many copies of IS*629* and IS*Ec8* are inserted in prophages, prophage-like integrative elements, and plasmids, which have carried these IS elements into the O157 genome and constitute a large proportion of O157-specific regions [13]. Given that IS elements (at least IS*629* and IS*Ec8*) are randomly inserted into the genome, it is most likely that clones with IS insertions into genes with essential functions have been selectively removed from the population, and those with insertions into regions that are less essential or nonessential for growth, such as many of the O157 Sakai-specific regions, have survived. Accordingly, IS insertion and IS-mediated genomic deletion have induced the inactivation or deletion of various foreign genes carried into the O157 lineage by MGEs, along with genes required for their mobility [22]. This suggests that IS insertion and IS-mediated genomic deletion generate various phenotypic differences among O157 strains, including differences in potential virulence (see Sect. 8.3.1).

8.2.3 IS-Printing: The IS Locus-Based Typing Method for EHEC O157

The highly variable distribution of IS*629* between O157 genomes was exploited to develop a multiplex PCR-based strain-typing method for O157, termed IS-printing [23]. The principle of the IS-printing method is schematically presented in Fig. 8.3. An outward universal primer for IS*629* and a set of outside primers are designed for the highly conserved region of IS*629* and for the adjacent regions of each IS*629* locus in the O157 genome, respectively. The positions of outside primers are designed to produce ladder patterns in the range of 100 bp to 1.0 kb with 50- to 100-bp spacing when multiplex PCR is performed using the outward primer and the outside primer set. Thus, O157 strains with different IS*629* insertion patterns (namely, different genomic locations of IS*629*) exhibit distinct ladder patterns (fingerprints). Using this method, 32 IS*629* loci were selected as the targets for

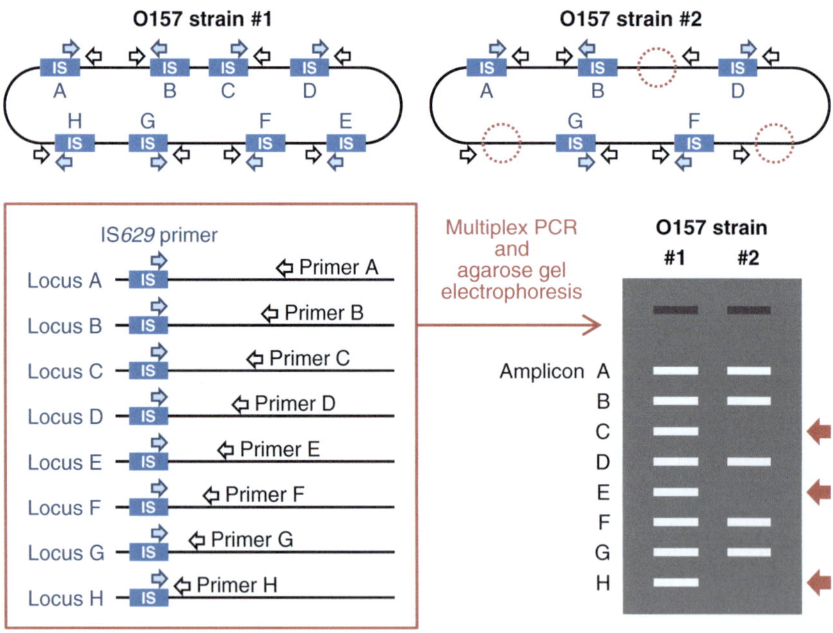

Fig. 8.3 Schematic of the IS-printing method. An outward universal primer for IS*629* (blue arrow) and the outside primers for the adjacent regions of each IS*629* locus in the O157 genome (open arrow) are designed such that the amplicons produce a ladder pattern. When comparing O157 strains #1 and #2, differences in the presence of IS*629* at loci C, E, and H are reflected by the observation of corresponding amplicons (red arrow)

amplification from 77 loci identified in eight WGPS-analyzed O157 strains [19] to maximize the discriminatory power [23].

Among the currently available methods for molecular typing of O157 strains, pulsed-field gel electrophoresis (PFGE) is most widely used for epidemiologic studies and the surveillance of O157 infections due to its high discriminatory power [24, 25]. However, PFGE analysis requires well-trained technical skills and several days to obtain results. It is also difficult to obtain consistently reproducible results among different laboratories, which hinders inter-laboratory data comparisons. On the other hand, the IS-printing method can be performed without the need for special equipment and techniques and can be completed within 2 h after colony picking. Furthermore, since banding patterns can be transformed into digital data (1 for presence and 0 for absence), the data are readily compared across different laboratories. Although the IS-printing method possesses these advantages, it has a lower discriminatory power than PFGE, for example, 201 O157 strains showing different XbaI digestion patterns by PFGE were classified into only 127 types by the IS-printing method [23]. Therefore, the IS-printing method may not be sufficiently discriminatory as a stand-alone typing method for O157 strains but instead can be used for initial screening to rapidly detect strains that are potentially associated with outbreaks as recently reported [23, 26, 27].

8.3 IEE-Promoted IS Excision and Genomic Deletion

8.3.1 Excision of IS*629* in EHEC O157

IS*629* insertion has been identified at various sites in *stx2* genes in O157 clinical isolates [17, 26, 28–31]. Although they occur at a very low frequency, such inactivated *stx2* genes were found to be reactivated by precise IS*629* excision (i.e., removal of IS*629* and a TSD sequence, which generates a wild-type *stx2* gene) in the *E. coli* laboratory strain K-12 [15]. Moreover, using a reporter plasmid-based assay, it was shown that precise IS*629* excision occurs much more frequently in O157 Sakai than in K-12, i.e., the excision frequencies of IS*629* carried on the reporter plasmid in O157 Sakai and in K-12 were 2.5×10^{-4} and 2.0×10^{-8}, respectively. Several other EHEC strains also exhibited a high frequency (within a range of 10^{-3} to 10^{-4}) of IS*629* excision [32]. These results suggested that some *E. coli* strains, including O157, contain a system that promotes IS excision and regenerates the donor DNA lacking the IS copy.

The clear difference between O157 Sakai and K-12 in terms of excision frequency further suggests that this system is encoded by some of the approximately 1600 O157 Sakai genes that are not present in K-12 [13].

8.3.2 IS-Excision Enhancer (IEE)

A genetic determinant that is responsible for the high frequency of IS*629* excision in O157 was identified by a combination of reporter plasmid-based IS*629*-excision assay and CGH analysis of *E. coli* strains and designated as IEE [33]. As shown in Table 8.1, no precise IS*629* excision in an IEE-deletion mutant of O157 Sakai

Table 8.1 The effects of *iee* on IS*629* excision frequency in O157 Sakai, K-12, and their derivatives

Strain	*iee*	Excision frequency
O157 Sakai	+	2.5×10^{-4}
GMSS401[a]	–	$<10^{-9}$
GMSS401(pIEE1)[b]	+	1.6×10^{-3}
GMEC101[c]	+	9.7×10^{-3}
K-12	–	2.0×10^{-8}
K-12(pIEE1)[b]	+	6.1×10^{-2}

[a]GMSS401 is an isogenic mutant of O157 Sakai in which *iee* was deleted from the chromosome
[b]pIEE1 is an IEE expression plasmid
[c]GMEC101 is an isogenic mutant of K-12 in which *iee* was inserted in the chromosome

(strain GMSS401), and the excision frequency of the mutant returned to the parental level by reintroducing IEE [GMSS401(pIEE1)]. Furthermore, K-12 derivatives containing IEE (chromosomally inserted or plasmid encoded) showed a high IS*629* excision rate comparable to that in O157 Sakai [GMEC101 and K-12(pIEE1)]. These results indicate that IEE enhances precise IS*629* excision in the genetic backgrounds of O157 and K-12. The gene encoding IEE (*iee*) is located on a large integrative element called SpLE1 [13], which is one of the O157 Sakai-specific genomic regions. The IEE-mediated enhancement of IS*629* excision requires the IS*629* TPase. The complete DDE motif in the TPase, which represents the active center of the IS TPase [2], was indispensable for this activity. Thus, IEE promotes IS*629* excision from the O157 genome in an IS TPase-dependent manner [33].

By using reporter plasmid-based IS-excision assays for two other IS*3* family members (IS*2* and IS*3*) and six IS elements each belonging to the IS*1*, IS*4*, IS*5*, IS*26*, IS*30*, and IS*621* families, IEE was found to enhance the excision of members of the IS*3*, IS*1*, and IS*30* families [33]. As the excision of IS*2* and IS*3* is promoted by IEE at the same level as that for IS*629*, IEE most likely acts on the entire IS*3* family. The excision frequencies of IS*1* and IS*30* also clearly increase in the presence of IEE, but to a lesser extent than that of the IS*3* family. The activity of IEE on IS*1* is much weaker than that on IS*3* family members and even IS*30*. This phenomenon is interesting because IS*1* appears to be able to use two pathways for transposition [34]: one involves the formation of a circular transposition intermediate (similar to the IS*3* family) and thus is expected to be IEE-sensitive, whereas the other involves cointegrate formation and thus may be insensitive to IEE.

8.3.3 IEE-Promoted Genomic Deletion

Using the IS-printing method, structural alterations were detected at the target IS*629* insertion sites in the O157 genome (that is, deletion of IS*629* and/or flanking regions) by inspecting the ladder patterns (Sect. 8.2.3). When O157 Sakai cells possessing both IS*629* TPase- and IEE-expressing plasmids (with IS*629* and *iee* left on the chromosome) were cultivated for 24 h in Luria-Bertani (LB) broth,

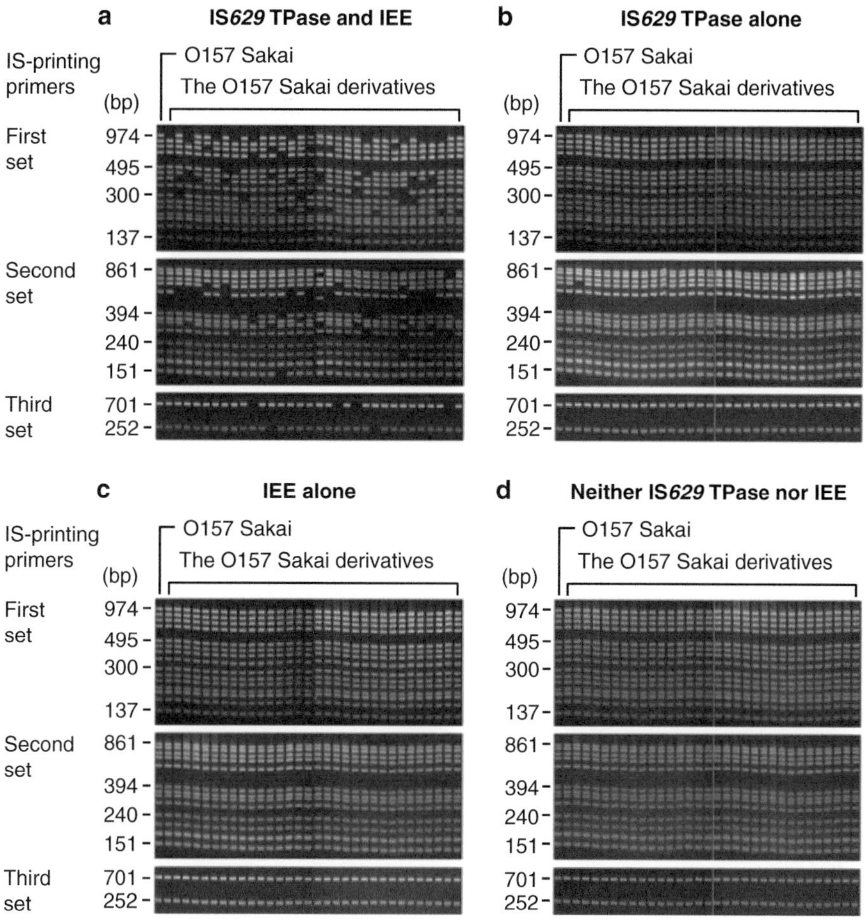

Fig. 8.4 IS*629* TPase- and IEE-induced structural alterations at IS*629* insertion sites in the O157 Sakai genome. The results of IS-printing analyses for four sets of O157 Sakai derivatives (32 clones each) that were obtained after overexpression of the IS*629* TPase and IEE together (**a**), IS*629* TPase alone (**b**), IEE alone (**c**), or neither IS*629* TPase nor IEE (**d**) are shown. IS-printing analysis was performed using three sets of multiplex PCR primers (the first to third sets). All 23 IS*629* insertion sites in the O157 Sakai genome were examined by multiplex PCR followed by agarose gel electrophoresis. The presence or absence of structural alterations at each IS*629* insertion site was determined according to the ladder band patterns observed for each clone

32 single colonies were randomly selected and analyzed by the IS-printing method, and alterations to the ladder pattern were detected in 29 clones (91%) in which 1–7 bands disappeared (Fig. 8.4a). No such alterations in the ladder pattern occurred in O157 Sakai cells containing the TPase- or the IEE-expressing plasmid alone (Fig. 8.4b–d). Thus, IEE, TPase, and an IS*629* sequence with complete TIRs recognized by TPase during transposition are all required for IEE-induced structural alterations [33].

Analysis of the structures of all IS*629* insertion sites that underwent structural alterations via the actions of IEE and IS*629* TPase (78 sites in 29 clones) provided somewhat expected findings; not only simple IS deletion but also various types genomic deletions were generated at the IS insertion sites [33]. As summarized in Fig. 8.5, these genomic deletions were categorized into four types: (I) deletion of IS*629* with an adjacent short sequence of 1–7 bp (mainly 3 or 4 bp); (II) deletion of IS*629* with an adjacent long sequence; (III) partial deletion of IS*629*; and (IV) deletion of a long sequence adjacent to IS*629* without IS deletion. In types I, II, and III, additional deletion of a 3–5 bp sequence (mostly 3 bp) which is located adjacent to the other end of IS*629* was also sometimes observed (types Ib, IIb, and IIIb in

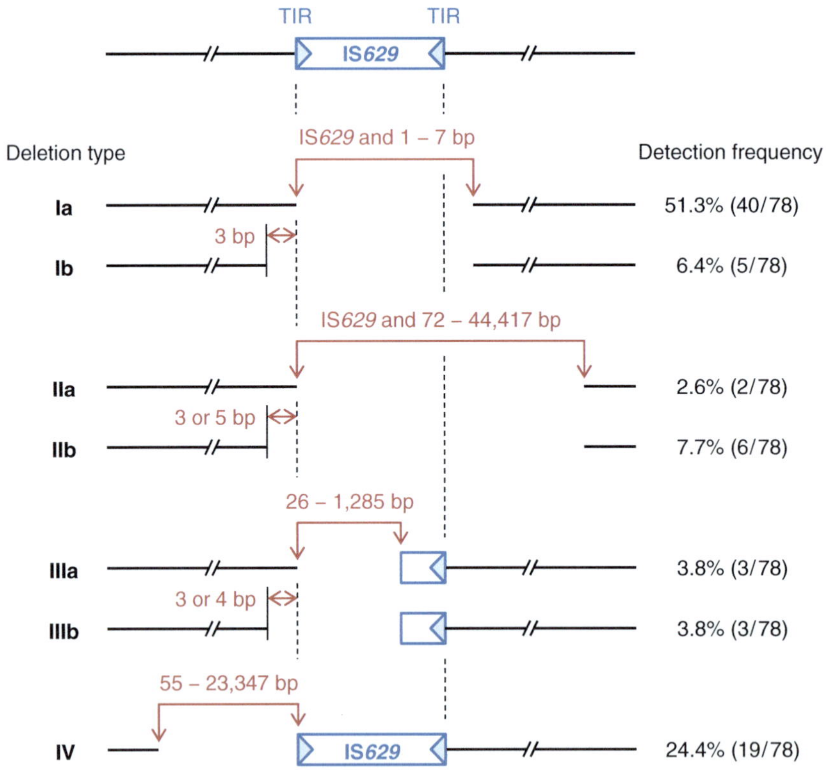

Fig. 8.5 Various genomic deletions generated by IS629 TPase and IEE. Structural changes at IS*629* insertion sites in the O157 Sakai genome that were induced by IEE in the presence of IS*629* TPase were classified into four types. In types I and II, the entire IS*629* sequence was deleted, together with short (1–7 bp, mainly 3 or 4 bp) and long (72–44,417 bp) adjacent sequences, respectively. In type III, portions (26–1285 bp) of the 1310-bp IS*629* sequence were deleted. Deletion occurred at one IS end in types Ia, IIa, and IIIa, and additional deletions of adjacent sequences occurred at the other IS end in types Ib, IIb, and IIIb. In type IV, only the genomic segments adjacent to IS*629* (55–23,347 bp) were deleted without deletion of the IS element. Blue triangles indicate TIRs of IS*629*. The detection frequency of each type is expressed as the percentage, and the number of clones obtained for each type is indicated in parenthesis

Fig. 8.5). Among the four types of deletions, the most frequently generated was type Ia (51.3%), which corresponds to precise or simple excision.

During the transposition process for IS*3* family members, TPases first cleave one DNA strand at the 3′ end of the IS element [7, 35] (Fig. 8.6a), and the liberated 3′ end is transferred to the same DNA strand at a position 3 or 4 bp away from the 5′ end of the IS element to generate a figure-eight intermediate (Fig. 8.6b, c). In the final step, DNA replication from the figure-eight junction generates a circular transposition intermediate and regenerates donor DNA that contains the original IS copy [2, 36]. Considering these processes and the proposed molecular mechanisms underlying each step, IEE most likely functions in the last step to guide the process toward deleting the IS*629* copy from the donor DNA (Fig. 8.6d). Indeed, the molecular mechanism underlying this phenomenon has yet to be clarified, but second-strand cleavage and resealing of donor DNA may be involved. In several cases, deletion of a 1-, 5-, 6-, or 7-bp sequence adjacent to the IS element was also observed (Fig. 8.5). These atypical deletions may have been generated by imprecise strand transfer. Other types of deletions (types II, III, and IV) were also formed only when IEE and IS*629* TPase were co-expressed (Fig. 8.4). This finding indicates that these deletions were also generated via a process coupled with IS excision and promoted by IEE and IS*629* TPase. The most plausible explanation at present is that the deletions were generated by aberrant strand transfer during the process promoted by IEE and IS TPase (Fig. 8.6e). Notably, all variant deletions were observed in naturally occurring O157 strains (see Sect. 8.2.2).

8.3.4 Distribution of IEE in Bacteria

Pathogenic *E. coli* strains other than EHEC are also important etiological agents of zoonotic or food-borne disease in humans and of colibacillosis in domestic animals [37, 38]. Among these, enterotoxigenic *E. coli* (ETEC) is an important cause of diarrhea in children, which is associated with high morbidity and mortality in non-industrialized countries. ETEC is also the main cause of diarrhea in travelers to these countries [37]. In swine, ETEC infections, which occur immediately after birth (neonatal diarrhea), and ETEC or Shiga toxin-producing *E. coli* infections, which occur after weaning (post-weaning diarrhea or edema disease), are responsible for significant economic losses due to diarrhea, growth retardation, and mortality [39, 40]. The majority of swine-pathogenic *E. coli* strains belong to a limited range of O serogroups, and O139 and O149 are among the most frequently reported serogroups worldwide [39, 41, 42].

Recently, these swine ETEC strains in serogroups O139 or O149 were found to possess the *iee* gene [43]. The *iee* gene is located on a large integrative element of O157 EHEC, SpLE1, and on SpLE1-like elements in O26, O111, and O103 EHECs. In the swine ETEC strains, IEE is distributed specifically among three distinct lineages, and the gene is encoded by integrative elements similar to SpLE1 [43]. SpLE1-like elements are highly conserved in their genomic structure among these ETEC lineages, and similar to SpLE1, they carry the *iha* gene and the *ure* operon,

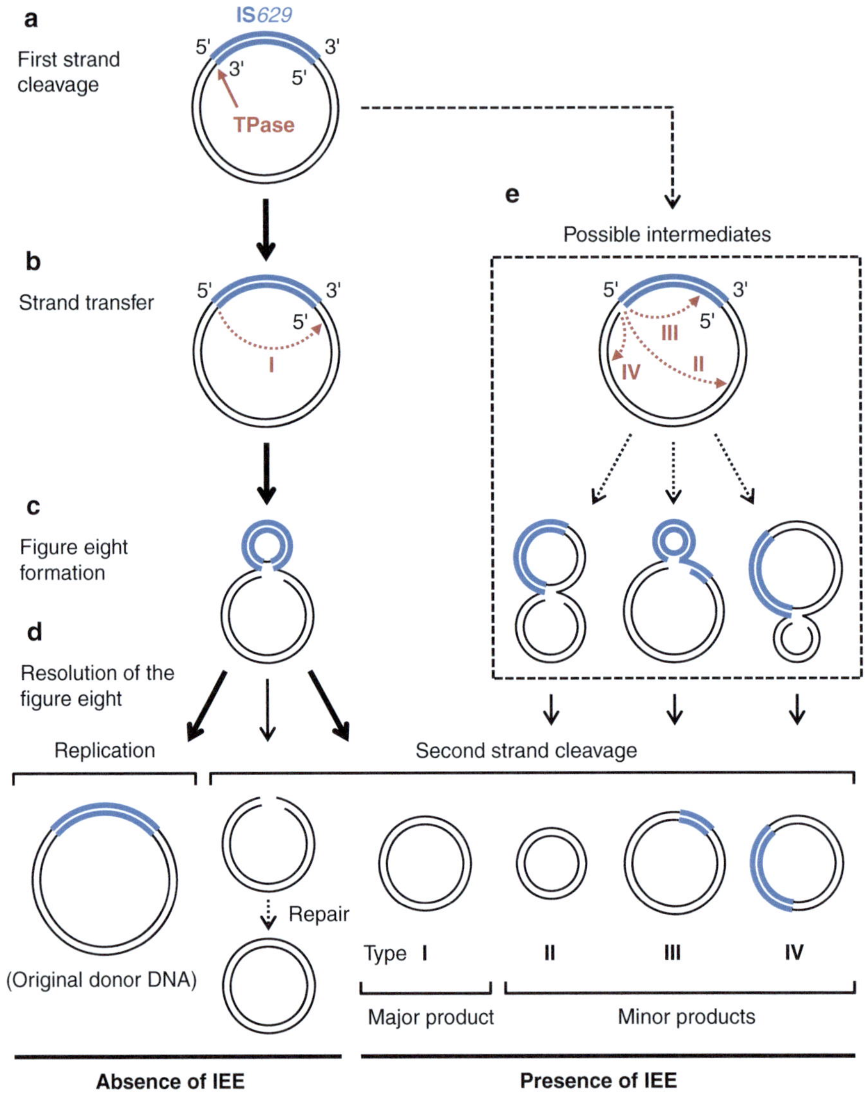

Fig. 8.6 Possible pathways for IS*629* excision and for the generation of various types of genomic deletions by IS excision. (**a**) TPase first cleaves one strand at the 3′ end of IS*629* (thick blue line) as indicated by the red arrow. (**b**) The liberated 3′ end is transferred to the proximity of the 5′ end of IS*629* (usually 3 or 4 bp away from the 5′ end) on the same strand, as indicated by the dotted red arrow I. (**c**) The strand transfer event generates a figure-eight structure in which the ends of IS*629* are joined by a single-stranded bridge. (**d**) DNA replication from the figure-eight junction generates a circular transposition intermediate and regenerates the donor DNA containing the IS copy. In the presence of IEE, various types of genomic deletions occur at IS insertion sites. The major product is a type I deletion representing simple excision, and other types of genomic deletions (types II, III, and IV) are minor products. (**e**) If aberrant strand transfer [to a genome position far away from the 5′ end of IS*629*, inside IS*629*, or very far from the 5′ end (i.e., near the 3′ end of IS*629*)] occurred, as indicated by dotted arrows II, III, and IV, these events would be intermediates for the formation of each variant type of genomic deletion

which are required for efficient colonization by O157 in the swine intestine [44]. Most of the structural variations observed between the elements of EHECs and those of ETEC are minor variations associated with IS insertion (or deletion) [43]. Considering that the three IEE-positive ETEC lineages are phylogenetically distantly related each other and that one lineage is closely related to the O26 and O111 EHECs, the SpLE1-like elements of ETEC strains may have been transferred from these EHEC lineages. Alternatively, SpLE1-like elements may circulate more widely in the *E. coli* population than recognized and transfer frequently between *E. coli* strains. Notably, all IEE-positive ETEC lineages also contain multiple copies of IS*629*, a preferred substrate of IEE, and their genomic locations vary significantly between strains [43]. Therefore, it is very likely that IS*629* actively moves in the genomes of these ETEC strains and, in combination with IEE, promotes the diversification of their genomes, as observed in EHEC O157.

Importantly, IEE homologs have been identified in a broad range of bacterial species [33]. According to a phylogenetic analysis of the IEE homologs, the homologs from different phyla often cluster together (Fig. 8.7). This finding indicates that the evolution of IEE homologs did not follow that of the host genomes. The distribution of IEE homologs suggests that they have spread in each species in a strain-specific manner, as observed in *E. coli*. For example, among the four fully sequenced *Legionella pneumophila* strains, IEE homologs were found in only two. In *Bacteroides fragilis*, *Desulfitobacterium hafniense*, *Lactobacillus rhamnosus*, and *Streptococcus pneumoniae*, IEE homologs were found only in one strain among the two or more sequenced strains. In these five species, genes for their IEE homologs were located in strain-specific regions or genomic islands. Furthermore, many IEE homologs are encoded in genomic regions exhibiting low GC content and/or containing genes related to MGEs [33]. Thus, IEE and its homologs appear to represent a novel protein family that has been widely spread in bacteria by horizontal gene transfer and has coevolved with MGEs, especially IS elements. It is possible that IEE family proteins play pivotal roles in the evolution of many bacterial species or strains by inducing IS excision and various types of genomic deletions.

8.4 Conclusion: Impact of IS Elements and IEE on Bacterial Genome Evolution

IEE promotes the excision of IS elements belonging to the IS*3* family, as well as those belonging to the IS*1* and IS*30* families, from bacterial genomes and generates various types of genomic deletions in cooperation with the IS TPase. Although the molecular mechanism of IEE action remains unknown, the identification of IEE is important from several perspectives. First, IEE is the first bacterial protein shown to promote IS excision in a TPase- or transposition-dependent manner. Several host factors, such as proteins for DNA repair and recombination, may also induce the deletion of TEs [45] but in a TPase-independent manner and at much lower efficiencies [33]. This function of IEE is particularly important for bacterial cells because,

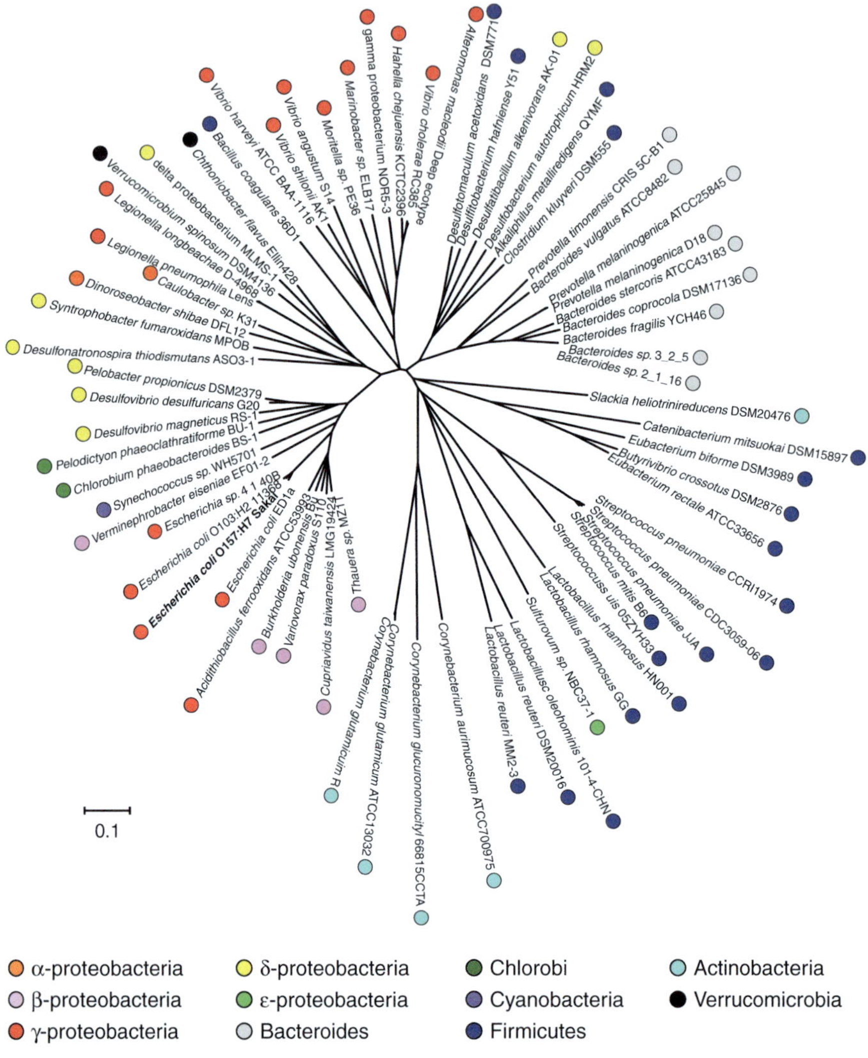

Fig. 8.7 Phylogeny of IEE homologs. The names of the bacterial species and strains in which each homolog was discovered are shown. The phylum (or class) of the species is indicated by colored circles, as shown below the phylogenetic tree. The size bar represents the number of substitutions per site. The IEE homologs from different phyla often cluster together, indicating that the evolution of the IEE homologs did not follow that of their host genomes

unlike eukaryotic cells, most bacterial species demonstrate no specific end-joining activity to regenerate and protect the donor genome after IS excision [8, 9]. Second, IEE-promoted IS excision generates a variety of genomic deletions. IS elements and their transpositions are generally regarded as one of the major driving forces generating various mutations and genomic structural changes. Such structural changes include genomic deletions. Many genomic deletions have been detected in sequenced bacterial genomes and are often suspected to have been generated by IS transposition-related events, but the precise mechanisms remain unknown. The action of IEE can explain how genomic deletions are generated upon IS transposition or excision. Third, the structural features of the donor DNA regenerated after IEE-promoted IS excision, particularly those associated with variant types of deletions, may provide further insights into the processes or molecular mechanisms of IS transposition. The processes leading to the generation of such variant types of genomic deletions may occur only when IEE is present. However, it is also possible that rare or aberrant intermediates that are unable to survive during the normal transposition process (that is, in the absence of IEE) may be trapped by the action of IEE. If so, the generation of these deletions may represent the inaccuracy present in the IS transposition mechanism. Finally, IEE can be regarded as a new type of genetic system that controls TE copy numbers in host genomes. The invasion and proliferation of TEs are important for long-term genome evolution, but their uncontrolled proliferation is sometimes detrimental to the host. Thus, various genetic systems to suppress or attenuate their transposition activities have coevolved. Importantly, however, the coexpression of IEE and the IS*629* TPase reduced the copy number of IS*629* in O157 [33], and this may also happen to other members of the IS*3* family and to members of the IS*1* and IS*30* families. This type of control mechanism, which actively reduces TE copy numbers, has not previously been identified, even in eukaryotes and archaea.

Acknowledgments This work was supported in part by the Research Project for Improving Animal Disease Prevention Technologies to Combat Antimicrobial Resistance 2017-2021 FY of the Ministry of Agriculture, Forestry, and Fisheries, Japan and by a Grant-in-Aid from the Japan Society for the Promotion of Science (JSPS) under grant number 15K08484 to M.K. This work was also supported in part by a Grant-in-Aid from Ministry of Health, Labor, and Welfare, Japan (H29-shokuhin-ippan-001) to T.H. and by the Japan Agency for Medical Research and Development (AMED) under grant number JP17fk0108308j0003 to T.H.

References

1. Sinzelle L, Izsvak Z, Ivics Z. Molecular domestication of transposable elements: from detrimental parasites to useful host genes. Cell Mol Life Sci. 2009;66:1073–93.
2. Chandler M, Mahillon J. Insertion sequences revisited. In: Craig NL, Craigie R, Gellert M, Lambowitz AM, editors. Mobile DNA II. Washington, DC: ASM Press; 2002. p. 305–66.
3. Kothapalli S, Nair S, Alokam S, Pang T, Khakhria R, Woodward D, Johnson W, Stocker BA, Sanderson KE, Liu SL. Diversity of genome structure in *Salmonella enterica* serovar Typhi populations. J Bacteriol. 2005;187:2638–50.
4. Wei J, Goldberg MB, Burland V, Venkatesan MM, Deng W, Fournier G, Mayhew GF, Plunkett G 3rd, Rose DJ, Darling A, Mau B, Perna NT, Payne SM, Runyen-Janecky LJ, Zhou S, Schwartz DC, Blattner FR. Complete genome sequence and comparative genomics of *Shigella flexneri* serotype 2a strain 2457T. Infect Immun. 2003;71:2775–86.
5. Siguier P, Perochon J, Lestrade L, Mahillon J, Chandler M. ISfinder: the reference centre for bacterial insertion sequences. Nucleic Acids Res. 2006;34:D32–6.
6. Mahillon J, Chandler M. Insertion sequences. Microbiol Mol Biol Rev. 1998;62:725–74.
7. Chandler M, Fayet O, Rousseau P, Hoang B, Duval-Valentin G. Copy-out-paste-in transposition of IS*911*: a major transposition pathway. In: Craig NL, Chandler M, Gellert M, Lambowitz

AM, Rice PA, Sandmeyer SB, editors. Mobile DNA III. Washington, DC: ASM Press; 2014. p. 591–607.
8. Synder L, Champness W. Molecular genetics of bacteria. 2nd ed. Washington, DC: ASM Press; 2003.
9. Stephanou NC, Gao F, Bongiorno P, Ehrt S, Schnappinger D, Shuman S, Glickman MS. Mycobacterial nonhomologous end joining mediates mutagenic repair of chromosomal double-strand DNA breaks. J Bacteriol. 2007;189:5237–46.
10. Foster TJ, Lundblad V, Hanley-Way S, Halling SM, Kleckner N. Three Tn10-associated excision events: relationship to transposition and role of direct and inverted repeats. Cell. 1981;23:215–27.
11. Nagel R, Chan A. Enhanced Tn10 and mini-Tn10 precise excision in DNA replication mutants of *Escherichia coli* K12. Mutat Res. 2000;459:275–84.
12. Mead PS, Griffin PM. *Escherichia coli* O157:H7. Lancet. 1998;352:1207–12.
13. Hayashi T, Makino K, Ohnishi M, Kurokawa K, Ishii K, Yokoyama K, Han CG, Ohtsubo E, Nakayama K, Murata T, Tanaka M, Tobe T, Iida T, Takami H, Honda T, Sasakawa C, Ogasawara N, Yasunaga T, Kuhara S, Shiba T, Hattori M, Shinagawa H. Complete genome sequence of enterohemorrhagic *Escherichia coli* O157:H7 and genomic comparison with a laboratory strain K-12. DNA Res. 2001;8:11–22.
14. Makino K, Ishii K, Yasunaga T, Hattori M, Yokoyama K, Yutsudo CH, Kubota Y, Yamaichi Y, Iida T, Yamamoto K, Honda T, Han CG, Ohtsubo E, Kasamatsu M, Hayashi T, Kuhara S, Shinagawa H. Complete nucleotide sequences of 93-kb and 3.3-kb plasmids of an enterohemorrhagic *Escherichia coli* O157:H7 derived from Sakai outbreak. DNA Res. 1998;5:1–9.
15. Kusumoto M, Nishiya Y, Kawamura Y. Reactivation of insertionally inactivated Shiga toxin 2 genes of *Escherichia coli* O157:H7 caused by nonreplicative transposition of the insertion sequence. Appl Environ Microbiol. 2000;66:1133–8.
16. Sekine Y, Eisaki N, Ohtsubo E. Translational control in production of transposase and in transposition of insertion sequence IS*3*. J Mol Biol. 1994;235:1406–20.
17. Kusumoto M, Nishiya Y, Kawamura Y, Shinagawa K. Identification of an insertion sequence, IS*1203* variant, in a Shiga toxin 2 gene of *Escherichia coli* O157:H7. J Biosci Bioeng. 1999;87:93–6.
18. Ohnishi M, Kurokawa K, Hayashi T. Diversification of *Escherichia coli* genomes: are bacteriophages the major contributors? Trends Microbiol. 2001;9:481–5.
19. Ohnishi M, Terajima J, Kurokawa K, Nakayama K, Murata T, Tamura K, Ogura Y, Watanabe H, Hayashi T. Genomic diversity of enterohemorrhagic *Escherichia coli* O157 revealed by whole genome PCR scanning. Proc Natl Acad Sci U S A. 2002;99:17043–8.
20. Ogura Y, Kurokawa K, Ooka T, Tashiro K, Tobe T, Ohnishi M, Nakayama K, Morimoto T, Terajima J, Watanabe H, Kuhara S, Hayashi T. Complexity of the genomic diversity in enterohemorrhagic *Escherichia coli* O157 revealed by the combinational use of the O157 Sakai OligoDNA microarray and the Whole Genome PCR scanning. DNA Res. 2006;13:3–14.
21. Iguchi A, Iyoda S, Terajima J, Watanabe H, Osawa R. Spontaneous recombination between homologous prophage regions causes large-scale inversions within the *Escherichia coli* O157:H7 chromosome. Gene. 2006;372:199–207.
22. Ooka T, Ogura Y, Asadulghani M, Ohnishi M, Nakayama K, Terajima J, Watanabe H, Hayashi T. Inference of the impact of insertion sequence (IS) elements on bacterial genome diversification through analysis of small-size structural polymorphisms in *Escherichia coli* O157 genomes. Genome Res. 2009a;19:1809–16.
23. Ooka T, Terajima J, Kusumoto M, Iguchi A, Kurokawa K, Ogura Y, Asadulghani M, Nakayama K, Murase K, Ohnishi M, Iyoda S, Watanabe H, Hayashi T. Development of a multiplex PCR-based rapid typing method for enterohemorrhagic *Escherichia coli* O157 strains. J Clin Microbiol. 2009b;47:2888–94.
24. Davis MA, Hancock DD, Besser TE, Call DR. Evaluation of pulsed-field gel electrophoresis as a tool for determining the degree of genetic relatedness between strains of *Escherichia coli* O157:H7. J Clin Microbiol. 2003;41:1843–9.
25. Terajima J, Izumiya H, Iyoda S, Mitobe J, Miura M, Watanabe H. Effectiveness of pulsed-field gel electrophoresis for the early detection of diffuse outbreaks due to Shiga toxin-producing Escherichia coli in Japan. Foodborne Pathog Dis. 2006;3:68–73.

26. Asano Y, Karasudani T, Tanaka H, Matsumoto J, Okada M, Nakamura K, Kondo H, Shinomiya H. Characterization of the *Escherichia coli* O157:H7 outbreak strain whose Shiga toxin 2 gene is inactivated by IS*1203*v insertion. Jpn J Infect Dis. 2013;66:201–6.
27. Hirai S, Yokoyama E, Etoh Y, Seto J, Ichihara S, Suzuki Y, Maeda E, Sera N, Horikawa K, Sato S, Yamamoto T. Putative classification of clades of enterohemorrhagic Escherichia coli O157 using an IS-printing system. Lett Appl Microbiol. 2015;61:267–73.
28. Ding H, Huang L, Mao X, Zou Q. Characterization of *stx2* and its variants in *Escherichia coli* O157:H7 isolated from patients and animals. Afr J Biotechnol. 2011;10:2991–8.
29. Harada T, Hirai Y, Itou T, Hayashida M, Seto K, Taguchi M, Kumeda Y. Laboratory investigation of an *Escherichia coli* O157:H7 strain possessing a *vtx2c* gene with an IS*1203* variant insertion sequence isolated from an asymptomatic food handler in Japan. Diagn Microbiol Infect Dis. 2013;77:176–8.
30. Jinneman KC, Weagant SD, Johnson JM, Abbott SL, Hill WE, Tenge BJ, Dang N-L, Ramsden R, Omiecinski CJ. A large insertion in the Shiga-like toxin 2 gene (stx_2) of an *Escherichia coli* O157:H7 clinical isolate. Int J Food Microbiol. 2000;57:115–24.
31. Okitsu T, Kusumoto M, Suzuki R, Sata S, Nishiya Y, Kawamura Y, Yamai S. Identification of Shiga toxin-producing *Escherichia coli* possessing insertionally inactivated Shiga toxin gene. Microbiol Immunol. 2001;45:319–22.
32. Kusumoto M, Suzuki R, Nishiya Y, Okitsu T, Oka M. Host-dependent activation of IS*1203*v excision in Shiga toxin-producing *Escherichia coli*. J Biosci Bioeng. 2004;97:406–11.
33. Kusumoto M, Ooka T, Nishiya Y, Ogura Y, Saito T, Sekine Y, Iwata T, Akiba M, Hayashi T. Insertion sequence-excision enhancer removes transposable elements from bacterial genomes and induces various genomic deletions. Nat Commun. 2011;2:152.
34. Turlan C, Chandler M. IS1-mediated intramolecular rearrangements: formation of excised transposon circles and replicative deletions. EMBO J. 1995;14:5410–21.
35. Sekine Y, Aihara K, Ohtsubo E. Linearization and transposition of circular molecules of insertion sequence IS*3*. J Mol Biol. 1999;294:21–34.
36. Duval-Valentin G, Marty-Cointin B, Chandler M. Requirement of IS911 replication before integration defines a new bacterial transposition pathway. EMBO J. 2004;23:3897–906.
37. Kaper JB, Nataro JP, Mobley HL. Pathogenic *Escherichia coli*. Nat Rev Microbiol. 2004;2:123–40.
38. Nataro JP, Kaper JB. Diarrheagenic *Escherichia coli*. Clin Microbiol Rev. 1998;11:142–201.
39. Frydendahl K. Prevalence of serogroups and virulence genes in *Escherichia coli* associated with postweaning diarrhoea and edema disease in pigs and a comparison of diagnostic approaches. Vet Microbiol. 2002;85:169–82.
40. Holland RE. Some infectious causes of diarrhea in young farm animals. Clin Microbiol Rev. 1990;3:345–75.
41. Fairbrother JM, Nadeau E, Gyles CL. *Escherichia coli* in postweaning diarrhea in pigs: an update on bacterial types, pathogenesis, and prevention strategies. Anim Health Res Rev. 2005;6:17–39.
42. Kusumoto M, Hikoda Y, Fujii Y, Murata M, Miyoshi H, Ogura Y, Gotoh Y, Iwata T, Hayashi T, Akiba M. Emergence of a multidrug-resistant shiga toxin-producing enterotoxigenic *Escherichia coli* lineage in diseased swine in Japan. J Clin Microbiol. 2016;54:1074–81.
43. Kusumoto M, Fukamizu D, Ogura Y, Yoshida E, Yamamoto F, Iwata T, Ooka T, Akiba M, Hayashi T. Lineage-specific distribution of insertion sequence excision enhancer in enterotoxigenic *Escherichia coli* isolated from swine. Appl Environ Microbiol. 2014;80:1394–402.
44. Yin X, Wheatcroft R, Chambers JR, Liu B, Zhu J, Gyles CL. Contributions of O island 48 to adherence of enterohemorrhagic *Escherichia coli* O157:H7 to epithelial cells in vitro and in ligated pig ileal loops. Appl Environ Microbiol. 2009;75:5779–86.
45. Lundblad V, Taylor AF, Smith GR, Kleckner N. Unusual alleles of recB and recC stimulate excision of inverted repeat transposons Tn10 and Tn5. Proc Natl Acad Sci U S A. 1984;81:824–8.

Mobile Genetic Elements Involved in the Evolution of Bacteria that Degrade Recalcitrant Xenobiotic Compounds

9

Yuji Nagata, Hiromi Kato, Yoshiyuki Ohtsubo, and Masataka Tsuda

Abstract

Bacterial strains capable of degrading man-made xenobiotic compounds are thought to have evolved to degrade such compounds within only several decades. Various sphingomonad strains belonging to *Alphaproteobacteria* were isolated that degrade highly recalcitrant compounds including man-made xenobiotics, indicating the versatility of this bacterial group. However, each strain degrades only a limited set of compounds, and specific genes for the degradation of a specific compound are found only in the genomes of its degraders, suggesting the plasticity of genomes of sphingomonads. Comparison of the complete genome sequences of four γ-hexachlorocyclohexane (γ-HCH)-degrading sphingomonad strains, *Sphingobium japonicum* UT26, *Sphingomonas* sp. MM-1, *Sphingobium* sp. MI1205, and *Sphingobium* sp. TKS, strongly suggested that the γ-HCH-degrading strains emerged through the recruitment of specific *lin* genes for the γ-HCH degradation into ancestral strains that had core functions of sphingomonads. Plasmids, most of which are specialized for sphingomonads, and IS*6100*, the most abundant insertion sequence in the four strains, seem to be involved in the recruitment of the *lin* genes and the diversification of the *lin*-flanking regions in the four strains. Here, we show concrete examples and discuss the important roles of such mobile genetic elements for the emergence and evolution of the bacterial strains degrading highly recalcitrant compounds.

Expertise: Environmental microbiology (Nagata Y.), Microbial ecology (Kato H.), Bacterial genomics (Ohtsubo Y.), Bacterial genetics (Tsuda M.).

Y. Nagata (✉) · H. Kato · Y. Ohtsubo · M. Tsuda
Graduate School of Life Sciences, Tohoku University, Sendai, Japan
e-mail: aynaga@ige.tohoku.ac.jp; katee@ige.tohoku.ac.jp; yohtsubo@ige.tohoku.ac.jp; mtsuda@ige.tohoku.ac.jp

H. Nishida, T. Oshima (eds.), *DNA Traffic in the Environment*,
https://doi.org/10.1007/978-981-13-3411-5_9

Keywords
Biodegradation · Mobile genetic element · Plasmid · Insertion sequence Organochlorine pesticide · Xenobiotics · Dehalogenase · Sphingomonads Genome

9.1 Introduction

Various chemical compounds have been released into the environment by human activities, and they often cause serious environmental problems, since most such compounds cannot be readily degraded in the environment and have harmful effects on the natural ecosystem, including on humans [1–3]. Some microorganisms that degrade environmental pollutants have been isolated and characterized for the purpose of bioremediation [4–7]. Such microorganisms have also attracted attention from the viewpoint of microbial adaptation and evolution toward chemical compounds that they have never or only rarely encountered before [4–7].

There are various environmental pollutants, and their degradability in the environment varies depending on the substance. Simple aromatic compounds, e.g., benzene, toluene, phenol, and naphthalene, are major environmental pollutants but relatively easily degraded by microorganisms [5]. In fact, many bacterial strains degrading such aromatic compounds have been isolated and studied in detail. In most cases, a series of genes encoding enzymes necessary for transformation of these compounds into TCA cycle intermediates constitute a gene cluster whose expression is regulated as an operon [8]. Moreover, such operons are usually located on mobile genetic elements, e.g., transposons, plasmids, and integrative and conjugative elements (ICEs), and can be transferred between bacterial cells as a set [8–13], and thus non-degrading bacterial cells can easily become degraders of aromatic compounds only by acquiring such "ready-made" gene clusters. In other words, the degradation system of simple aromatic compounds has already been well established in nature, and the gene clusters necessary for the degradation can be distributed among bacterial cells in the environments contaminated with these compounds, where cells having the ability to assimilate the compounds have a survival advantage. The fact that systems degrading simple aromatic compounds have been well established in nature is not surprising, since most of such compounds are not man-made but natural products and have existed for a long time in the environment, although their distribution and abundance on the surface of the earth have been broadened and increased, respectively, by human activities. On the other hand, anthropogenic compounds that were chemically synthesized or industrially produced are usually highly recalcitrant, because microorganisms have never or rarely encountered such chemical compounds and have not fully established systems to degrade them [4–7]. However, bacteria that can degrade even anthropogenic chemicals have been isolated, and surprisingly, most aerobic xenobiotics-degrading bacteria can use such chemicals as their sole sources of carbon and energy. Since it has been proposed that the pathways for aerobic degradation of man-made xenobiotic compounds evolved relatively quickly (within several decades) after the release of such compounds into the environment, the bacterial strains capable of degrading man-made xenobiotic compounds are excellent models for studying the adaptation and evolution of bacteria in the environment [4–7]. However, with the exception of a few speculative examples, the evolutionary processes of these bacterial strains remain largely unknown. Recent genome analyses of such strains have strongly suggested that they indeed emerged relatively recently by gathering genes for the degradation of xenobiotic compounds, and that mobile genetic elements played important roles for the recruitment of the genes [6, 14, 15].

In this chapter, we will avoid a simple enumeration of examples of mobile genetic elements involved in the degradation of highly recalcitrant compounds. Instead, we will explain in detail our hypothesis for the emergence and evolution of the degraders of highly recalcitrant xenobiotic compounds and the roles of mobile genetic elements on the process by showing selected typical examples. For this purpose, we will mainly use γ-hexachlorocyclohexane (γ-HCH)-degrading sphingomonad strains, since the γ-HCH degradation system in aerobic bacteria is an excellent model for investigating fundamental issues in microbial and molecular evolution [6, 16].

9.2 Sphingomonads, a Bacterial Group Containing Various Strains Degrading Highly Recalcitrant Compounds

Microorganisms have developed new metabolic pathways for various chemical compounds, including xenobiotic compounds, in order to exploit new carbon sources and to detoxify harmful compounds [5]. The mechanisms by which they adapt to xenobiotics are of great interest. Many xenobiotics-degrading bacterial strains belonging to various taxonomical classification have been isolated [14, 15], and "sphingomonads" are one of the most important bacterial groups for the degradation of recalcitrant hydrophobic compounds [7, 17–19]. Sphingomonads is a collective category comprising *Sphingomonas*, *Sphingobium*, *Novosphingobium*, *Sphingopyxis*, and their related genera belonging to *Alphaproteobacteria* [20], and various sphingomonad strains have been isolated that degrade highly recalcitrant hydrophobic compounds, e.g., γ-HCH [16, 21], pentachlorophenol [22], dioxin-related compounds [23], lignin-related compounds [24], polyaromatic hydrocarbons [25], polyvinyl alcohol [26], polyethylene/polypropylene glycol [27–30], and organophosphate [31]. Of course, many strains belonging to other bacterial groups are also known that can degrade highly recalcitrant compounds, but no other bacterial groups have members degrading such various highly recalcitrant compounds including man-made ones at least among bacteria that are widely distributed in the environment and can be easily cultured under laboratory conditions, suggesting that sphingomonads can adapt more quickly or more efficiently to the degradation of new compounds in the environment than members of the other bacterial genera. However, it should be noted that each sphingomonad strain degrades only a limited set of compounds.

For example, γ-HCH degraders cannot degrade any of the other "special," i.e., highly recalcitrant and anthropogenic, compounds. In addition, there are also sphingomonad strains that degrade no "special" compound, and such strains seem to be one of the major bacteria in the environment [32]. In other words, most sphingomonads are "ordinary," but have the potential to be specialists for the degradation of highly recalcitrant hydrophobic compounds.

The genome sequences of a vast number of bacterial strains have already been determined, and the complete genome sequences of xenobiotics-degrading bacterial strains are now available. Thus, it has become possible to discuss the emergence and evolution of bacterial strains that degrade xenobiotics on the basis of their genomic information. Comparison of genome sequences of sphingomonad strains including degraders of highly recalcitrant compounds supports the idea that sphingomonads are a versatile group because of the plasticity of their genomes [33, 34]. The organization and coding potential of the sphingomonad genomes appear to be highly variable, and the specific genes necessary for the degradation of specific compounds are found only in the genomes of the degraders of the compounds. It is strongly suggested that plasmid-mediated gene transfer and chromosome-plasmid recombination, together with prophage and transposon-mediated rearrangements, play prominent roles in the genome evolution of sphingomonads. In some cases, the gene organizations seem to be edited by using insertion sequences. These points will be explained in greater detail in the following sections by using aerobic γ-HCH-degrading sphingomonads as an example.

9.3 γ-HCH-Degrading Sphingomonad Strains

γ-HCH (also known as γ-BHC or lindane) is a completely man-made chlorinated pesticide that has caused serious environmental problems due to its toxicity and long persistence in upland soils [18, 35, 36]. Although the use of γ-HCH is now banned in many countries, this compound still remains in various environments and causes serious environmental problems [36]. γ-HCH is chemically synthesized by the process of photochlorination of benzene. The synthesized product is called technical-HCH (t-HCH) and consists mainly of five isomers, α- (60–70%), γ- (12–16%), β- (10–12%), δ- (6–10%), and ε-HCH (3–4%) [36]. Among t-HCH isomers, only γ-HCH has insecticidal activity, and it is used after purification as the insecticide lindane (> 99% purity). The remaining isomers were in many cases improperly deposited and also caused serious environmental problems. In addition to γ-HCH, α- and β-HCH isomers were also included as additional POPs, persistent organic pollutants that must be controlled under international agreement, at the Stockholm Convention [36]. Among the HCH isomers, β-HCH is the most recalcitrant; it is usually the predominant isomer remaining in contaminated soils and in animal tissues and fluids [37].

Only 60 years after the first release of γ-HCH into the environment, a number of bacterial strains that aerobically degrade γ-HCH have been isolated from geographically dispersed locations [18]. The archetypal γ-HCH-degrading bacterium *Sphingobium japonicum* UT26 was isolated from an upland experimental field to which γ-HCH had been applied once a year for 12 years, and the aerobic degradation pathway of γ-HCH in this strain was intensively studied [38] (Fig. 9.1). In strain UT26, γ-HCH is converted to β-ketoadipate via reactions catalyzed by dehydrochlorinase (LinA), haloalkane dehalogenase (LinB), dehydrogenase (LinC), reductive dechlorinase (LinD), ring-cleavage dioxygenase (LinE), and maleylacetate reductase (LinF); β-ketoadipate is further converted to succinyl-CoA and acetyl-CoA by succinyl-CoA:3-oxoadipate CoA transferase (LinGH) and β-ketoadipyl CoA thiolase (LinJ), respectively [34, 38]. The last two compounds are metabolized in the TCA cycle. In the early degradation process, two dead-end products, 1,2,4-trichlorobenzene (1,2,4-TCB) and 2,5-dichlorophenol (2,5-DCP), which are not degraded by strain UT26 are produced (Fig. 9.1). The *linA*, *linB*, and *linC* genes are constitutively expressed at a relatively high level in UT26, while the *linD* and *linE* genes constitute an operon, and their expression is regulated by an LysR-type transcriptional regulator (LinR) [39]. Since the β-ketoadipate pathway is often used by environmental bacterial strains [40], the *lin* genes for the conversion of γ-HCH to β-ketoadipate (Fig. 9.1: *linA* to *linF*) are peculiar to the γ-HCH-degrading pathway. In particular, the *linA* gene is unique, because it does not show significant similarity to any sequences in the databases except for the almost identical *linA* genes (> 90% identical) from other bacterial strains and metagenomes [18, 38]. LinA and LinB are also important targets from the viewpoint of protein evolution, since the LinA and LinB variants show different levels of enzymatic activity toward different HCH isomers and their metabolites [18, 38, 41].

In addition to catabolic enzymes, a putative ABC-type transporter system is necessary for the γ-HCH utilization in UT26. This putative ABC transporter system consists of four components (Fig. 9.1): permease, ATPase, periplasmic protein, and lipoprotein, encoded by *linK*, *linL*, *linM*, and *linN*, respectively, which constitute a cluster. Mutation and complementation analysis indicated that all the *linKLMN* genes are required as a set for the γ-HCH utilization in UT26 [42]. One of the dead-end products, 2,5-DCP, is toxic to cells [43]. The LinKLMN system is involved in γ-HCH utilization by conferring tolerance toward this toxic metabolite to the cells [42]. However, the LinKLMN system is not a simple efflux pump of the toxic compound, but seems to be involved in the integrity of the outer membrane. The mutant cells defective in the LinKLMN system showed higher γ-HCH degradation activity and greater accumulation of the toxic dead-end product 2,5-DCP, higher sensitivity of the cells to 2,5-DCP itself, and higher permeability of hydrophobic compounds than the wild-type cells [42]. It still remains unknown how the LinKLMN system is involved in the integrity of the outer membrane, but the periplasmic protein LinM has a mammalian cell entry (Mce) domain [44], which is necessary for the lipid binding [45], and thus it is speculated that the LinKLMN system transports lipid-related compounds, e.g., sphingolipid, for the integrity of the outer membrane.

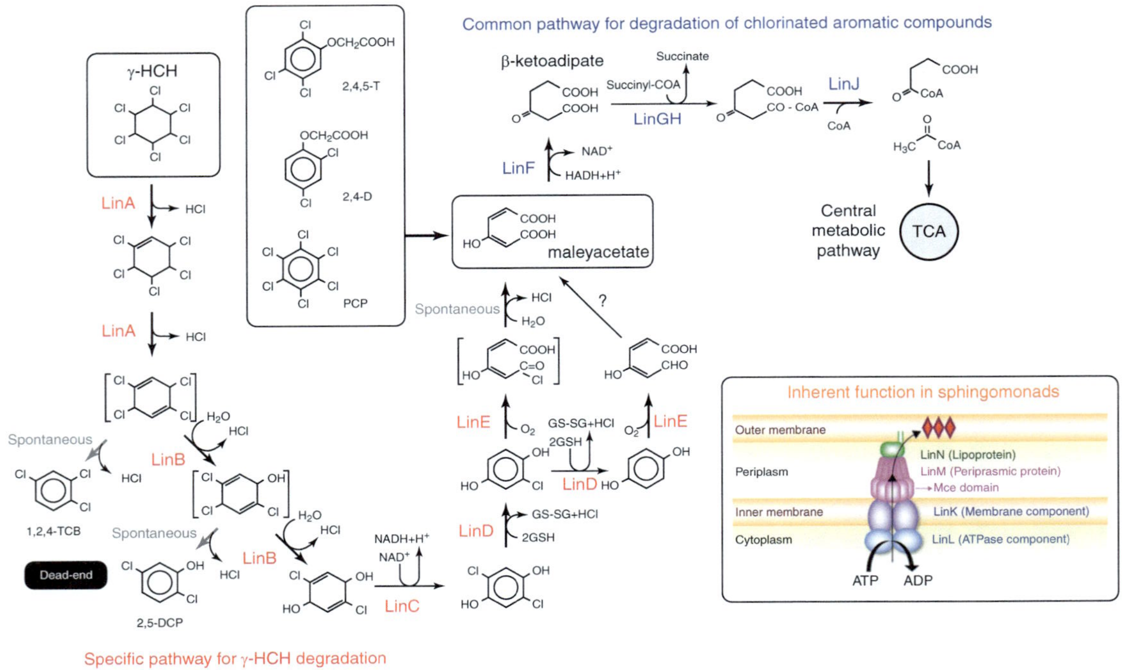

Fig. 9.1 Functions necessary for γ-HCH utilization in sphingomonads. Enzymes for the specific pathway for γ-HCH degradation (only "specific" *lin* genes are used for the functions) and for a common pathway for degradation of chlorinated aromatic compounds, e.g., 2,4,5-T, 2,4-D, and PCP (more than one genes have been found for each function), are shown in red and blue, respectively. Genes for these functions are often located on plasmids and may be obtained by horizontal gene transfer. Note that β-ketoadipate pathway is common pathway for the degradation of aromatic compounds and is widely distributed among environmental bacteria. 1,2,4-TCB and 2,5-DCP are dead-end products, and 2,5-DCP has toxic effect for the cells. The LinKLNM-type ABC transporter, which is an inherent function in sphingomonads (genes for the function are located on chromosomes and their divergence reflects the phylogenetic relationship of their host), is involved in the tolerance for the toxic effect of 2,5-DCP. See text for details

9.4 Genomes of γ-HCH-Degrading Sphingomonad Strains

Sphingobium japonicum UT26 is an archetypal γ-HCH-degrading strain, and its complete genome sequence was the first determined among γ-HCH degraders [34, 46]. The *S. japonicum* UT26 genome consists of five replicons, two circular chromosomes (Chr), Chr1 (3.5 Mb), and Chr2 (682 kb), and three circular plasmids, pCHQ1 (191 kb), pUT1 (32 kb), and pUT2 (5 kb) ([34, 46]; Table 9.1). The *lin* genes for the γ-HCH degradation are dispersed on Chr 1 (*linA*, *linB*, *linC*, and *linKLMN*), Chr 2 (*linF* and *linGHIJ*), and pCHQ1 (*linRED*) [34] (Table 9.1). The comparison of the genome of UT26 with those of other sphingomonad strains revealed that the UT26 genome consists of regions conserved among these sphingomonad strains as well as specific regions unique to UT26 [34]. All of the 196 bacterial essential genes proposed by Gil et al. [47] are dispersed on the conserved regions of Chr1. The G + C contents of the specific regions of Chr1 are lower than those of conserved regions. Another important point is that the *linKLMN* and *linGHIJ* genes are located on conserved regions among sphingomonads, while the *linA*, *linB*, *linC*, *linRED*, and *linF* genes are located within unique regions of UT26 [34].

Aerobic γ-HCH-degrading bacterial strains other than UT26 have been isolated around the world, and most of such strains—particularly those that have been intensively analyzed—are sphingomonads [17, 18]. To gain more insight into the emergence and evolution of γ-HCH-degrading sphingomonad strains, we determined the complete genome sequences of three other γ-HCH degraders: *Sphingobium* sp. MI1205 from Miyagi, Japan [48, 49], *Sphingomonas* sp. MM-1 from India [50, 51], and *Sphingobium* sp. TKS from Kyushu, Japan [52]. These three γ-HCH-degrading strains and UT26 are phylogenetically dispersed on the basis of 16S rRNA gene analysis among closely related sphingomonad strains. In addition, comparison of genome sequences of the four strains revealed that their gene repertories (each strain has 4128–5248 ORF clusters among total 10,325 ORF clusters of the four strains) are quite different from each other (only 1288 ORF clusters are shared) and that the organization of replicons is completely different among the four strains [16] (Table 9.1). These results clearly indicated that the four γ-HCH degraders are phylogenetically divergent, and it was strongly suggested that the four γ-HCH degraders independently acquired γ-HCH degradation ability. At least, it is unlikely that the four strains were derived from a single ancestral γ-HCH degrader.

γ-HCH is degraded in MI1205, MM-1, and TKS by the same pathway as in UT26 (Fig. 9.1). All four strains carry almost identical *linA* to *linE* genes for the conversion of γ-HCH to maleylacetate, and MI1205 and MM-1 also carry almost identical *linF* and *linGHIJ* genes for the metabolism of maleylacetate, while different genes that show no significant similarity to *linF* and *linGHIJ* genes at DNA level (*linFb* and *linGHIJ* homologues) are used for the latter conversion steps in TKS. The *linKLMN* genes for the putative ABC transporter necessary for γ-HCH utilization exhibit structural divergence, which reflects the phylogenetic relationship of their hosts [16], but the *linKLMN* homologues of MM-1, which is phylogenetically the most distant strain from UT26, could complement the *linKLMN* function in UT26 [16]. Moreover, as mentioned above, the *linKLMN* homologues were found not only

Table 9.1 Genome organization of four γ-HCH-degrading sphingomonad strains

Strain name	Isolated site	Replicon	Length (bp)	Number of *rrn* operon	Number of ORF	Number of IS*6100*	*lin* genes[a]
Sphingobium japonicum UT26	Tokyo, Japan	Chromosome 1	35,14,822	1	3529	5	*linA, linB, linC, linKLMN*
		Chromosome 2	6,81,892	2	589	2	*linF, linGHIJ, linEb*
	Soil artificially polluted with γ-HCH	pCHQ1	1,90,974	0	224	4	*linRED*
		pUT1	31,776	0	44	2	–
		pUT2	5398	0	8	0	–
		Total	44,24,862	3	4394	13	
Sphingomonas sp. MM-1	Lucknow, India	Chromosome	40,54,833	2	3801	0	*linKbLbMbNb*
		pISP0	2,75,840	0	251	1	*linF, linGHIJ*
	Soil polluted with HCH isomers	pISP1[b]	1,72,140	0	174	7	*linA, linC, linF′*
		pISP2	53,841	0	52	2	–
		pISP3	43,776	0	44	1	*linRED*
		pISP4	33,183	0	39	4	*linB, linC, linF′*
		Total	46,33,613	2	4361	15	
Sphingobium sp. TKS	Kyushu, Japan	Chromosome 1[b]	42,49,857	1	4172	7	*linB, linC, linF′, linFb, linKLMN*
		Chromosome 2	9,89,120	2	843	0	*linGHIJ*_homologue
	Sediment polluted with HCH isomers	pTK1[b]	5,20,614	0	470	6	–
		pTK2	1,95,308	0	182	1	–
		pTK3[b]	87,635	0	92	8	*linB, linC, linF′*
		pTK4[b]	75,938	0	86	6	*linA, linC*
		pTK5	53,908	0	76	0	–
		pTK6	34,300	0	35	1	*linRED*
		pTK7	9585	0	12	0	–
		pTK8	7223	0	11	0	–
		pTK9	5391	0	8	0	–
		Total	62,28,879	3	5987	29	

Sphingobium sp. MI1205	Miyagi, Japan	Chromosome 1	33,51,250	1	3285	0	*linKLMN*
		Chromosome 2	5,67,154	1	516	0	–
	Soil polluted with HCH isomers	pMI1	2,92,135	0	299	7	*linB, linC, linRED, linEb, linF, linF″, linGHIJ*
		pMI2[b]	2,87,488	0	323	7	*linA, linRD*
		pMI3[b]	88,374	0	102	7	*linB, linC, linF″*
		pMI4[b]	32,974	0	45	3	*linRED*
		Total	46,19,375	2	4570	24	

[a]*linF′* and *linF″* are probably pseudogenes
[b]Replicons having more than one *rep* genes

in γ-HCH degraders but also in non-γ-HCH-degrading sphingomonad strains [34]. These findings strongly suggest that the *linKLMN* system is one of the inherent functions necessary for γ-HCH utilization in sphingomonads. In summary, it can be concluded that the *lin* genes for the utilization of γ-HCH consist of three types of genes for (1) the "specific" pathway for γ-HCH degradation, (2) a common pathway for the degradation of chlorinated aromatic compounds (more than one genes have been found for the function), and (3) inherent function(s) in sphingomonads (Fig. 9.1).

In addition to the genes for γ-HCH degradation, putative genes for the degradation of aromatic compounds were found in the genomes of the four strains [16]. The numbers of ORFs potentially involved in the degradation of aromatic compounds in the four strains (62, 46, 27, and 25 for TKS, UT26, MI1205, and MM-1, respectively) are much smaller than those in versatile recalcitrant pollutant degraders, *Cupriavidus necator* JMP134 [53, 54] and *Burkholderia xenovorans* LB400 [55, 56] (149 and 135 for JMP134 and LB400, respectively). In particular, those in UT26, MI1205 and MM-1 are even smaller than those in the typical metabolically versatile soil bacterial strains *Burkholderia multivorans* ATCC 17616 [57–60] and *Pseudomonas putida* KT2440 [61] (73 and 62 for KT2440 and ATCC 17616, respectively). In addition, no gene for the degradation of other highly recalcitrant compounds was found in their genomes. These results support our hypothesis that the four sphingomonad strains are "specialists" for γ-HCH degradation, but not "generalists" for the degradation of many recalcitrant compounds.

9.5 Plasmids in Sphingomonads

It is generally accepted that horizontal gene transfer (HGT) is an important mechanism of microbial adaptation and genomic evolution [8, 9, 11–13]. HGT between bacteria in natural habitats is largely mediated by mobile genetic elements (MGEs), e.g., self-transmissible plasmids, transposons, integrons, IS elements, ICEs, and bacteriophages. Among such known MGEs, plasmids are particularly important for the rapid adaptation of bacteria toward xenobiotics [9, 19, 62, 63], and genes for the degradation of recalcitrant compounds are also often located on plasmids [19, 64, 65].

After publication of the UT26 genome, many draft genome sequences of other HCH (including not only γ-HCH but also other HCH isomers) degraders and their related but non-HCH-degrading strains were determined, and their comparative analyses have been conducted [66, 67]. These studies provided us some important primary information on the evolution of HCH degraders with the involvement of plasmids and insertion sequences. In addition, our analysis of the complete genomes of the four γ-HCH degraders clearly indicated how plasmids/insertion sequences are involved in the emergence and evolution of HCH degraders [16]. As described above, all four of the strains carry almost identical specific *lin* genes, suggesting they acquired such genes by HGT. However, the specific *lin* genes are dispersed on multiple replicons in the four strains (Table 9.1). In UT26, *linA* to *linC* are located on Chr1, and only the *linRED* cluster is located on a plasmid. On the other hand, all the specific *lin* genes are dispersed on multiple plasmids in various combinations in the other three strains, although additional copies of *linB* and *linC* are also located on Chr1 in TKS (Table 9.1). Finally, there are various replicon types of such plasmids carrying the specific *lin* genes (Table 9.2). In other words, no plasmid carrying a whole set of the specific *lin* genes has been found. These observations indicate that these four strains did not simply acquire all the specific *lin* genes at once as a cluster. This contrasts with other aromatic compound-degrading strains, which may have acquired a whole set of responsible genes at once by the simple conjugative transfer of plasmids and/or ICEs [10, 62, 63].

The putative replication origins (*oriC*s) of the main chromosomes (Chr1s) of the four γ-HCH degraders are of α-proteobacterial-chromosome type [79, 80]; these *oriC*s were located upstream of the uroporphyrinogen decarboxylase gene (*hemE*) with multiple DnaA boxes [TT(A/T)TNCACA] [81]. On the other hand, the Chr2s of UT26, TKS, and MI1205 have the plasmid-type replication and active partition systems [16, 34]. These three plasmid-type chromosomes, the 21 plasmids in the four strains and the plasmids from other sphingomonads, were classified into 18 types based on the similarities of their RepA (DNA replication initiator) proteins (Table 9.2). The RepA proteins of plasmids in sphingomonads show a very low level of similarity to those of well-studied plasmids (e.g., IncP-1, F, IincP-7, and IncP-9 plasmids), and thus we compared only plasmids in sphingomonads. Since the RepA proteins of the 18 types are very divergent, they were further categorized into four major groups, in each of which the RepA proteins exhibit 22–60% identity: (1) the $Chr2_{UT26}$- and pCHQ1-types, (2) the pUT1-, pISP3-, pTK3_1-, pTK3_2-, and pNIC1-types, (3) the pISP4-, pTK2-, pTK7-, pTK8-, pSA2-, pSM103mini-, and pDE6-types, and (4) the pLB1 and pDE3-types. Based on our BLASTP analysis, the RepA proteins of pUT2 and pNL1 did not show similarity to those of any of the other types of plasmids listed in Table 9.2, although the RepA of pUT2 was similar to those of the IncP-9 family of plasmids.

The sizes and gene contents of the same type plasmids, even ones having the identical *repA* gene, are highly divergent (Table 9.2), suggesting that plasmids in sphingomonads underwent dynamic rearrangements. It was clearly indicated that the replicons having highly conserved replication/partition genes are distributed among sphingomonad strains with frequent recombination events including replicon fusion [17]. Active DNA rearrangements between plasmids and host chromosomes might have caused the "mosaic" genetic structures of sphingomonads [63]. Interestingly, six pISP4-type plasmids carry identical *repA* and *parA* genes, and five of them also have other types of *repA* genes (Table 9.2), suggesting a prevalent fusion event of replicons in the pISP4-type plasmids. It is noteworthy that all six pISP4-type plasmids carrying identical *repA* and *parA* genes contain the *lin* genes (Table 9.2), indicating that this type of plasmid plays an important role in dissemination of the *lin* genes.

The genes for conjugal transfer consist of those encoding proteins involved in mating pair formation (Mpf) and DNA transfer and replication (Dtr) [82]. The *mpf* genes encode proteins that assemble in a large macromolecular structure called the

Table 9.2 Classification of sphingomonad plasmids and plasmid-type replicons

Replicon[a]	Size (bp)	Host	Host feature	Identity (%) of RepA protein to representative[a]	Gene cluster for conjugation	Genes for degradation	Reference
UT26_Chr2	6,81,892	Sphingobium japonicum UT26	γ-HCH degradation	–		*linF, linGHIJ*	[34]
pSRL2	1,08,430	Sphingobium indicum B90A	HCH degradation	99 (398/400)			[21]
L-1_Chr2	13,68,670	Sphingobium chlorophenolicum L-1	PCP degradation	89 (359/400)		*pcpBDR, pcpEMAC*	[22]
TKS_Chr2	9,89,120	Sphingobium sp. TKS	γ-HCH degradation	89 (360/401)		*linGHIJ_* homologues	[16]
pMSHV	9,59,196	Sphingobium herbicidovorans MH	Chlorophenoxy herbicide degradation	84 (338/402)	F	*spdA, cadABCD*	[68]
MI_Chr2	5,67,154	Sphingobium sp. MI1205	γ-HCH degradation	83 (338/405)			[16]
pNIC2	1,77,851	Sphingomonas sp. NIC1	Nicotine degradation	76 (308/405)			[69]
pNL2	4,87,268	Novosphingobium aromaticivorans DSM 12444	Aromatic compounds degradation	70 (284/401)			Unpublished

Lpl	1,92,103	Novosphingobium sp. PP1Y	Aromatic compounds degradation	67 (288/425)			[25]
pISP1[b]	1,72,140	Sphingomonas sp. MM-1	γ-HCH degradation	55 (201/363)	F	*linA, linC, linF*	[16]
TKS_Chr1[b]	42,49,857	Sphingobium sp. TKS	γ-HCH degradation	55 (200/363)	F	*linB, linC, linF, linFb, linKLMN*	[16]
pMI2[b]	2,87,488	Sphingobium sp. MI1205	γ-HCH degradation	55 (200/363)	F	linA, linRD	[16]
pSA3	3,54,886	Novosphingobium resinovorum SA1	Sulfanilic acid degradation	52 (198/378)	Ti	*sad* and *sca* gene clusters	[70]
p203_2	1,51,240	Sphingopyxis macrogoltabida 203 (NBRC 15033)	PEG degradation[c]	43 (183/422)	F		[27]
p203N_2	1,51,240	Sphingopyxis macrogoltabida 203 N (NBRC 111659)	PEG degradation	43 (183/422)	F		[30]
pMI1	2,92,135	Sphingobium sp. MI1205	γ-HCH degradation	42 (174/409)	F	*linB, linC, linRED, linEb, linF, linF'', linGHIJ*	[49]
pTK1[b]	5,20,614	Sphingobium sp. TKS	γ-HCH degradation	42 (171/401)	Ti, F		[52]
pCHQ1	1,90,974	Sphingobium japonicum UT26	γ-HCH degradation	–	Ti, F	*linRED*	[34]
pTK1[b]	5,20,614	Sphingobium sp. TKS	γ-HCH degradation	Identical	Ti, F		[16]
pSLGP	1,48,801	Sphingobium sp. SYK-6	Lignin degradation	Identical	Ti		[71]

(continued)

Table 9.2 (continued)

Replicon[a]	Size (bp)	Host	Host feature	Identity (%) of RepA protein to representative[a]	Gene cluster for conjugation	Genes for degradation	Reference
pSPHCH01	1,23,733	Sphingobium chlorophenolicum L-1	PCP degradation	Identical	Ti		[22]
pSRL1	1,39,218	Sphingobium indicum B90A	HCH degradation	Identical		*linA1*, *linC*	[21]
pISP0	2,75,840	Sphingomonas sp. MM-1	γ-HCH degradation	98 (381/387)	Ti	*linF*, *linGHIJ*	[16]
pMSHV1	3,12,167	Sphingobium herbicidovorans MH	Chlorophenoxy herbicide degradation	98 (356/362)	F		[68]
pDE1	2,29,716	Sphingobium baderi DE-13	Alkyl-substituted aniline degradation	98 (356/362)	F		[72]
pEY-1_1	1,96,952	Sphingopyxis macrogoltabida EY-1	PPG and PEG degradation	76 (289/378)			[28]
pUT1	31,776	Sphingobium japonicum UT26	γ-HCH degradation	–			[34]
pMI2[b]	2,87,488	Sphingobium sp. MI1205	γ-HCH degradation	Identical	F	*linA*, *linRD*	[16]
pISP2	53,841	Sphingomonas sp. MM-1	γ-HCH degradation	Identical			[16]
pTK4[b]	75,938	Sphingobium sp. TKS	γ-HCH degradation	Identical		*linA*, *linC*	[16]
pDE4	43,676	Sphingobium baderi DE-13	Alkyl-substituted aniline degradation	Identical			[72]
pISP3	43,776	Sphingomonas sp. MM-1	γ-HCH degradation	–		*linRED*	[16]

pMI4[b]	32,974	Sphingobium sp. MI1205	γ-HCH degradation	Identical		*linRED*	[16]
pTK6	34,300	Sphingobium sp. TKS	γ-HCH degradation	Identical		*linRED*	[16]
pSRL3	43,761	Sphingobium indicum B90A	HCH degradation	Identical		*linRED*	[21]
pSA4	67,284	Novosphingobium resinovorum SA1	Sulfanilic acid degradation	Identical			[70]
pEY-1_3	34,474	Sphingopyxis macrogoltabida EY-1	PPG and PEG degradation	Identical			[28]
pDC03_1[b]	1,15,498	Sphingomonas sp. DC-6	Chloroacetanilide herbicides degradation	95 (355/370)	Ti		[73]
pTK3_1[b]	87,635	Sphingobium sp. TKS	γ-HCH degradation	–		*linB, linC, linF'*	[16]
pDE5	21,405	Sphingobium baderi DE-13	Alkyl-substituted aniline degradation	90 (288/317)			[72]
pEY-1_4	30,496	Sphingopyxis macrogoltabida EY-1	PPG and PEG degradation	75 (237/314)			[28]
pEY-1_5	19,860	Sphingopyxis macrogoltabida EY-1	PPG and PEG degradation	75 (237/314)			[28]
pTK3_2[b]	87,635	Sphingobium sp. TKS	γ-HCH degradation	–		*linB, linC, linF'*	[16]
pNIC1	74,167	Sphingomonas sp. NIC1	Nicotine degradation	–	Ti		[69]
p4	79,937	Sphingobium hydrophobicum C1		88 (253/286)			Unpublished
pBSY17_3	63,550	Sphingobium sp. RAC03		80 (230/286)			Unpublished

(continued)

Table 9.2 (continued)

Replicon[a]	Size (bp)	Host	Host feature	Identity (%) of RepA protein to representative[a]	Gene cluster for conjugation	Genes for degradation	Reference
pDE2	85,174	Sphingobium baderi DE-13	Alkyl-substituted aniline degradation	70 (201/284)	Ti		[72]
pISP4	33,183	Sphingomonas sp. MM-1	γ-HCH degradation	–		*linB, linC, linF′*	[16]
pISP1[b]	1,72,140	Sphingomonas sp. MM-1	γ-HCH degradation	Identical	F	*linA, linC, linF′*	[16]
pTK3[b]	87,635	Sphingobium sp. TKS	γ-HCH degradation	Identical		*linB, linC, linF′*	[16]
pTK4[b]	75,938	Sphingobium sp. TKS	γ-HCH degradation	Identical		*linA, linC*	[16]
pMI3[b]	88,374	Sphingobium sp. MI1205	γ-HCH degradation	Identical	Ti	*linB, linC, linF″*	[16]
pMI4[b]	32,974	Sphingobium sp. MI1205	γ-HCH degradation	Identical		*linRED*	[16]
pMSHV2	39,633	Sphingobium herbicidovorans MH	Chlorophenoxy herbicide degradation	90 (268/296)			[68]
pTK2	1,95,308	Sphingobium sp. TKS	γ-HCH degradation	–	Ti		[16]
p113P3	2,43,437	Sphingopyxis sp. 113P3 (NBRC 111507)	PVA degradation	88 (334/376)			[26]
pTK7	9585	Sphingobium sp. TKS	γ-HCH degradation	–			[16]
pTK8	7223	Sphingobium sp. TKS	γ-HCH degradation	–			[16]
pSA2	9,60,805	Novosphingobium resinovorum SA1	Sulfanilic acid degradation	–	F		[70]
Unidentified	Contig	Novosphingobium sp. LH128	PAH degradation	99 (455/456)			[74]
pSM103mini	4328	Sphingopyxis macrogoltabida 103	PEG degradation	–			[75]

p203-1	4328	Sphingopyxis terrae 203-1 (NBRC 111660)	PEG degradation	99 (242/243)			[29]
pDE6	10,566	Sphingobium baderi DE-13	Alkyl-substituted aniline degradation	–			[72]
pDE7	6108	Sphingobium baderi DE-13	Alkyl-substituted aniline degradation	82 (231/280)			[72]
pDE8	5818	Sphingobium baderi DE-13	Alkyl-substituted aniline degradation	52 (142/269)			[72]
pLB1	65,998	unidentified soil bacterium (S. japonicum UT26)	γ-HCH degradation	–	Ti	*linB*	[76]
pMI3[b]	88,374	Sphingobium sp. MI1205	γ-HCH degradation	Identical	Ti	*linB, linC, linF'*	[16]
pLA2	62,341	Novosphingobium pentaromativorans US6-1	Benzo(a)pyrene degradation	98 (256/260)	Ti		[77]
pEY-1_2	58,892	Sphingopyxis macrogoltabida EY-1	PPG and PEG degradation	98 (256/260)	Ti		[28]
pTK5	53,908	Sphingobium sp. TKS	γ-HCH degradation	97 (253/260)	Ti		[16]
pDE3	73,561	Sphingobium baderi DE-13	Alkyl-substituted aniline degradation	–			[72]
pDC03_2[b]	1,15,498	Sphingomonas sp. DC-6	Chloroacetanilide herbicides degradation	99 (273/275)	Ti		[73]

(continued)

Table 9.2 (continued)

Replicon[a]	Size (bp)	Host	Host feature	Identity (%) of RepA protein to representative[a]	Gene cluster for conjugation	Genes for degradation	Reference
pUT2	5398	Sphingobium japonicum UT26	γ-HCH degradation	–			[34]
pTK9	5391	Sphingobium sp. TKS	γ-HCH degradation	Identical			[16]
pMSHV3	9968	Sphingobium herbicidovorans MH	Chlorophenoxy herbicide degradation	65 (134/206)			[68]
pNL1	1,84,462	Novosphingobium aromaticivorans DSM 12444	Aromatic compounds degradation	–			Unpublished
pCAR3	2,54,797	Novosphingobium sp. KA1	Carbazole degradation	91 (398/433)		Carbazole-degradative gene clusters	[78]
Mpl	11,61,602	Novosphingobium sp. PP1Y	Aromatic compounds degradation	85 (372/434)			[25]
pSA1	17,56,808	Novosphingobium resinovorum SA1	Sulfanilic acid degradation	82 (361/437)	Ti	*pob/pca/ben/cat* gene clusters	[70]
p203_1	4,22,455	Sphingopyxis macrogoltabida 203 (NBRC 15033)	PEG degradation[c]	82 (359/435)	F		[27]
p203N_1	4,21,025	Sphingopyxis macrogoltabida 203 N (NBRC 111659)	PEG degradation	82 (359/435)	F		[30]
pSWIT02	2,22,757	Sphingomonas wittichii RW1	Dioxin degradation	82 (356/429)			[23]

[a]The representative replicons of each type ones are indicated in bold in the first column
[b]Replicons having more than one *rep* genes
[c]No degradation gene was found in the genome

Type IV secretion system (T4SS), whereas the *dtr* genes encode proteins that bind to the DNA at the origin of transfer region, *oriT*, forming a structure called a relaxosome. This modular gene organization is shared by most conjugative systems, and indicates a high degree of gene synteny conservation. Among the sphingomonad plasmids listed in Table 9.2, the conjugal transferability of pCHQ1 and pLB1 has been experimentally confirmed [76, 83]. Moreover, using the exogenous plasmid isolation technique, we successfully isolated from HCH-contaminated soil a conjugative plasmid, pLB1, which carries two copies of *linB* [76]. Metagenomic analysis also suggested the importance of the horizontal transfer of the specific *lin* genes by plasmids for HCH-degradation in the environment [84]. These facts strongly suggest that conjugative plasmids play important roles in the distribution of the specific *lin* genes under environmental conditions. Some sphingomonad plasmids encoding transfer machineries may be self-transmissible (Table 9.2) [63]. In particular, the pCHQ1- and pLB1-type plasmids, whose conjugal transferability has been experimentally proven [76, 83], seem to be representative self-transmissible plasmids in sphingomonads. These plasmids are also suggested to have a narrow-host range property [76, 83].

9.6 Genome "Editing" Role of IS*6100* in γ-HCH Degraders

Many putative transposable elements including insertion sequence (IS) elements and Tn*3*-type transposons were found in the genomes of the four γ-HCH degraders [16, 34]. Although most of the IS elements are present as a single-copy form in the four strains, IS*6100* is the most abundant element in the UT26, MM-1, TKS, and MI1205 genomes (13, 15, 29, and 24 copies, respectively: Table 9.1). This suggests that IS*6100* can transpose and increase its copy number in these γ-HCH degraders. The IS entrapment experiments indeed detected the transposition of IS*6100* in γ-HCH degraders [16].

IS*6100*, which belongs to the IS*6* family, is 880 bp long and carries a transposase gene and 14-bp terminal inverted repeats (IR) at both ends. Since IS*6100* is a "replicative" IS element [85], its transposition without apparent preference of target specificity causes the duplication of IS*6100* with an 8-bp duplication of the target sequence. Therefore, the IS*6100* transposition can generate three types of DNA rearrangements (Fig. 9.2): intra-molecular transposition with a deletion/resolution (intra-replicon 1) or inversion (intra-replicon 2) event, and inter-molecular transposition with a fusion (inter-replicon) event. On the basis of this IS*6100* transposition mechanism (Fig. 9.2), the most plausible past events caused by transposition of IS*6100* can be inferred by comparison of the regions just upstream and downstream of copies of IS*6100* [16]. Here, we show the case in TKS as an example. We found seven pairs of 8-bp sequences among the regions just upstream and downstream of the 29 copies of IS*6100* in the TKS genome (Table 9.3), and inferred the most plausible past events caused by transposition of IS*6100* (Fig. 9.3); it is indicated that not only simple transposition with inversion but also transposition accompanied with the fusion and resolution of replicons must have occurred. TKS was isolated from

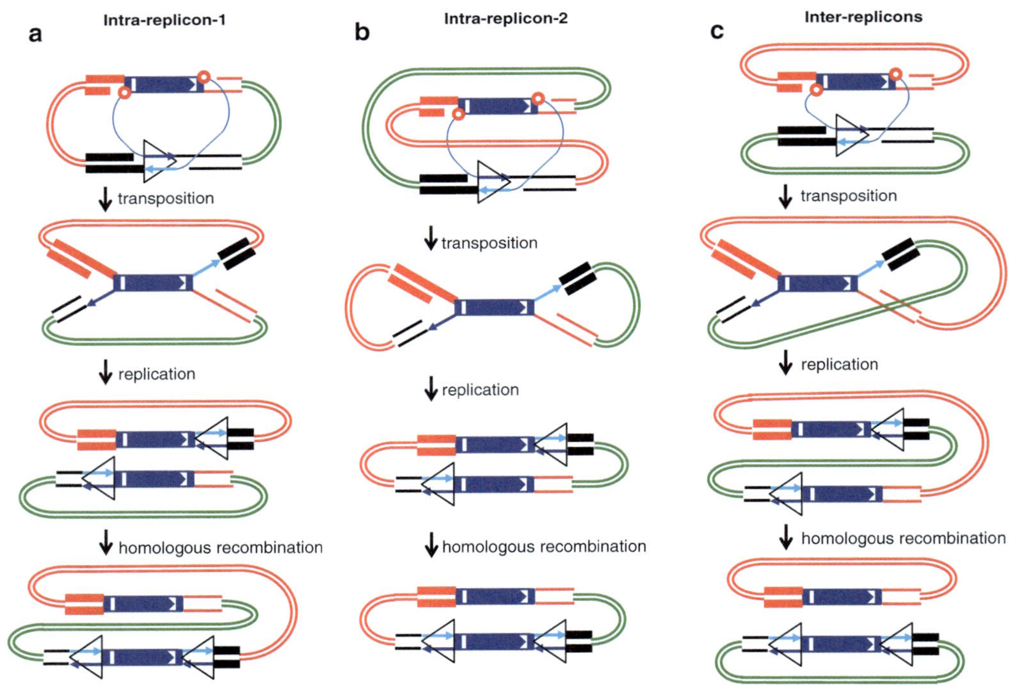

Fig. 9.2 Patterns for transposition of IS*6100*. Transposition into the same replicon with deletion/resolution (**a**. intra-replicon 1) or inversion (**b**. intra-replicon 2) and transposition between two replicons with fusion (**c**. inter-replicons) are schematically shown. IS*6100* (880 bp) is shown as blue pentagon and squares on the both sides showing direction of the *tnpA* gene (795 bp) and the 14-bp inverted repeats, respectively. All these transpositions cause duplication of IS*6100* and generate 8-bp target (blue and cyan arrows with triangle) duplication at the upstream and downstream of IS*6100*. Note that sequences shown in Table 9.3 are cyan strands of 8-bp targets in this figure and the directions of triangles are the same as in Fig. 9.3. The 3-'OH groups at IS*6100* ends are shown by red circles. The 3-'OH groups then attack the target sites as shown by curved thin blue lines

Table 9.3 Flanking 8-bp sequences of IS*6100* in TKS genome

Location of IS*6100* [a]	Flanking 8-bp sequence [b]	
	Upstream	Downstream
Chr1_229152_230031_c	CTCGCGGA	CCCATTTG (a)
Chr1_231415_232294_c	GGGCGTCA	TTTAATGT
Chr1_464035_464914	TATCACCG	GCAGCGCG
Chr1_467435_468314	GGCGTAAA	GCAGCGCG
Chr1_471676_472555_c	GTCGATAT	GCAGCGCG
Chr1_479503_480382_c	GGGCGGGC (b)	CGACGGAT
Chr1_4083212_4084091	CCCATTTG (a)	GGGCGGGC (b)
pTK1_10576_11455_c	CGATGGCC	TTATAAAG
pTK1_41506_42385_c	CGCCTCGC	TCATAGAT (c)
pTK1_381386_382265_c	CCGGTGAG	GGATTTCG
pTK1_395485_396364	GCTTCATG	GAGCCTGT
pTK1_401181_402060_c	GGCCTGCG (d)	GGGACAGC
pTK1_453367_454246	TCATAGAT (c)	GGCCTGCG (d)
pTK2_43892_44771_c	ATCATGGT	GGCGGGAG
pTK3_17346_18225_c	CGCTGTAG	TTCTATTA
pTK3_33765_34644_c	CGCGCGCC	GATCGACC (e)
pTK3_35948_36827_c	GATCGACC (e)	AAAAATGT
pTK3_42069_42944	GTCAACAC	ATGCGGCG
pTK3_48377_49256	TGTTCGCC (f)	GTTCGCTG (g)
pTK3_73465_74344	GTTCGCTG (g)	ACGCCAAG
pTK3_76693_77572_c	GGCGTAAA	GATGCTGG
pTK3_79935_80814_c	GAGTAGCG	TTTCTGAC
pTK4_6762_7641	GCGCAAGG	TGTTCGCC (f)
pTK4_20604_21483_c	GCTGAAAT	GTTGGGAT
pTK4_23754_24633_c	GAGTAGCG	CCAGCATC
pTK4_32697_33576	ATTGCCGC	CGCACCAC
pTK4_36402_37281_c	AAACGGTA	CTCCCAGT
pTK4_58412_59291	CGCTGTAG	TTCTATTA
pTK6_23365_24244	AGCCTGTC	CTTGCGCC

[a] c, encoding on complementary strand

b Pairs of the same sequences are marked with the same colors and alphabets in parentheses (see Fig. 9.3). Note that sequences shown in this table are cyan strands of 8-bp targets in Fig. 9.2.

γ-HCH-enriched liquid cultivation of a microbial community from a sediment sample contaminated with HCH isomers [16]. Recently we found three predicted previous structures in metagenome sequence of the enrichment culture from which TKS was isolated (Fig. 9.3; unpublished data), suggesting such events occurred during the repeated single-colony isolation processes on the solid medium. This fact shows that rapid genome evolution is occurring and suggests that the genome

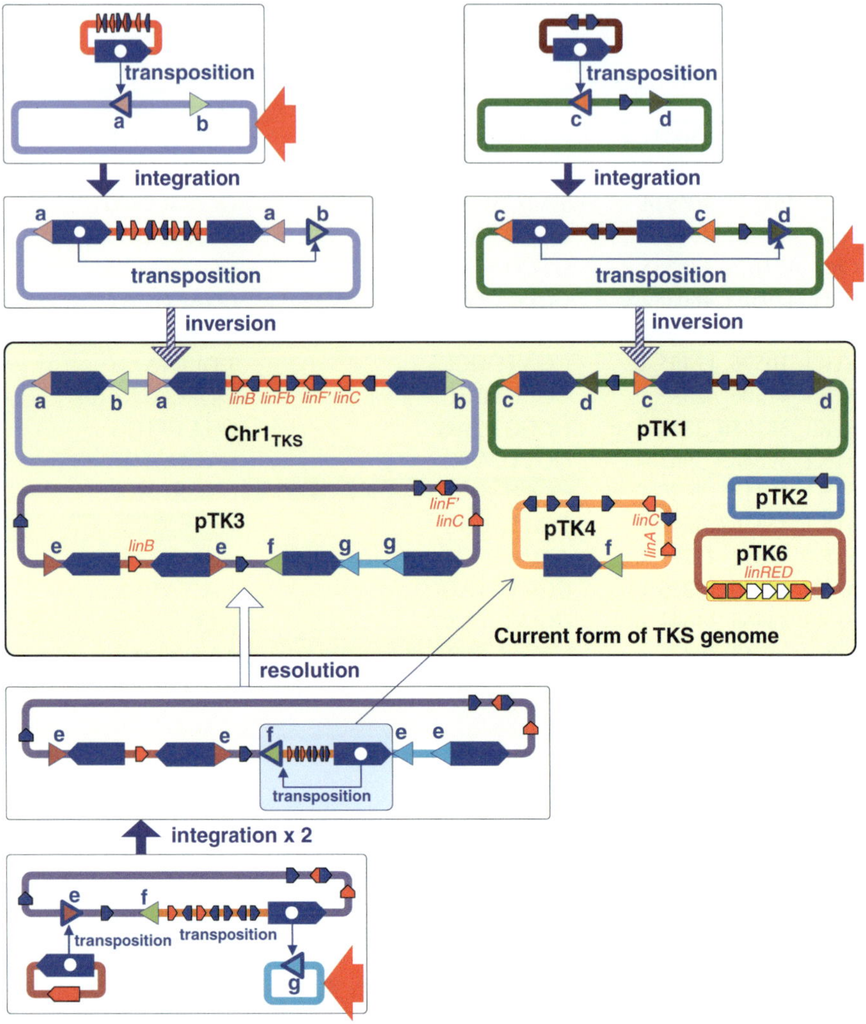

Fig. 9.3 Inference of the past genome rearrangements via IS*6100* in *Sphingobium* sp. strain TKS. Blue pentagons, triangles with alphabet, and red pentagons indicate IS*6100*, 8-bp target sites, and *lin* genes, respectively. Triangles with the same alphabet mean identical sequence and direction (see Table 9.3 and Fig. 9.2 for details: note that sequences shown in Table 9.3 are cyan strands of 8-bp targets in Fig. 9.2). Blue pentagons marked with internal white circle indicate the IS*6100* element which transposed. IS*6100* is a "replicative" IS element, and it increases its copy number with the transposition (Fig. 9.2). Only replicons carrying IS*6100* are illustrated, and relative positions and directions of IS*6100* and *lin* genes in each replicon are schematically shown. The IS*6100* elements involved in the proposed past genome rearrangements are shown in larger size. Three predicted previous structures found in metagenome sequences of the γ-HCH-enrichment culture from which TKS was isolated are shown in red arrows

structure of the bacterial strain isolated in the laboratory may be different from the ancestral strain inhabiting the environment.

Copies of IS*6100* are often found in close proximity to *lin* genes in HCH-degrading strains and the metagenomic sequences from HCH-contaminated sites [17, 18, 86, 87]. A plasmid pLB1 that carries an IS*6100*-composite transposon containing two copies of *linB* was isolated by the exogenous plasmid isolation technique [76]. These facts suggest that IS*6100* plays an important role in the recruitment of the specific *lin* genes. Comparison of the specific *lin*-flanking regions in the four strains revealed that not only the *lin* genes themselves but also their flanking regions are highly conserved (Fig. 9.4). Interestingly, such conserved regions are located very close to IS*6100* and the distances between the IS*6100* copies and the *lin* genes are vary [16] (Fig. 9.4). This means that IS*6100* is likely to play a crucial editing

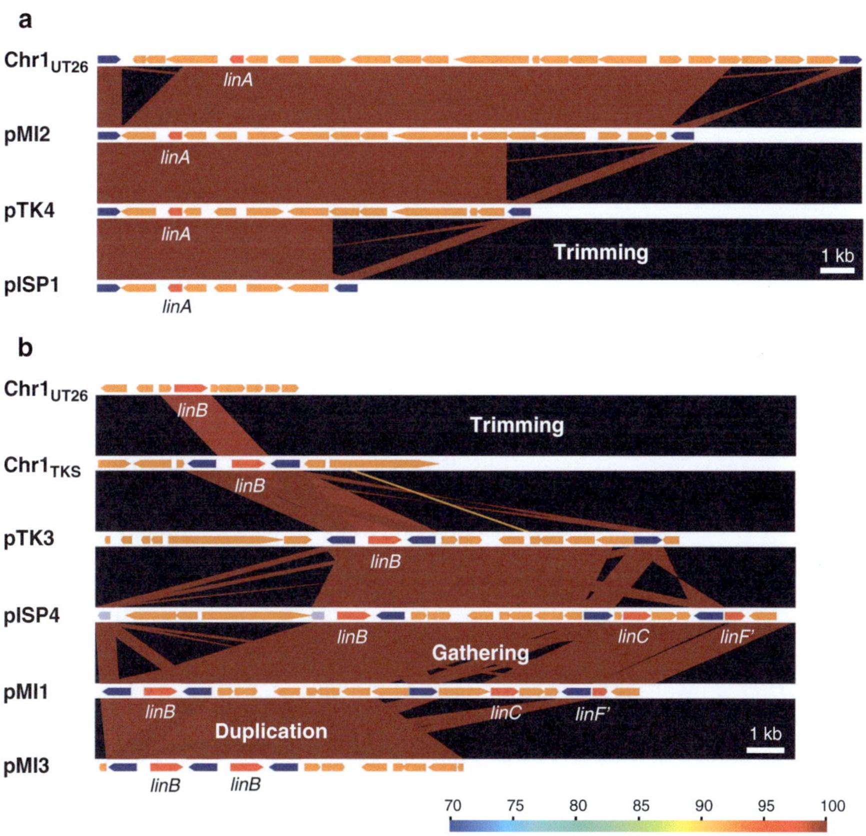

Fig. 9.4 Comparison of regions containing the *linA* (**a**) and *linB* (**b**) genes in the four γ-HCH-degrading sphingomonad strains. The regions homologous to each other were colored in the gradient depending on the level of similarity as shown in explanatory note. The *lin* genes, transposase gene of IS*6100*, and other ORFs were shown by pentagons in red, blue, and orange, respectively. Drawn by GenomeMatcher [88]

role in the "trimming" the regions unnecessary for HCH utilization and the "gathering" of the specific *lin* genes. This observation supports the "Selfish Operon model," in which HGT allows genes to cluster into an operon by a series of approximations [89, 90]. At least, the most plausible explanation is that the transposition of IS*6100* led to the diversification of the distribution and organization of the *lin* genes in the genomes. The distance between IS*6100* and *linA* is the longest in UT26 (Fig. 9.4a), and the *linB* gene in UT26 has no IS*6100* element in its flanking regions (Fig. 9.4b). Moreover, IS*6100* is located at only one side of *linC* and the *linRED* cluster in UT26 [16, 34]. These results suggested that UT26 is the closest to the prototype of the γ-HCH degrader, at least among the four strains. In addition, IS*6100* seems to be involved in the genetic instability of the specific *lin* genes. The *linA*, *linC*, and *linRED* genes in UT26 are genetically unstable, i.e., spontaneous deletion mutants of the regions containing these genes could easily be obtained, and these deletion processes in the mutants can be most simply explained by the involvement of IS*6100* [34]. As in the case of IS*6100*, IS*1071*, a member of the Tn*3* family, is also often associated with the genes for the degradation of xenobiotics, including atrazine [15] and 2,4-D [9], suggesting that IS*1071* also has the functions like those of IS*6100*. Generally, diverse IS family transposase genes are associated with genes for the degradation of xenobiotics [9]. It will be of great interest to learn how such combinations between IS elements and degradative genes have arisen.

9.7 Emergence and Evolution of γ-HCH-Degrading Bacteria

Comparison of the genomes of γ-HCH-degrading and non-γ-HCH-degrading sphingomonad strains strongly suggested that the γ-HCH-degrading bacteria emerged through the recruitment of the specific *lin* genes into an ancestral strain that had core functions of sphingomonads, such as the LinKLMN-type ABC transporter system and the β-ketoadipate pathway (Fig. 9.5). Other unknown factors may also be involved in the core functions. The important conclusion at present is that at least four γ-HCH degraders emerged independently and in parallel around the world. Multiple plasmids whose replication/partition machineries are highly conserved in sphingomonads might have played important roles in the recruitment of the specific *lin* genes by their HGT. In addition, IS*6100* likely plays a crucial "editing" role in the distribution and organization of the *lin* genes in genomes. It is also speculated that IS*6100* is involved in the recruitment of the *lin* genes from an environmental gene pool whose details are still obscure. Recently, the complete genome sequence of *Sphingobium indicum* B90A was published [21], which is a HCH degrader phylogenetically closely related to UT26. Since most of the genomic regions of B90A and UT26 are highly conserved [21], B90A may be a kind of UT26-derived strain emerged from a common ancestral γ-HCH degrader via transpositions of IS*6100*. However, the replicon organizations of these two strains are different from each other (Table 9.2), and thus acquisition and/or loss of some plasmids seems to have occurred during the diversification process of the two strains. In the future, our hypothesis may be confirmed in experiments using the HCH degraders and their

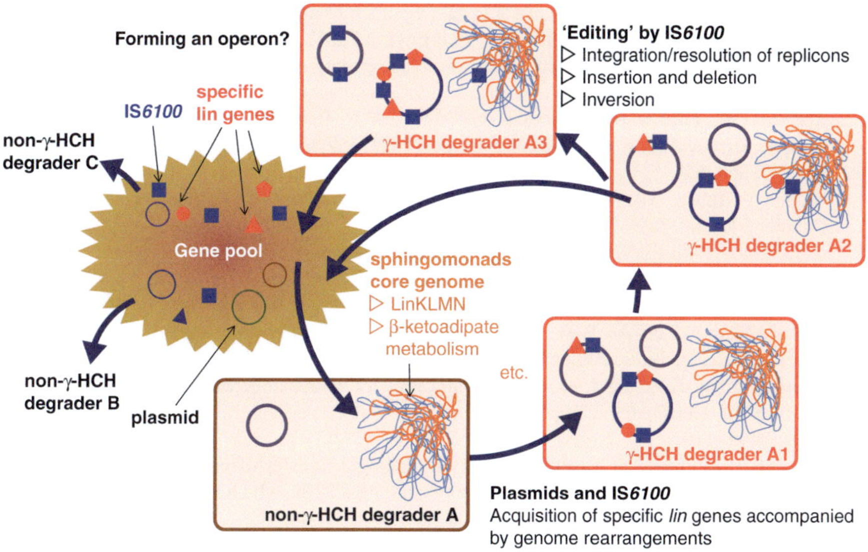

Fig. 9.5 Proposed model for the emergence and evolution of γ-HCH-degrading sphingomonad strains. Ancestral "non-γ-HCH-degrader A" having core functions of sphingomonads, e.g., LinKLMN-type ABC transporter and β-ketoadipate pathway, turns to be "γ-HCH-degrader A1" by acquisition of specific *lin* genes by using sphingomonads-specific plasmids and IS*6100* accompanied by genome rearrangements. The organization of *lin* genes is "edited" by IS*6100* and acquisition/loss of plasmids, resulting in the derivative γ-HCH degraders A2 and A3. "The *lin* operon" and/or a plasmid carrying all the specific *lin* genes may be formed in the future. It is noteworthy that these events can occur also in phylogenetically different sphingomonad strains having the core functions ("non-γ-HCH-degrader B and C"). See text for details

related but non-HCH-degrading and/or IS*6100*-free sphingomonad strains. Further, the *lin* system may still be evolving toward one or more optimal states by using the combination of plasmids and IS*6100*, e.g., by forming an operon of *lin* genes (Fig. 9.5). It is noteworthy that IS*6100* is also involved in the loss of the *lin* genes [34], and thus can contribute to the adaptation for other conditions under which the *lin* genes are no longer necessary.

Acknowledgements This work was supported by Grants-in-Aid from the Ministry of Education, Culture, Sports, Science and Technology of Japan, and the Institute for Fermentation, Osaka (IFO), Japan.

References

1. El-Shahawi MS, Hamza A, Bashammakh AS, Al-Saggaf WT. An overview on the accumulation, distribution, transformations, toxicity and analytical methods for the monitoring of persistent organic pollutants. Talanta. 2010;80:1587–97.
2. Ogata Y, Takada H, Mizukawa K, et al. International pellet watch: global monitoring of persistent organic pollutants (POPs) in coastal waters. 1. Initial phase data on PCBs, DDTs, and HCHs. Mar Pollut Bull. 2009;58:1437–46.

3. Tarcau D, Cucu-Man S, Boruvkova J, et al. Organochlorine pesticides in soil, moss and tree-bark from North-Eastern Romania. Sci Total Environ. 2013;456–457:317–24.
4. Copley SD. Evolution of efficient pathways for degradation of anthropogenic chemicals. Nat Chem Biol. 2009;5:559–66.
5. Janssen DB, Dinkla IJ, Poelarends GJ, Terpstra P. Bacterial degradation of xenobiotic compounds: evolution and distribution of novel enzyme activities. Environ Microbiol. 2005;7:1868–82.
6. Nagata Y, Tabata M, Ohtsubo Y, Tsuda M. Biodegradation of organochlorine pesticides. Chapter 512. In: Yates M, Nakatsu C, Miller R, Pillai S, editors. Manual of environmental microbiology. 4th ed. Washington, DC: ASM Press; 2016. p. 1–30.
7. Stolz A. Molecular characteristics of xenobiotic-degrading sphingomonads. Appl Microbiol Biotechnol. 2009;81:793–811.
8. van der Meer JR, de Vos WM, Harayama S, Zehnder AJ. Molecular mechanisms of genetic adaptation to xenobiotic compounds. Microbiol Rev. 1992;56:677–94.
9. Liang B, Jiang J, Zhang J, et al. Horizontal transfer of dehalogenase genes involved in the catalysis of chlorinated compounds: evidence and ecological role. Crit Rev Microbiol. 2012;38:95–110.
10. Ohtsubo Y, Ishibashi Y, Naganawa H, et al. Conjugal transfer of polychlorinated biphenyl/biphenyl degradation genes in *Acidovorax* sp. strain KKS102, which are located on an integrative and conjugative element. J Bacteriol. 2012;194:4237–48.
11. Springael D, Top EM. Horizontal gene transfer and microbial adaptation to xenobiotics: new types of mobile genetic elements and lessons from ecological studies. Trends Microbiol. 2004;12:53–8.
12. Top EM, Springael D. The role of mobile genetic elements in bacterial adaptation to xenobiotic organic compounds. Curr Opin Biotechnol. 2003;14:262–9.
13. Tsuda M, Tan HM, Nishi A, Furukawa K. Mobile catabolic genes in bacteria. J Biosci Bioeng. 1999;87:401–10.
14. Satola B, Wubbeler JH, Steinbuchel A. Metabolic characteristics of the species *Variovorax paradoxus*. Appl Microbiol Biotechnol. 2013;97:541–60.
15. Udikovic-Kolic N, Scott C, Martin-Laurent F. Evolution of atrazine-degrading capabilities in the environment. Appl Microbiol Biotechnol. 2012;96:1175–89.
16. Tabata M, Ohhata S, Nikawadori Y, et al. Comparison of the complete genome sequences of four gamma-hexachlorocyclohexane-degrading bacterial strains: insights into the evolution of bacteria able to degrade a recalcitrant man-made pesticide. DNA Res. 2016;23:581–99.
17. Lal R, Dogra C, Malhotra S, Sharma P, Pal R. Diversity, distribution and divergence of lin genes in hexachlorocyclohexane-degrading sphingomonads. Trends Biotechnol. 2006;24:121–30.
18. Lal R, Pandey G, Sharma P, et al. Biochemistry of microbial degradation of hexachlorocyclohexane and prospects for bioremediation. Microbiol Mol Biol Rev. 2010;74:58–80.
19. Stolz A. Degradative plasmids from sphingomonads. FEMS Microbiol Lett. 2014;350:9–19.
20. Yabuuchi E, Kosako Y, Fujiwara N, et al. Emendation of the genus *Sphingomonas* Yabuuchi et al. 1990 and junior objective synonymy of the species of three genera, *Sphingobium*, *Novosphingobium* and *Sphingopyxis*, in conjunction with *Blastomonas ursincola*. Int J Syst Evol Microbiol. 2002;52:1485–96.
21. Verma H, Bajaj A, Kumar R, et al. Genome organization of *Sphingobium indicum* B90A: an archetypal hexachlorocyclohexane (HCH) degrading genotype. Genome Biol Evol. 2017;9:2191–7.
22. Copley SD, Rokicki J, Turner P, et al. The whole genome sequence of *Sphingobium chlorophenolicum* L-1: insights into the evolution of the pentachlorophenol degradation pathway. Genome Biol Evol. 2011;4:184–98.
23. Miller TR, Delcher AL, Salzberg SL, et al. Genome sequence of the dioxin-mineralizing bacterium *Sphingomonas wittichii* RW1. J Bacteriol. 2010;192:6101–2.
24. Masai E, Katayama Y, Fukuda M. Genetic and biochemical investigations on bacterial catabolic pathways for lignin-derived aromatic compounds. Biosci Biotechnol Biochem. 2007;71:1–15.

25. D'Argenio V, Petrillo M, Cantiello P, et al. De novo sequencing and assembly of the whole genome of *Novosphingobium* sp. strain PP1Y. J Bacteriol. 2011;193:4296.
26. Ohtsubo Y, Nagata Y, Numata M, et al. Complete genome sequence of polyvinyl alcohol-degrading strain *Sphingopyxis* sp. 113P3 (NBRC 111507). Genome Announc. 2015;3:e01169–15.
27. Ohtsubo Y, Nagata Y, Numata M, et al. Complete genome sequence of *Sphingopyxis macrogoltabida* type strain NBRC 15033, originally isolated as a polyethylene glycol degrader. Genome Announc. 2015;3:e01401–15.
28. Ohtsubo Y, Nagata Y, Numata M, et al. Complete genome sequence of polypropylene glycol- and polyethylene glycol-degrading *Sphingopyxis macrogoltabida* strain EY-1. Genome Announc. 2015;3:e01399–15.
29. Ohtsubo Y, Nonoyama S, Nagata Y, et al. Complete genome sequence of *Sphingopyxis terrae* strain 203-1 (NBRC 111660), a polyethylene glycol degrader. Genome Announc. 2016;4:e00530–16.
30. Ohtsubo Y, Nonoyama S, Nagata Y, et al. Complete genome sequence of *Sphingopyxis macrogoltabida* strain 203N (NBRC 111659), a polyethylene glycol degrader. Genome Announc. 2016b;4:e00529–16.
31. Parthasarathy S, Azam S, Lakshman SA, et al. Genome-guided insights reveal organophosphate-degrading *Brevundimonas diminuta* as *Sphingopyxis wildii* and define its versatile metabolic capabilities and environmental adaptations. Genome Biol Evol. 2017;9:77–81.
32. Lauro FM, McDougald D, Thomas T, et al. The genomic basis of trophic strategy in marine bacteria. Proc Natl Acad Sci U S A. 2009;106:15527–33.
33. Aylward FO, McDonald BR, Adams SM, et al. Comparison of 26 sphingomonad genomes reveals diverse environmental adaptations and biodegradative capabilities. Appl Environ Microbiol. 2013;79:3724–33.
34. Nagata Y, Natsui S, Endo R, et al. Genomic organization and genomic structural rearrangements of *Sphingobium japonicum* UT26, an archetypal γ-hexachlorocyclohexane-degrading bacterium. Enzyme Microb Technol. 2011;49:499–508.
35. Phillips TM, Seech AG, Lee H, Trevors JT. Biodegradation of hexachlorocyclohexane (HCH) by microorganisms. Biodegradation. 2005;16:363–92.
36. Vijgen J, Abhilash PC, Li YF, et al. Hexachlorocyclohexane (HCH) as new Stockholm Convention POPs-a global perspective on the management of lindane and its waste isomers. Environ Sci Pollut Res Int. 2011;18:152–62.
37. Willett K, Ulrich E, Hites R. Differential toxicity and environmental fates of hexachlorocyclohexane isomers. Environ Sci Technol. 1998;32:2197–207.
38. Nagata Y, Endo R, Ito M, et al. Aerobic degradation of lindane (γ-hexachlorocyclohexane) in bacteria and its biochemical and molecular basis. Appl Microbiol Biotechnol. 2007;76:741–52.
39. Miyauchi K, Lee HS, Fukuda M, et al. Cloning and characterization of *linR*, involved in regulation of the downstream pathway for γ-hexachlorocyclohexane degradation in *Sphingomonas paucimobilis* UT26. Appl Environ Microbiol. 2002;68:1803–7.
40. Harwood CS, Parales RE. The β-ketoadipate pathway and the biology of self-identity. Annu Rev Microbiol. 1996;50:553–90.
41. Nagata Y, Ohtsubo Y, Tsuda M. Properties and biotechnological applications of natural and engineered haloalkane dehalogenases. Appl Microbiol Biotechnol. 2015;99:9865–81.
42. Endo R, Ohtsubo Y, Tsuda M, Nagata Y. Identification and characterization of genes encoding a putative ABC-type transporter essential for utilization of γ-hexachlorocyclohexane in *Sphingobium japonicum* UT26. J Bacteriol. 2007;189:3712–20.
43. Endo R, Ohtsubo Y, Tsuda M, Nagata Y. Growth inhibition by metabolites of γ-hexachlorocyclohexane in *Sphingobium japonicum* UT26. Biosci Biotechnol Biochem. 2006;70:1029–32.
44. Casali N, Riley LW. A phylogenomic analysis of the Actinomycetales *mce* operons. BMC Genomics. 2007;8:60.

45. Awai K, Xu C, Tamot B, Benning C. A phosphatidic acid-binding protein of the chloroplast inner envelope membrane involved in lipid trafficking. Proc Natl Acad Sci U S A. 2006;103:10817–22.
46. Nagata Y, Ohtsubo Y, Endo R, et al. Complete genome sequence of the representative γ-hexachlorocyclohexane-degrading bacterium *Sphingobium japonicum* UT26. J Bacteriol. 2010;192:5852–3.
47. Gil R, Silva FJ, Pereto J, Moya A. Determination of the core of a minimal bacterial gene set. Microbiol Mol Biol Rev. 2004;68:518–37.
48. Ito M, Prokop Z, Klvana M, et al. Degradation of β-hexachlorocyclohexane by haloalkane dehalogenase LinB from γ-hexachlorocyclohexane-utilizing bacterium *Sphingobium* sp. MI1205. Arch Microbiol. 2007;188:313–25.
49. Tabata M, Ohhata S, Nikawadori Y, et al. Complete genome sequence of a γ-hexachlorocyclohexane-degrading bacterium, *Sphingobium* sp. strain MI1205. Genome Announc. 2016;4:e00246–16.
50. Tabata M, Endo R, Ito M, et al. The *lin* genes for γ-hexachlorocyclohexane degradation in *Sphingomonas* sp. MM-1 proved to be dispersed across multiple plasmids. Biosci Biotechnol Biochem. 2011;75:466–72.
51. Tabata M, Ohtsubo Y, Ohhata S, et al. Complete genome sequence of the γ-hexachlorocyclohexane-degrading bacterium *Sphingomonas* sp. strain MM-1. Genome Announc. 2013;1:e00247–13.
52. Tabata M, Ohhata S, Kawasumi T, et al. Complete genome sequence of a γ-hexachlorocyclohexane degrader, *Sphingobium* sp. strain TKS, isolated from a γ-hexachlorocyclohexane-degrading microbial community. Genome Announc. 2016;4:e00247–16.
53. Lykidis A, Perez-Pantoja D, Ledger T, et al. The complete multipartite genome sequence of *Cupriavidus necator* JMP134, a versatile pollutant degrader. PLoS One. 2010;5:e9729.
54. Perez-Pantoja D, De la Iglesia R, Pieper DH, Gonzalez B. Metabolic reconstruction of aromatic compounds degradation from the genome of the amazing pollutant-degrading bacterium *Cupriavidus necator* JMP134. FEMS Microbiol Rev. 2008;32:736–94.
55. Chain PS, Denef VJ, Konstantinidis KT, et al. *Burkholderia xenovorans* LB400 harbors a multi-replicon, 9.73-Mbp genome shaped for versatility. Proc Natl Acad Sci U S A. 2006;103:15280–7.
56. Romero-Silva MJ, Mendez V, Agullo L, Seeger M. Genomic and functional analyses of the gentisate and protocatechuate ring-cleavage pathways and related 3-hydroxybenzoate and 4-hydroxybenzoate peripheral pathways in *Burkholderia xenovorans* LB400. PLoS One. 2013;8:e56038.
57. Nagata Y, Senbongi J, Ishibashi Y, et al. Identification of *Burkholderia multivorans* ATCC 17616 genetic determinants for fitness in soil by using signature-tagged mutagenesis. Microbiology. 2014;160:883–91.
58. Perez-Pantoja D, Donoso R, Agullo L, et al. Genomic analysis of the potential for aromatic compounds biodegradation in Burkholderiales. Environ Microbiol. 2012;14:1091–117.
59. Stanier RY, Palleroni NJ, Doudoroff M. The aerobic pseudomonads: a taxonomic study. J Gen Microbiol. 1966;43:159–271.
60. Yuhara S, Komatsu H, Goto H, et al. Pleiotropic roles of iron-responsive transcriptional regulator Fur in *Burkholderia multivorans*. Microbiology. 2008;154:1763–74.
61. Nelson KE, Weinel C, Paulsen IT, et al. Complete genome sequence and comparative analysis of the metabolically versatile *Pseudomonas putida* KT2440. Environ Microbiol. 2002;4:799–808.
62. Davison J. Genetic exchange between bacteria in the environment. Plasmid. 1999;42:73–91.
63. Shintani M, Nojiri H. Mobile genetic elements (MGEs) carrying catabolic genes. In: Malik A, Grohmann E, Alves M, editors. Management of microbial resources in the environment. Dordrecht: Springer; 2013.
64. Martinez B, Tomkins J, Wackett LP, et al. Complete nucleotide sequence and organization of the atrazine catabolic plasmid pADP-1 from *Pseudomonas* sp. strain ADP. J Bacteriol. 2001;183:5684–97.

65. Trefault N, De la Iglesia R, Molina AM, et al. Genetic organization of the catabolic plasmid pJP4 from *Ralstonia eutropha* JMP134 (pJP4) reveals mechanisms of adaptation to chloroaromatic pollutants and evolution of specialized chloroaromatic degradation pathways. Environ Microbiol. 2004;6:655–68.
66. Pearce SL, Oakeshott JG, Pandey G. Insights into ongoing evolution of the hexachlorocyclohexane catabolic pathway from comparative genomics of ten Sphingomonadaceae strains. G3 (Bethesda). 2015;5:1081–94.
67. Verma H, Kumar R, Oldach P, et al. Comparative genomic analysis of nine *Sphingobium* strains: insights into their evolution and hexachlorocyclohexane (HCH) degradation pathways. BMC Genomics. 2014;15:1014.
68. Nielsen TK, Rasmussen M, Demaneche S, et al. Evolution of sphingomonad gene clusters related to pesticide catabolism revealed by genome sequence and mobilomics of *Sphingobium herbicidovorans* MH. Genome Biol Evol. 2017;9:2477–90.
69. Zhu X, Wang W, Xu P, Tang H. Complete genome sequence of *Sphingomonas* sp. strain NIC1, an efficient nicotine-degrading bacterium. Genome Announc. 2016;4:e00666–16.
70. Hegedus B, Kos PB, Balint B, et al. Complete genome sequence of *Novosphingobium resinovorum* SA1, a versatile xenobiotic-degrading bacterium capable of utilizing sulfanilic acid. J Biotechnol. 2017;241:76–80.
71. Masai E, Kamimura N, Kasai D, et al. Complete genome sequence of *Sphingobium* sp. strain SYK-6, a degrader of lignin-derived biaryls and monoaryls. J Bacteriol. 2012;194:534–5.
72. Cheng M, Li Y, Ma Y, et al. Complete genome sequence of *Sphingobium baderi* DE-13, an alkyl-substituted aniline-mineralizing bacterium. Curr Microbiol. 2018;75:27–31.
73. Chen Q, Wang CH, Deng SK, et al. Novel three-component Rieske non-heme iron oxygenase system catalyzing the *N*-dealkylation of chloroacetanilide herbicides in sphingomonads DC-6 and DC-2. Appl Environ Microbiol. 2014;80:5078–85.
74. Fida TT, Breugelmans P, Lavigne R, et al. Identification of *opsA*, a gene involved in solute stress mitigation and survival in soil, in the polycyclic aromatic hydrocarbon-degrading bacterium *Novosphingobium* sp. strain LH128. Appl Environ Microbiol. 2014;80:3350–61.
75. Tani A, Tanaka A, Minami T, et al. Characterization of a cryptic plasmid, pSM103mini, from polyethylene-glycol degrading *Sphingopyxis macrogoltabida* strain 103. Biosci Biotechnol Biochem. 2011;75:295–8.
76. Miyazaki R, Sato Y, Ito M, et al. Complete nucleotide sequence of an exogenously isolated plasmid, pLB1, involved in γ-hexachlorocyclohexane degradation. Appl Environ Microbiol. 2006;72:6923–33.
77. Luo YR, Kang SG, Kim SJ, et al. Genome sequence of benzo(*a*)pyrene-degrading bacterium *Novosphingobium pentaromativorans* US6-1. J Bacteriol. 2012;194:907.
78. Shintani M, Urata M, Inoue K, et al. The *Sphingomonas* plasmid pCAR3 is involved in complete mineralization of carbazole. J Bacteriol. 2007;189:2007–20.
79. Brassinga AK, Marczynski GT. Replication intermediate analysis confirms that chromosomal replication origin initiates from an unusual intergenic region in *Caulobacter crescentus*. Nucleic Acids Res. 2001;29:4441–51.
80. Sibley CD, MacLellan SR, Finan T. The *Sinorhizobium meliloti* chromosomal origin of replication. Microbiology. 2006;152:443–55.
81. Schaper S, Messer W. Interaction of the initiator protein DnaA of *Escherichia coli* with its DNA target. J Biol Chem. 1995;270:17622–6.
82. Cabezon E, Ripoll-Rozada J, Pena A, et al. Towards an integrated model of bacterial conjugation. FEMS Microbiol Rev. 2015;39:81–95.
83. Nagata Y, Kamakura M, Endo R, et al. Distribution of γ-hexachlorocyclohexane-degrading genes on three replicons in *Sphingobium japonicum* UT26. FEMS Microbiol Lett. 2006;256:112–8.
84. Sangwan N, Lata P, Dwivedi V, et al. Comparative metagenomic analysis of soil microbial communities across three hexachlorocyclohexane contamination levels. PLoS One. 2012;7:e46219.
85. Mahillon J, Chandler M. Insertion sequences. Microbiol Mol Biol Rev. 1998;62:725–74.

86. Boltner D, Moreno-Morillas S, Ramos JL. 16S rDNA phylogeny and distribution of *lin* genes in novel hexachlorocyclohexane-degrading *Sphingomonas* strains. Environ Microbiol. 2005;7:1329–38.
87. Fuchu G, Ohtsubo Y, Ito M, et al. Insertion sequence-based cassette PCR: cultivation-independent isolation of γ-hexachlorocyclohexane-degrading genes from soil DNA. Appl Microbiol Biotechnol. 2008;79:627–32.
88. Ohtsubo Y, Ikeda-Ohtsubo W, Nagata Y, Tsuda M. GenomeMatcher: a graphical user interface for DNA sequence comparison. BMC Bioinformatics. 2008;9:376.
89. Lawrence J. Selfish operons: the evolutionary impact of gene clustering in prokaryotes and eukaryotes. Curr Opin Genet Dev. 1999;9:642–8.
90. Lawrence JG, Roth JR. Selfish operons: horizontal transfer may drive the evolution of gene clusters. Genetics. 1996;143:1843–60.

DNA Traffic in the Environment and Antimicrobial Resistance

10

Steven P. Hooton, Andrew D. Millard, Michelle Baker, Dov J. Stekel, and Jon L. Hobman

Abstract

The seemingly insurmountable problem of antimicrobial resistance (AMR) in clinical, food, and agricultural environments requires considerable efforts to be made in order to mitigate associated risks. Understanding the dynamics of the multitude of processes contributing towards AMR development, spread, and persistence in microbial populations will prove paramount in resolving these problems. Mobile genetic elements (MGEs) such as plasmids, transposons/insertion sequences, and bacteriophages contribute towards horizontal gene transfer of antimicrobial resistance genes (ARGs) in the environment. ARGs can be transferred from naturally resistant, ubiquitously distributed microbial populations acting as reservoirs for these genes. When ARGs are introduced into pathogens or opportunistic pathogens, these microorganisms subsequently become problematic when introduced into human/animal populations.

The role of MGEs in the evolution and emergence of pathogens of significant clinical and veterinary importance is well-documented. From a microbiological perspective, improving our knowledge of MGE-mediated AMR transmission by the application of traditional microbial culture techniques, molecular biology methods, and genomic/metagenomic/transcriptomic approaches will enhance our understanding of the flow of genetic information in bacteria. Mathematical modelling will prove to be integral to developing testable hypotheses regarding gene transfer rates, the consequences of positive selection, persistence in the absence of selection, and the fitness cost of gaining/losing resistance.

S. P. Hooton · M. Baker · D. J. Stekel · J. L. Hobman (✉)
University of Nottingham, School of Biosciences, Leicestershire, UK
e-mail: steven.hooton@nottingham.ac.uk; michelle.baker@nottingham.ac.uk; dov.stekel@nottingham.ac.uk; jon.hobman@nottingham.ac.uk

A. D. Millard
Department of Immunity, Infection, and Inflammation, University of Leicester, Leicester, UK
e-mail: adm39@leicester.ac.uk

H. Nishida, T. Oshima (eds.), *DNA Traffic in the Environment*,
https://doi.org/10.1007/978-981-13-3411-5_10

Horizontal gene transfer of AMR genes has led to the emergence of significant globally distributed enterobacterial pathogens such as *Escherichia coli*, *Salmonella* spp., and *Klebsiella* spp. The consequences of the emergence of these pathogens pose significant risks for humans and veterinary medicine, in what is a highly convoluted and at present, a seemingly intractable problem.

Keywords
Antimicrobial resistance · Mobile genetic elements · Horizontal gene transfer Plasmids · Transposable elements · Bacteriophages · Mathematical modelling

10.1 Introduction

Factors governing the emergence, establishment, and persistence of multidrug-resistant (MDR) bacteria can largely be attributed to a number of natural processes—mutation/evolution, selection, and DNA transfer in vivo and in the environment. The flow of DNA can be vertical (mother-to-daughter) or horizontal (between bacterial cells) and forms a complex continuum of interactions in microbial populations [1]. Beneficial traits have the potential to spread throughout bacterial populations via horizontal transmission and become established in lineages following vertical transmission. Although positive selection exerts a major influence over this process, such genetic traits can persist in the absence of selection [2]. With information flowing in multiple directions, mixed microbial communities can benefit and evolve from the loss (e.g. insertional inactivation) or acquisition of genes that increase fitness of the recipient. Horizontal gene transfer (HGT) constitutes the major route of DNA traffic in many environments, although there are roles for bacteriophage (phage)-mediated gene transfer, and transformation of bacteria by DNA in the environment. A wide range of mechanisms underpin HGT in nature [3], most notably mobile genetic elements (MGEs). MGEs such as plasmids and associated mobile elements including transposons, insertion sequences, and class I integrons have played pivotal roles in the emergence of many human and animal bacterial pathogens [4, 5]. Phages also contribute towards HGT via the processes of generalised and specialised transduction [6–8]. As vectors of genetic information, the significant role MGEs perform in driving bacterial evolution is well-documented [9]. From a microbial perspective, the potential to acquire novel traits such as antimicrobial resistance genes (ARGs) conferring the ability to survive in the presence of lethal compounds including antibiotics of clinical and veterinary importance, toxic metals, and disinfectants is evolutionarily advantageous in environments where lethal levels of antimicrobial compounds are present. Reservoirs of ARGs can be found in environments harbouring both pathogenic and non-pathogenic bacteria. Significantly, habitats such as the human and animal gut harbour high levels of Gram-negative *Enterobacteriaceae*, and therefore higher levels of ARGs [10]. It is readily accepted that genetic exchanges occur frequently between pathogens/non-pathogens in the environment. Consequently, the development and spread of antimicrobial resistance (AMR) is a seemingly foregone conclusion. Delineating the important factors driving the global problem of AMR requires multidisciplinary approaches capable of drawing meaningful conclusions from often highly convoluted microbial interactions. Traditional microbial culture-based analysis, functional and comparative genomics, and mathematical modelling will all play crucial roles in the study of AMR.

The cumulative effect of mobilising and disseminating ARGs in the environment is the spread of multiple resistance mechanisms capable of mitigating the effects of antimicrobial compounds, rendering treatment ineffective. Masking or mutation of the target site of the antimicrobial compound can render it ineffective. Resistance to macrolides, lincosamides, and streptogramins (MLS) is conferred by the activity of genes encoded by the *erm* (erythromycin ribosome methylase) family of methylases. Methylation of 16S rRNA mediated by *erm* alters the cognate target site of erythromycin and similar compounds. Actively pumping out antimicrobial compounds via the acquisition of genes encoding specific/non-specific efflux pumps reduces the efficacy of many antimicrobial agents including antibiotics, metals, and disinfectants. Active modification of antimicrobial compounds into non-/less-toxic forms also leads to a loss of function and may ultimately lead to treatment failure [11]. Target site masking/mutation, efflux pumps, detoxification, and other associated resistance mechanisms are nearly always found to be acting in synergy. The combined effect leads to higher levels of resistance following positive selection for more resilient microbes.

10.2 Horizontal Gene Transfer

10.2.1 Plasmids

Plasmids are closed-circular autonomously replicating DNA molecules encoding genes for the propagation/retention of the plasmid and can also contain genes encoding various functions associated with increasing bacterial host fitness. Many plasmids are self-transmissible as they encode the necessary conjugative apparatus for transfer to recipient bacteria. Others, whilst notably lacking the ability to self-transmit due to the absence of conjugation-associated genes, are capable of being transferred in the presence of self-transmissible plasmids [12, 13]. Plasmid-mediated horizontal gene transfer is a major driving force behind the dissemination of AMR in bacterial populations [14]. Large AMR/virulence plasmids can simultaneously encode resistance to multiple antibiotic classes (e.g. β-lactams/cephalosporins, macrolides, tetracyclines, aminoglycosides, and sulphonamides), a wide range of antimicrobial metal resistances (e.g. copper, silver, zinc, and mercury), and resistance to detergents/disinfectants (e.g. quaternary ammonium compound resistance (*qacE*)). Often co-located with AMR genes are virulence operons encoding diverse pathogenicity functions [15, 16]. Siderophore systems involved in iron acquisition, fimbriae/pili for cell-to-cell contact, and toxins are all common features of large virulence plasmids [13]. Smaller plasmids encoding single or a few resistances can often be found co-existing in the same bacterial cell with large plasmids. Plasmids can readily be transferred between bacterial cells during the process of conjugation, allowing the dissemination of AMR genes throughout microbial communities, and small plasmids lacking the apparatus for conjugation can be mobilised with large self-transmissible plasmids. Recombination events between plasmids take place frequently and evolution has generated a diverse array of hybrid/mosaic mobile elements [12, 13, 17]. The activity of composite and complex transposons moving between plasmids, chromosomes, and genomic islands allows expansion of resistance gene assemblages within and between plasmids, adding a further layer of complexity to the spread of AMR. Genes encoding resistances to a wide range of antibiotics including bla_{TEM} [16], $bla_{\mathrm{CTX\text{-}M\text{-}15}}$ [18], and $bla_{\mathrm{NDM\text{-}1}}$ [19] are often found linked to or within insertion sequences in composite transposons, whilst complex transposons such as Tn*3* family transposons Tn*5*, Tn*7*, and Tn*10* which all carry antimicrobial resistances have been found on plasmids. Other genetic elements such as class I integrons have been found to be plasmid encoded and play a major role in the capture of AMR-associated gene cassettes allowing the generation of novel resistance phenotypes [1]. The constant interplay of mobile elements, recombination events, and bacterial conjugation leads to the formation of a complex mobile genetic landscape—the plasmidome [20–22]. The mobile nature of the plasmidome drives the emergence of resistance in the environmental resistome [10, 23]. Understanding the intricate relationships between bacteria and mobile genetic elements in the environment is central to combatting AMR.

Plasmids represent a highly diverse pool of MGEs that are grouped according to the type of replicon complexes (PCR-based replicon typing (PBRT)), mobilisation genes (*mob*-encoded relaxases), and incompatibility (Inc) groups. Plasmid incompatibility is a phenomenon arising from the fact that two plasmids of an identical Inc group cannot stably co-exist in the same bacterial cell [13, 24, 25]. In silico detection of plasmid-incompatibility groups is possible using the freely available PlasmidFinder—https://cge.cbs.dtu.dk/services/PlasmidFinder [26]. Factors governing incompatibility include essential functional plasmid components undergoing competitive/antagonistic interactions, thus abrogating plasmid establishment. Plasmid partitioning apparatus, DNA replication control elements, and other important replication/establishment/maintenance functions can all be affected. It should be noted that a single plasmid can contain multiple replicons, e.g. IncFI/IncFIA/IncFII, and IncHI/IncHI2. Further, it is reported that incompatibility can arise between plasmids bearing multi-replicons when trying to occupy the same bacterial cell. IncFIA/IncHI incompatibility arises due to the presence of a *repFIA* replicon in IncHI plasmids. However IncHI2 and IncFIA plasmids can stably co-exist in the same cell together as the IncFIA/*repFIA* replicon is absent in IncHI2 plasmids [17]. Multi-replicon plasmids such as those observed in the IncF/IncH partnership can evolve following fusion of the two plasmid DNA elements [27]. It has also been shown that an IncHI plasmid pEQ1 (239,151 bp) can undergo recombination with a 48.5 kb IncXI plasmid generating a novel large virulence/resistance plasmid in *Escherichia coli* of equine origin. The IncXI plasmid encoded a progenitor β-lactamase gene $bla_{\mathrm{TEM\text{-}1}}$ and quinolone-resistance determinant *qnrS1*, and the resultant recombinant plasmid pEQ2 (287,616 bp) had successfully acquired these ARGs [28]. Generation of large hybrid plasmids via fusion events instantly pools the encoded antimicrobial resistance and virulence determinants of the two elements.

10.2.2 Plasmid Conjugation

The process of IncF plasmid transmission (bacterial conjugation) in Gram-negative bacteria has been intensively studied with attention being focused on *Enterobacteriaceae* including *E. coli* and *Salmonella* spp. [29]. Conjugation is the process by which self-transmissible plasmids are transferred from a donor-host to a recipient bacterium via apparatus such as Type IV secretion systems (T4SS) [30, 31]. Conjugation machinery constitutes a subfamily of T4SSs that are found to be related to secretion systems that deliver effector proteins to target prokaryotic/ eukaryotic cells. For both conjugative and effector secretion machinery, it is a necessity that intimate cell-to-cell contact is maintained throughout the substrate translocation process [32]. Conjugation systems are capable of driving the translocation of DNA (e.g. via rolling-circle replication/ssDNA transfer) into recipient cells [33]. Secreted effector systems constitute a virulence-associated mechanism wherein protein substrates (e.g. toxins and cytoskeletal modulators) can be injected into target cells [32]. The IncF plasmid T4SS constitutes a dynamic pilus structure capable of extending and retracting to guide cell-to-cell interactions and ultimately transfer plasmid DNA during conjugation. The genes associated with conjugal transfer transcriptional regulation, pilus-assembly, molecular chaperones, surface-entry/exclusion, and mating-pair stabilisation are encoded in the *tra* regions of IncF-like plasmids [29, 34, 35]. The ~34-Kb *tra* operon has been the focus of many studies investigating bacterial conjugation due to the association with the spread of AMR. A range of proteins associated with the formation and function of the T4SS pilus complex are produced during conjugation. Fertility inhibition is the process which controls expression of the *tra* operon. Transcriptional regulation of *traJ* (plasmid transcription factor) is achieved via the *finOP* two-component system. FinP is an antisense non-coding RNA that interacts with transcribed *traJ* mRNA with the overall effect of *tra* gene repression [29]. FinO protein is recognised as stabilising the *finP*/*traJ* mRNA complex; however, IS3-mediated insertional inactivation of *finO* can derepress control of the *tra* operon resulting in high-frequency transfer phenotypes [36, 37].

Expression of the *tra* operon leads to the production of the T4SS pilus complex of IncF plasmids, which is also a target for F-specific phages [38]. The major structural proteins of the T4SS serve to form three major substructures and contribute towards three important processes of the complex. Expression of pilus is required to successfully mediate cell-to-cell contact/adhesion via interactions with proteins such as the multi-subunit pilin protein TraA [39]. Transportation of plasmid DNA across the Gram-negative cell membrane into a recipient bacterium requires the formation of a pore channel. Pore formation following the assembly of proteins

TraB/TraK creates a structure observed to span the inner membrane, periplasm, and outer membrane. Pilus extension proceeds through the membrane-spanning pore and the internal tubular structure serves as a conduit for the secretion of DNA substrates. Translocation of single-stranded DNA (ssDNA) substrates generated through rolling-circle replication requires attachment of a type 4 coupling protein (T4CP) to *oriT*-nicked DNA substrate. A large protein complex (the relaxosome) is involved in the generation of the ssDNA molecule via relaxase-mediated substrate nicking, and the ssDNA molecule is transferred to the recipient cell [39–41]. Stabilisation of the mating-pair complex requires the activity of two proteins encoded by *traN*/*traG*. A stable association between the plasmid donor and recipient bacterial cell is mediated via tight-junction formation arising through interactions with the outer membrane protein TraN, inner membrane protein TraG, and the pilus complex [34, 35].

10.2.3 Plasmid Sequencing and ARG Identification/Annotation

Whole plasmid sequencing provides significantly more power towards establishing the true genetic relationship between different plasmid types/groups [25]. Plasmids have highly mosaic genetic structures that are under constant selective pressure to adapt and evolve. The application and cost-effective use of next-generation short-read sequencing technologies such as Illumina MiSeq has resulted in a huge increase in the number of microbial chromosomal/plasmid DNA sequences present in nucleotide databases such as NCBI. The DNA sequences obtained from short-read assemblies can often be quite fragmented in terms of multiple contigs [42]. Genetic recombination, repetitive regions (e.g. IS elements), fusion events, and large-scale deletions can all be identified from short-read assemblies. However, the true location and orientation of multiple contigs can be complex issues to resolve. Improved assembly software such as Plasmid SPAdes [43] and ReScaf allow for greater confidence in short-read plasmid assemblies—although gaps still exist and resolution may require multiple rounds of PCR. In recent years, long-read DNA sequencing via the PacBio platform is allowing previously unseen resolution of closed-circular plasmid molecules. In-depth inferences can be made as to the overall structure at the nucleotide level, corrections can be made to preexisting plasmid sequences generated from short-read assemblies, and stronger phylogenetic relationships between plasmids can be observed using complete closed-circular sequences obtained with long-read technologies. Over time, the costs associated with long-read technologies will become more comparable to those of short-reads. The positive effect on our ability to obtain highly accurate plasmid sequences will prove invaluable to our understanding of the flow of genetic information in the environment.

Multiple programs and databases are available for the identification and annotation of ARGs present in plasmid, chromosomal, and metagenomic sequence reads and assemblies. Simple Blast-based programs can rapidly search and annotate nucleotide fasta files, identify functional protein domains, and generate user-friendly outputs for examination. For the identification of ARGs from DNA sequences, ResFinder (https://cge.cbs.dtu.dk/services/ResFinder) and PATRIC (https://www.patricbrc.org/) can prove to be useful tools [44, 45]. ResFinder is capable of identifying acquired ARGs covering all major classes of antibiotics. Identification of chromosomal point mutations in genes associated with the development of antibiotic resistance in a number of pathogens including *Campylobacter* spp., *E. coli*, *Salmonella* spp., *Neisseria gonorrhoeae*, and *Mycobacterium tuberculosis* is possible [45]. The BacMet database (http://bacmet.biomedicine.gu.se) offers rapid identification of genes and proteins associated with microbial resistance to metals and disinfectant/biocides [46]. Annotation of shotgun metagenomic DNA sequences is possible using the popular freely available MG-RAST (https://www.mg-rast.org/) [47]. Resistance databases such as Comprehensive Antibiotic Resistance Database—CARD (https://card.mcmaster.ca/)—are powerful tools for identifying intrinsic, acquired, and mutational antibiotic resistances in microbial genomic datasets [48]. It is possible to identify most ARGs using general annotation programs such as PROKKA [49]; however, manual curation of predicted genes is recommended as some ARGs can be overlooked during the initial annotation.

10.2.4 Extended-Spectrum β-Lactamases

From a human perspective, the emergence and dissemination of antimicrobial resistance (AMR) in the clinic, agriculture, communities, and beyond poses significant threats to modern day civilisation. MDR extended-spectrum β-lactamase-producing (MDR-ESBL) Gram-negative *Enterobacteriaceae* encoding resistance towards several classes of antibiotics including third-/fourth-generation cephalosporins are of global concern [10]. Human and animal pathogens including MDR-ESBL *Salmonella* spp., *Klebsiella* spp., and *E. coli* therefore constitute considerable public and veterinary health risks [50–54]. MDR-ESBL pathogens can readily be isolated from people and associated environments (homes, hospitals, companion animals, etc.), livestock (farms, manures, slurries, etc.), and retail foods (dairy, meats, fruits, and vegetables). Of critical concern are emerging carbapenem-resistant ESBL-producing *Enterobacteriaceae* encoding the gene $bla_{\mathrm{NDM-1}}$. Production of the enzyme carbapenemase encoded by $bla_{\mathrm{NDM-1}}$ provides resistance against carbapenems essential for the treatment of MDR Gram-negative bacilli. Surveillance of ESBL-producing *Enterobacteriaceae* is therefore essential if prevention, containment, and treatment strategies are to be developed to control AMR [55, 56]. Abrogating the risks associated with AMR can only be achieved by enhancing knowledge of the varied factors influencing resistance in the environmental resistome. It is important to remember that environmental reservoirs of ARGs are observed in non-pathogenic bacteria—functional metagenomics studies can play an important role in studying antimicrobial resistance [57].

10.2.5 Class I Integrons

Integrons represent an environmental pool of ARGs that can be found in areas influenced by human activity. The pool is continually enriched with functional resistance determinants that reflect human use of antibiotics, metals, and compounds such as disinfectants and detergents [58]. Mobile gene cassettes encoding ARGs can be acquired by integrons through the activity of integrase (*intI*—a site-specific tyrosine recombinase) and subsequently expressed from integron-specific promoters (P_C/P_{ANT} and P2). DNA/protein sequence analysis of integron-associated integrase genes highlights several distinct classes, e.g. classes I, II, and III [59]. Class I integrons are recognised as being a major contributing factor towards the development and persistence of AMR in Gram-negative *Enterobacteriaceae*. Dispersal of class I integrons throughout microbial populations in human-associated environments including agriculture/food production, sewage/waterways/river catchments, soils, and the clinic is well-documented. The highly diverse and seemingly endless array of genetic configurations found in class I integrons can be attributed to >130 identified gene cassettes [60]. A recent study proposed an elegant mechanism for the origin and subsequent global dissemination of class I integrons. A series of nested transposons were found embedded in an IncP plasmid-associated class I integron harboured by an environmental *Enterobacter cloacae* isolate. The authors hypothesise that the progenitor clinical class I integron was mobilised from a β-proteobacterial chromosome as part of a Tn*5090* transposon which subsequently generated a Tn*402* transposable element. The newly formed plasmid-integrated Tn*402*-element gained the ability to mobilise between microbial populations via plasmid conjugation. Transposition of Tn*402* into a mercury resistance (*mer* operon) Tn*501*-like transposon formed the hallmark Tn*21* transposon of the Tn*3* family [61]. Pre-antibiotic era Tn*21*-like mercury resistances have been found not to contain the integron element [62, 63]. From an initially non-mobilisable genetic element, this chain of events allowed the evolution of one of the most pivotal antimicrobial resistance determinants within the microbial gene pool. The clinical class I integron was therefore well equipped to disseminate throughout globally distributed microbial populations.

Prior to the introduction of antibiotics as antimicrobial agents, the selective pressures driving the evolution of class I integrons are likely to have been disinfectants, and other small molecules with antibacterial activity. As antibiotic usage became common place in humans, agriculture, and horticulture, the necessary components of the class I integron were already in place allowing gene cassette expansion of novel determinants associated with antibiotic resistance. The first integron isolated in Japan in the mid-1950s was carried on the mercury resistance transposon Tn*21* and carried quaternary ammonium ion, sulphonamide, and streptomycin resistance [64]. Due to its widespread environmental distribution, the class I integron is considered to be a useful marker for anthropogenic pollution. Following the global dissemination of class I integrons, it is now well documented that the element is present in both pathogenic and non-pathogenic bacteria. Coupled with our knowledge of anthropogenically derived selective pressures such as the profligate use of

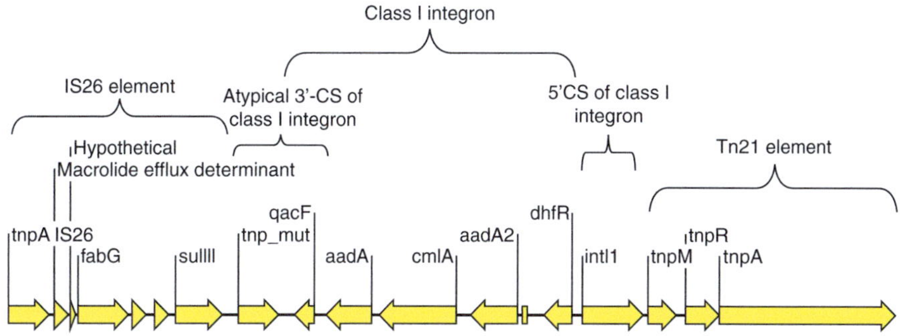

Fig. 10.1 Atypical class I integron of *Salmonella* Typhimurium U288 (pSTU288–1)—adapted from Hooton et al. [16]. An IS26-mediated transposition event has introduced *sulIII* into the 3′-end of the integron disrupting the *qac/sulI* fusion of typical class I integrons

antibiotics, metals, and disinfectants, utilising the class I integron (and the Tn*21*-family mercury resistance transposon that frequently carry the integrons) as a representation of environmental pollution will prove important in our attempts to understand AMR [58, 65].

The capture, integration, and excision of gene cassettes by class I integrons is a well-understood biological process. IntI1 facilitates the integration of gene cassettes via RecA-independent site-specific recombination between integron *attI* and gene cassette *attC* [58, 61]. Transcription of integrated ARGs is driven from P_C allowing expression of the encoded functional determinants [66]. Gene cassettes located proximal to *intI1* are observed to have higher levels of expression when compared to cassettes positioned distal to *intI1*. Over time, class I integrons can evolve through the loss or capture of specific ARGs, expansion of the number of integrated cassettes, or via disruption of gene synteny following transposition events [58]. The 3′-conserved segment (3′-CS) of class I integrons is recognisable by the presence of a sulphonamide resistance gene (*sulI*) and quaternary ammonium compounds (e.g. *qacEΔ1*). Interruption of *sul/qac* following IS26 transposition (see Fig. 10.1) such as observed in the large virulence plasmid of *S.* Typhimurium U288 (pSTU288-1-148,711 bp) creates an atypical 3′-CS [16, 66]. It has recently been reported that class I integrons containing atypical 3′-CS are associated with higher degrees of resistance when compared to typical 3′-CS [67].

10.2.6 Microbial Metal Resistance

The presence of specific gene operons capable of conferring resistance towards metal compounds is well documented in a wide range of bacterial species. Many natural and human-influenced environments contain levels of metals that require specific and effective mechanisms for microorganisms to survive (and sometimes prosper). Microbial-encoded resistance to metals can be conferred by diverse gene collections such as *cus/pco* (copper resistance), *czc* system (cobalt/zinc/cadmium

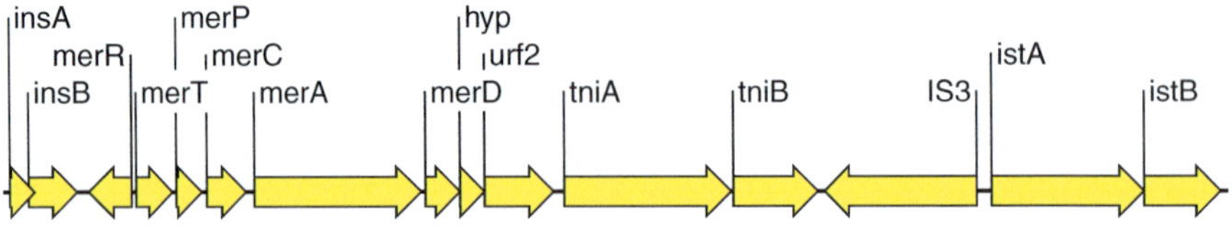

Fig. 10.2 *mer* operon of multidrug-resistant *S.* Typhimurium T000240 (DT12)—accession no. AP011957 [69]

resistance), *sil* operon (silver resistance), *mer* operon (mercury resistance), *ter* operon (tellurium resistance), and *ars* operon (arsenic resistance). Mobility of metal resistance determinants can be influenced via the activity of transposable elements. A striking example of the mobility of metal resistances are the mercury resistance (*mer*) operons found in Gram-negative bacteria, of which Tn*21* and Tn*501* are exemplars [64]. The *mer* operon is often linked to a Tn*21* element found flanking the 5′-CS of class I integron arrays. A striking example of metal resistance mobility is observed in the *sil*/*pco* region of *S.* Typhimurium metal resistance plasmid pMG101. The *sil* region is flanked by the Tn7 transposable element (Tns*ABCDE*) which provides mobility of the *pco*/*sil* operons. Interestingly, when pMG101 is conjugated into *E. coli* J53, the Tn7/*pco*/*sil* unit is observed to transpose into the chromosome of the recipient cell [68].

Mercury resistance is associated with genes encoded within the *mer* operon. The *mer* operon of multidrug-resistant *S.* Typhimurium T000240 (DT12) is flanked either side by IS1 transposable elements (*insA*/*insB*), transposon-associated genes (*tniA*/t*niB*), and an IS3 element (*istA*/*istB*). Mobility of the element is actually driven by its physical location as it is present along with a class I integron, ARGs, and virulence factors that form part of a larger mobile genomic island (SGI-DT12) discussed in detail below (also see Fig. 10.2). The IS1-elements flanking the *mer* operon precede the terminal regions of the integrated genomic island and may constitute an important factor in the spread of mobile genomic islands [69].

10.2.7 Co-selection of Antibiotic and Metal Resistance

Many metal resistance determinants are located on plasmids that also encode resistance to multiple antibiotics. Co-selection/co-occurrence and maintenance of AMR is particularly worrying due to the fact that even following removal of a selective pressure such as an antibiotic, resistance may be maintained due to co-localisation with genes that are still undergoing selection [46, 70]. The presence of copper and zinc in many human-dominated environments such as agricultural livestock production has allowed the proliferation of co-selected genetic elements in both Gram-positive and Gram-negative bacteria. It is reported that the use of various metals in food production environments can lead to proliferation and persistence of a range of pathogenic microorganisms. The use of copper in the pig-rearing process is linked to the development of MDR *Salmonella*, zinc in feedstuffs can select for methicillin-resistant *Staphylococcus aureus* [71], and zinc was observed to select for co-resistance to the antibiotics sulfamethoxazole and ciprofloxacin in *E. coli* [72]. The addition of metals such as copper and zinc to pig feed at concentrations well above normal physiological requirements (Zn 2–3000 ppm and Cu 125–250 ppm) results in improved growth post-weaning and also reduces the incidence of scour [73]. With such practices commonplace in agriculture, co-selection for simultaneous resistance towards numerous metal and antimicrobial compounds will undoubtedly persist. Even in the absence of selection, it is likely that such resistances will persist in the environmental resistome indefinitely.

10.2.8 Mobile Genomic Islands

Genomic islands (GIs) are large DNA elements that have the potential to excise from the bacterial chromosome and move via secondary interactions with conjugative plasmids. This process is exemplified in *Enterobacteriaceae* that can mobilise genomic islands in conjunction with IncA/C plasmids [74]. The IncA/C group constitutes a diverse collection of large multidrug-resistant, conjugative, broad-host range plasmids disseminated throughout clinical and environmental microbial populations [75]. Studies involving the MDR-conferring 43 kb *Salmonella* genomic island-1 (SGI-1) provide a fascinating insight into factors driving the dissemination of this integrative mobile element (IME). Resistances encoded on SGI-1 include the class I integron-borne (IN104) cassettes conferring resistance to ampicillin, chloramphenicol, streptomycin, sulphonamides, and tetracycline. SGI-1 has not only successfully spread through a number of *S. enterica* serovars, but it has also been reported in clinical *Proteus mirabilis* isolates. Chromosomal insertion of SGI-1 occurs in a site-specific manner at an 18-bp site situated at the 3′-end of *trmE* [74, 75]. The exact mechanism of mobilisation involves *in trans* interactions via a conjugative IncA/C following chromosomal excision of SGI-1. Interestingly, a mobile resistance-associated island found in *Vibrio cholerae* (MGI*Vchol*Hai6) is also mobilised via IncA/C plasmids, integrates in the *trmE* loci, and is associated with contributing towards the multidrug resistance phenotype of the host [76]. The adoption of whole genome sequencing has allowed for the identification of a highly diverse collection of IMEs in bacterial chromosomes. Originally identified in a case of human gastroenteritis in Japan, in 2000, the 82-kb genomic island SGI-DT12 of fluoroquinolone-resistant *S.* Typhimurium T000240 (DT12) provides an interesting example of pathogen evolution [69]. Encoded within SGI-DT12 are a range of genes providing virulence-associated functions such as siderophores (aerobactin IucABCD system) and high-affinity Fe^{2+}-binding proteins. Further determinants providing protection against a range of antibiotics including the transposon-linked tetracycline resistance genes and chloramphenicol acetyltransferase gene are also present. SGI-DT12 further encodes a class I integron structure flanked by a Tn*3*/Tn*21* transposable element and the classical mercury resistance operon (*mer*). The class I integron is observed to contain a β-lactamase-encoding $bla_{OXA\text{-}30}$ gene cassette that confers resistance to ampicillin and cefepime. SGI-DT12 was identified as having

IS-1 family transposons at both ends of the chromosomal insertion site. The presence of such elements indicates the likely mechanism of HGT that drove the integration of SGI-DT12 into the *S.* Typhimurium T000240 chromosome [69]. Diverse resistance phenotypes are capable of being acquired following IME acquisition and such elements are therefore major contributors towards bacterial evolution. Freely available online programs such as IslandViewer 4 can be used to interrogate genomic DNA sequences for the presence of genomic islands [77].

10.3 Bacteriophages

As obligate intracellular parasites of bacteria, bacteriophages (phages) have exclusive access to their microbial hosts' intracellular environment. As part of their natural lifecycles, phages can either integrate into the bacterial host chromosome during the process of lysogeny (temperate pathway) or by default actively replicate inside the bacterium thereby increasing virion copy number prior to lysis of the host cell (lytic pathway). Phages and bacteria have co-existed and interacted over millennia in countless predator/prey or host/parasite interactions. This shared pathway driven by lytic phage predation, temperate phage integration, and bacterial resistance mechanisms has resulted in a co-evolutionary arms race. Phages are therefore recognised as playing a pivotal role in the past and continued evolution of prokaryotic genomes [78]. A striking example of the role phages have performed in bacterial evolution is the emergence of Shiga-toxin (Stx)-producing *E. coli* (STEC). STEC (and STEC-like pathogens) are well documented as having emerged following the integration of multiple temperate phages encoding the *stx* genes necessary for Stx production—a process known as lysogenic conversion. *E. coli* O157:H7 is recognised as being a significant pathogen of humans capable of causing numerous disease sequelae including haemolytic uraemic syndrome, haemorrhagic colitis, and thrombotic thrombocytopenic purpura. Further outbreaks of Stx-associated disease arising from human infections with non-O157:H7 serotypes are a direct result of Stx phage disseminating throughout distinct *E. coli* lineages and other *Enterobacteriaceae*, e.g. *E. cloacae* [79]. The toxicity of Shiga toxin towards eukaryotic cells is well-documented and understood. It is believed that a potential beneficial function for *E. coli* carrying Stx-producing phages is protection afforded towards the avoidance of predation by grazing single-cell eukaryotes in the bovine intestine where *E. coli* O157:H7 is commonly found [80, 81].

10.3.1 Bacteriophages and HGT

The role of phages in the transfer of AMR genes in the environment still remains unclear. Undoubtedly, phages will have a role in the transfer of AMR genes, but how significant this role is still remains to be quantified. Phages in general can mediate the transfer of genes through four mechanisms: generalised, specialised, auto-transduction, and phage-mediated plasmid transformation.

Transduction is the transfer of DNA from one bacterial host to another via a phage particle, and this process is further divided into generalised and specialised transduction. Generalised transduction can occur in both lytic and temperate phages, whereby a mispackaging event causes host DNA to be incorporated within the phage virion. Upon lysis of the host, the transducing particle is released, and therefore capable of infecting another bacterial cell to transfer the host DNA incorporated within the virion particle. As generalised transduction is a result of the random mispackaging of host DNA, it provides a mechanism to transfer any part of the host genetic material to another bacterial host.

Specialised transduction was first discovered in the 1950s [82] and can only be carried out by temperate prophages. Specialised transduction occurs as a result of the inexact excision of a prophage from the host chromosome, resulting in host genetic material being incorporated into a virion particle along with phage DNA. Upon release of phage particles, the specialised transducing particles are capable of infecting and replicating in their next host, unlike generalised transducing particles which merely transfer host genetic material. Unlike generalised transduction, only genes that are localised next to prophage insertion sites can be transferred.

The third mechanism of phage-mediated transduction has been termed autotransduction and has so far only been demonstrated in *S. aureus* [83]. In autotransduction, a lysogenic population of bacteria releases phage particles that are capable of infecting a susceptible population of bacterial cells. These infected cells subsequently produce more phage virions and occasionally generalised transducing particles. The original lysogenic population is immune to the resulting phage particles owing to prophage homo-immunity, a system that prevents infection by the same phage type. However, the transducing particles are able to transfer DNA into the lysogenic population of bacteria. In the case of *S. aureus*, the process of autotransduction provided a mechanism for cells to quickly acquire resistance to antibiotics [83].

Recently, a fourth method of phage-mediated gene transfer has been described, which is distinct from transduction and is described as phage-mediated plasmid transformation. Upon infection of the host with what has become known as a "superspreader", an increased amount of plasmid DNA is released then transferred via the process of transformation into other cells. The underlying mechanisms that allow this are not fully understood but are thought to involve absence of endonucleases in the "superspreader" phages, which consequently do not degrade the plasmid DNA [84].

10.3.2 Bacteriophage Genomes and ARGs

Despite the mechanisms of generalised and specialised transduction long being known, how significant phage are in the transfer of ARGs in the environment is still unknown and largely understudied. A recent analysis of 1181 phage genomes revealed that only three phages carried an antibiotic resistance gene within their genomes [85], suggesting that phage rarely encode ARGs. Whilst this may be correct, further work is needed to determine the role of phages in the transfer of ARGs as well as the carriage of ARGs. The study of Enault and colleagues that examined complete phage genomes was based on the relatively small dataset that was available at the time that was dominated by lytic phages so may be a sampling bias and not representative of all phages. Our analysis of 8263 phage genomes using conservative parameters identified only 24 phage that carry ARGs (~0.23%), suggesting as Enault and Kleinheinz have that ARGs encoded in phage genomes are rare, despite a near sevenfold increase in the database (A. Millard—Unpublished data).

The rare occurrence of ARGs in lytic phage genomes is possibly not surprising as currently there is no clear selection pressure that can explain why they would be maintained within lytic phage genomes, as the very nature of lytic phages means that their host will die shortly after infection. Despite this, there are rare occurrences of antimicrobial resistance genes in lytic phages, such as a tellurite resistance gene found in a phage infecting *E. coli* isolated from pig faeces [86]. The presence of this gene suggests that it could be transferred to other bacteria only if the lytic infection is unsuccessful, thus providing an opportunity for recombination with the host bacterial DNA. For temperate prophages, the presence of ARGs can easily be seen to be advantageous, as the stable association of the prophage within the host chromosome for many generations provides the opportunity for carriage of ARGs to be selected for. Analysis of 1442 prophages found that ~1% carried one or more ARGs, in comparison to ~0.2% of 1642 lytic phages, suggesting that prophages are far more likely to carry ARGs than lytic phages [87].

Whilst the carriage of ARGs within the genome of a phage provides direct means to allow the transfer of ARGs, it is likely that the process of transduction is of more importance in the dissemination of ARGs by phages. However, there are relatively few reports of the direct assessment of transduction. The large-scale isolation of phages and subsequent testing of their ability to transduce ARGs has demonstrated that phage can transduce ARGs. The isolation of 243 phages from chickens reveals that ~25% of phage could transduce single/multiple ARGs, with considerable variation in the resistance genes that could be transduced, with kanamycin transduction occurring ~10 times more often than tetracycline [88]. Efforts have been made to directly study transduction in the environment using indigenous populations of marine bacteria. In a study using marine bacteria from Tampa Bay, it was found that the frequency of transduction was 1.5×10^{-8}–3.7×10^{-8} transductants/PFU [89]. This data was then incorporated into a model along with estimates of phage and bacterial abundance, and volume in Tampa Bay to estimate up to 1.3×10^{14} transductants per year occurred in the Tampa Bay Estuary [89]. Whilst there are limitations to the model used, it clearly demonstrates that phage could have an important impact on the movement of genes in the environment.

The role of the recently described phage-mediated plasmid transformation to transfer ARGs in the environment has also been investigated. A "superspreader" and "normal" phage were spiked into identical natural populations of soil bacteria, which resulted in >1000-fold increase in resistant colonies with addition of the "superspreader" compared to the "normal" phage [84]. In addition, it was found that

E. coli-specific superspreader phage SUSP2 mediated the transfer of plasmid DNA from the Gram-negative *E. coli* to a Gram-positive *Bacillus* spp. [84]. Whilst it is known that plasmids can replicate in both *E. coli* and *Bacillus* spp. [90], it adds further complexity to our understanding of the role of phages in the transfer of ARGs.

10.3.3 Environmental Bacteriophage Genomics

In addition to culture-based techniques used to assess the ability of phages to mediate the transfer of ARGs, the molecular methods of qPCR and metagenomics have been used. Metagenomics allows for a greater understanding of the total viral community and any ARGs within it to be assessed, without the need for isolation of phages. Since its first use in 2002 [91], there have been some important developments in viral metagenomics. The early years of viral metagenomics required a number of innovative steps for the sequencing of viral metagenomes compared to bacterial metagenomics [92]. These steps were largely to overcome the toxic nature of cloning bacteriophage genes into *E. coli* prior to Sanger sequencing. With the advent of next-generation sequencing, cloning of viral DNA became redundant as sequencing technology developed. With the development of second-generation sequencing, a previously unimaginable depth of coverage for viral genomes has become possible. This has allowed viral metagenomics to move from answering what is present towards a quantitative measurement of how much is present, with optimisation of extraction of the viral fraction and library preparation for sequencing [93].

With a move towards quantitative viral metagenomics, it has been possible to determine the abundance of ARGs within viral metagenomes from a variety of different environments [94, 95]. What is not in doubt is that viral metagenomes contain ARGs [85, 94, 95]; however, what is still to be determined is how important this is as a reservoir and how it relates to the abundance of ARGs in the bacterial fraction. A virome can contain bacterial genes for three reasons: (1) phage can incorporate them into their genomes, (2) phage are capable of transduction, therefore will package host genes within their capsids, and (3) insufficient removal of host DNA during purification. To understand the role in the transfer of ARGs, it is essential to minimise the effect of contaminating host DNA. There has been much work to optimise the production [96] and analysis of viral metagenomes [97, 98] to minimise and identify contaminating host DNA. However, recent work has identified that contaminating bacterial DNA and low-similarity matches have overestimated the abundance of ARGs in previously reported viromes [85]. Despite the possible pitfalls associated with metagenomics, it is possible to gather useful information on the abundance of ARGs within viromes if samples that are high in contaminating host DNA are excluded, and stringent values for predicting ARGs are used. A study that reanalysed 33 viromes found ARGs in diverse environments including freshwater, marine, sewage, swine, and human faeces [94, 95], with ARGs least abundant in human-associated viromes, and most abundant in viromes associated with pig faeces and freshwater habitats. In particular, genes conferring resistance to β-lactam antibiotics were the most abundant class of ARG found in any virome [94, 95]. Thus, viral metagenomics clearly demonstrates that phages can harbour ARGs in a range of environments. The use of viral metagenomics alone, without metagenomics of the bacterial fraction from the sample, does not give any indication of the direction of the flow of genes, from phage to host or host to phage.

The use of qPCR that targets a small number of known genes provides the opportunity to do this, allowing accurate quantification of ARGs in the bacterial and phage fraction, at a fraction of the cost of metagenomics. The analysis of water samples collected downstream of wastewater treatment plant revealed the differential abundance of ARGs in bacterial chromosomal DNA, phages, and plasmids between upstream and downstream sites [94]. Whilst ARG abundance was higher in the bacterial fraction, the differences in abundance between upstream and downstream sites were often greatest for the phage or plasmid fraction, suggesting phages and plasmids are important vectors in the dissemination of ARGs in river water [94, 95].

A higher abundance of ARGs in the bacterial fraction compared to the phage fraction has also been demonstrated in other environments. In a soil microcosm study that focused on soil contaminated with tetracycline and monitored the abundance of six tetracycline resistance genes, it was found that the bacterial fraction contained genes at two orders of magnitude higher than the phage fraction [99]. There was also a correlation in the abundance of each gene between the bacterial and phage fractions, with decreasing abundances in both fractions as the tetracycline concentration decreased over time [99].

In a study of the River Llobregat, the abundance of six ARGs was studied in both the bacterial and viral fraction of sediment and river water over an annual cycle. In common with other studies, ARGs were higher in the bacterial fraction in both sediment and water samples [100]. A clear seasonal variation was observed in the abundance of ARGs in the phage fraction, although the reasons behind this are unknown. Furthermore, it was found that ARGs in both phages and bacteria are persistent in sediment for over 20 days, suggesting that both phage and bacteria are a reservoir of resistance genes in the environment [100]. In another study of the Funan river in China using qPCR to target 15 ARGs, it was found that sites located near to a wastewater treatment facility and a hospital contained the highest abundance of ARGs in the phage fraction [101]. Thus, providing further evidence that phages are a reservoir of ARGs, and the increased abundance is likely a result of anthropogenic inputs. A recent study of agricultural soil and fresh cut vegetables (lettuce, cucumber, and spinach) provides further evidence that phages are an important reservoir of ARGs [102]. Using qPCR to detect 10 ARGs, it was found that there were variable levels of ARGs in the phage fraction extracted from soil and vegetables. Critically within this study, it was shown the phage fraction containing ARGs was able to propagate in *E. coli* [102]. Thus, not only does the phage fraction contain ARGs, they are not merely transducing particles but are capable of active replication and further spread of ARGs.

Through a combination of culture-based studies and molecular methods, it is becoming clear that phages have an important role in both the transfer of ARGs within the environment and acting as a reservoir of ARGs. However, there is still much to be understood as to how much different methods of phage-mediated gene transfer contribute to the overall transfer of ARGs.

10.4 Genomics/Metagenomics

Deep-sequencing an environmental metagenome provides an indication as to microbial population structures and genes present at a given moment in time [1, 103]. Sequence-based read mapping is extremely useful but it is limited by the fact that genes of interest cannot necessarily be directly linked to taxonomic distribution. Identifying AMR genes in metagenomic datasets also provides comprehensive information as to the range of resistances present. However, the ownership of AMR genes in environmental samples often remains elusive and currently cannot be investigated using deep-sequencing strategies alone. To overcome this, emulsion, paired-isolation, and concatenation PCR (epicPCR) can be used to link genes of interest with 16S rRNA markers (see Fig. 10.3). A suspension of bacterial cells in oil emulsion/acrylamide is polymerised allowing capture of single cells in a solid matrix. In situ lysis of bacterial cells in a porous matrix provides a sufficient platform for downstream PCR steps. Enzymes and reagents can readily diffuse into the bead allowing access to the embedded genomic and plasmid DNA. Combining 16S PCR with amplification of a target such as ARGs will provide insights into environmental distribution of antimicrobial resistance determinants. Adapting approaches such as epicPCR for the study of antimicrobial resistance will also allow for

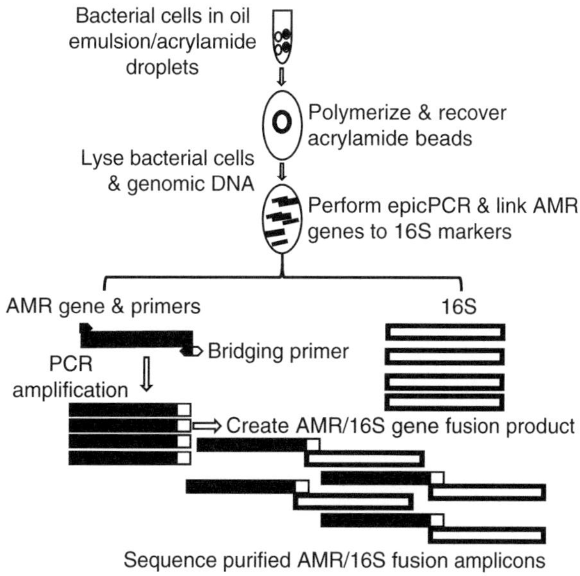

Fig. 10.3 Schematic overview of epicPCR technique (adapted from [104])

high-throughput screening of environmental samples. Combining metagenomic-based approaches with epicPCR also has the potential to identify previously unobservable AMR relationships. High-throughput sequencing of epicPCR libraries produces unique data by linking environmental genes with bacterial species-level identification [104]. Environmental metagenomes may further contain "hidden" information about currently uncharacterised or unidentified ARGs [105, 106]. Identification of novel environmental resistance genes remains an underinvestigated field of AMR research.

Functional metagenomic DNA libraries can be used to screen for the presence of known/unknown ARGs in the environment. Dependent on the insert size used during library preparation, single genes, operons, or large resistance clusters may be identified. Appropriate selection of resistant clones using clinically relevant antibiotics allows for the identification of functional AMR-associated genetic markers. High-throughput sequencing of DNA inserts obtained from resistant clones provides genetic data based on phenotypic resistances to the chosen panel of selective agents [105–107]. Data obtained from functional metagenomic studies has the potential to guide the design of epicPCR experiments [104]. Inverse PCR can also be integrated as a platform for high-throughput environmental studies. In combination with third-generation long-read sequencing technologies, inverse PCR can be used to study large amplicons generated using AMR-specific primers. PacBio sequencing of inverse PCR products generates high-quality data relating to ARGs and immediate sequence environments [108]. Establishing new roles for traditional molecular biology techniques in the context of high-throughput sequencing has provided novel avenues of investigation for AMR research [104–106, 108]. Knowledge of the prevalence and phylogenetic distribution of environmental variants of known/novel ARGs utilising metagenomic approaches such as epicPCR and inverse PCR will further our understanding of environmental AMR.

10.5 Transformation with Exogenous DNA

Transformation of competent microbes with exogenous sources of DNA constitutes another route for the flow of genetic information [109]. The full importance of transformation in the spread of AMR is not fully understood—most reports of transformation by exogenous AMR gene DNA involve Gram-positive bacteria such as *Streptococcus pneumoniae* and *N. gonorrhoeae* being transformed. Other bacterial strains have been reported to be naturally competent, e.g. *Bacillus* spp. and *Ralstonia* spp., and there have been reports that Gram-negative bacteria such as *Haemophilus influenza*e, *Haemophilus parainfluenzae*, and *Haemophilus suis*, *Pseudomonas* sp., and *E. coli* can be transformed by exogenous DNA under natural environmental conditions (reviewed in [9]). Natural competence in bacteria can also be dependent upon environmental factors—bacterial cells may need to be metabolically active, but have shifted to unbalanced growth [109, 110], presence of metal ions such as Ca^{2+}, and Mg^{2+} [110, 111], and that natural waters containing Ca^{2+} at levels as low as 1–2 mM are sufficient to induce transformation in *E. coli* at levels equivalent to those reported for other naturally competent bacteria [112].

In addition to the cellular conditions required for natural competence, transformation requires a number of other conditions to occur: The DNA can be damaged or fragmented, but can still transform bacterial cells, but needs to be in intimate contact with them, and once taken up the DNA need not be restricted by the host's own restriction/modification systems, and then must be integrated into cellular DNA that replicates, or carries all of the machinery that allows it to replicate autonomously [9, 113, 114]. DNA is known to be released by bacterial species such as *Pseudomonas*, *Bacillus*, and *Neisseria*, and this released extracellular DNA not only plays a role in transformation but may also play a role in DNA damage repair, and is known to be important in the formation of microbial biofilm communities [115]. There also appears to be a potential role in natural transformation for lytic bacteriophage, as "superspreaders" of AMR genes, by causing the release of bacterial plasmid DNA after cell lysis [84].

10.6 Mathematical Modelling of Spread of Resistance Through HGT

Mathematical models for spread of antimicrobial resistance can be useful in a number of ways: to identify those factors to which spread of resistance is most sensitive; to make predictions that can help inform policy; and to make predictions in the face of specific outbreaks [116].

Mathematical models can be built in a number of ways. The most common approach is to use ordinary differential equations [117–119]. These typically consider populations of sensitive and resistant bacteria, in a homogeneous environment, although they can be extended to more than one compartment [120]. Differential equation models are rapid to simulate and have a wide range of analytical techniques available. For example, sensitivity analyses have highlighted the importance of horizontal gene transfer to spread of resistance, for example in a slurry tank system [117] or between compartments, such as animals and humans [118].

Other approaches include the use of stochastic models that can include randomness [120, 121]. Stochastic effects have been shown to be important in the way that plasmids control the genes for their replicative and conjugative apparatus [122–124]. Individual-based models treat bacteria as single cells acting in a spatial domain [125]. Individual-based models can only model small populations of bacteria, but have the advantage of being more realistic. They also allow simulations of spatial processes, for example demonstrating that growth-dependent conjugation rates can explain how biofilms can resist plasmid invasion [126].

To illustrate how a differential equation model works, we describe a simple ordinary differential model for horizontal transfer of resistance. The model we show contains two populations: bacteria that are sensitive to antibiotic (S), and bacteria that are resistant to the antibiotic (R). This model can be used to describe bacteriostatic or bactericidal antibiotics, as well as resistance that can be transmitted horizontally (i.e. through plasmids) or only vertically (i.e. chromosomal).

The model equations are

$$\frac{dS}{dt} = r\left(1 - \frac{R+S}{K}\right)E_S S + \varepsilon r(1-\alpha)\left(1 - \frac{R+S}{K}\right)E_R R - \frac{\beta SR}{S+R} - \delta G_S S$$

$$\frac{dR}{dt} = r(1-\varepsilon)(1-\alpha)\left(1 - \frac{R+S}{K}\right)E_R R + \frac{\beta SR}{S+R} - \delta G_R R$$

These equations contain the following processes:

- Bacterial growth; this has maximal growth rate r and maximum population size K.
- Fitness cost α of carrying the antibiotic resistance genes.
- Impact of a bacteriostatic antibiotic on growth of sensitive bacteria (E_S) and resistant bacteria (E_R). Fully resistant bacteria would have $E_R = 1$. In the absence of bacteriostatic antibiotic, E_S would also be equal to 1.
- Loss of resistance due to plasmid segregation (in the horizontal gene transfer case) or simply mutation (in the vertical transfer case) with rate ε.
- Horizontal gene transfer at rate β. If we are considering only vertical transmission, then $\beta = 0$.
- Death of bacteria at rate δ. Previous models have not included death terms [117, 119], but they are essential for three reasons. First, bacteria do die, especially from phage or protozoal predation [127], so it is more realistic. Second, it is required for inclusion of bactericidal antibiotics. Third, and more important, it is impossible to implement the effect of bacteriostatic antibiotics correctly in a logistic growth model without death; without death, the logistic model will operate near to its carrying capacity, where bacteriostatic effects will be minimal. Alternatively, the model would be required to be run far from equilibrium, in order to observe bacteriostatic effects, which is also unrealistic. For these reasons, it is essential to include death terms into models for resistance.
- Impact of a bactericidal antibiotic on death of sensitive bacteria (G_S) or resistant bacteria (G_R), fully resistant bacteria would have $G_R = 1$. In the absence of bactericidal antibiotic, G_S would also be equal to 1.

This model could be used in many ways, for example to consider spread of resistance under exposure to either or both of a bacteriostatic or bactericidal antibiotic. It could be subjected to sensitivity analyses to determine factors to which it is most sensitive. Or it could form the basis of a more complex model that might include multiple compartments (whether patients or animals, locations, or spatial environments), many different antibiotics, different resistances, different gene transfer agents, or different bacterial species. Indeed, the complexity of antimicrobial resistance in real settings makes developing realistic models a considerable challenge [116].

10.7 Conclusions

The role of horizontal gene transfer in the dissemination of AMR genes amongst bacteria has been known for over 60 years, but some of the mechanisms, their importance, and impact on acquisition of resistance are still not quantified. Horizontal gene transfer arises from a multiplicity of events driven by mobile genetic elements—understanding their true influence will require the application of multidisciplinary approaches. Plasmids, phages, transposons, etc. all serve to drive the continued evolution of bacteria, some of which come to our attention due to their pathogenicity. The continued use of antibiotics, metals, and disinfectants/biocides will positively select for bacteria that are capable of acquiring resistance determinants and adapting to potentially toxic environments. Even if anthropogenic selection was withdrawn, it is highly likely resistance will be maintained throughout the prokaryotic kingdom. The use of microbiology culture methods, metagenomics, functional metagenomics, and molecular methods will prove essential for enhancing our understanding of AMR at the genetic and ecological level. Predicting the spread of resistance genes via mobile genetic elements is possible using mathematical models. Developing simple ordinary differential equations capable of describing horizontal transfer rates of resistance genes in microbial populations provides useful information that can be tested experimentally. Furthering our understanding of the drivers for acquisition and retention of resistance genes and the importance of DNA trafficking in the environment will be central to controlling the AMR problem.

Acknowledgement The authors acknowledge funding from the UK Natural Environment Research Council through grant NE/N019881/1.

References

1. Gillings MR, Paulsen IT, Tetu SG. Genomics and the evolution of antibiotic resistance. Ann N Y Acad Sci. 2017;1388:92–107.
2. Hooton SP, Timms AR, Rowsell J, Wilson R, Connerton IF. Salmonella Typhimurium-specific bacteriophage PhiSH19 and the origins of species specificity in the Vi01-like phage family. Virol J. 2011;8:498.
3. Soucy SM, Huang J, Gogarten JP. Horizontal gene transfer: building the web of life. Nat Rev Genet. 2015;16:472–82.
4. Bennett PM. Plasmid encoded antibiotic resistance: acquisition and transfer of antibiotic resistance genes in bacteria. Br J Pharmacol. 2008;153(Suppl 1):S347–57.
5. Hudson JA, Frewer LJ, Jones G, Brereton PA, Whittingham MJ, Stewart G. The agri-food chain and antimicrobial resistance: a review. Trends Food Sci Technol. 2017;69:131–47.
6. Day A, Ahn J, Fang X, Salmond GPC. Environmental bacteriophages of the emerging enterobacterial phytopathogen, Dickeya solani, show genomic conservation and capacity for horizontal gene transfer between their bacterial hosts. Front Microbiol. 2017;8:1654.
7. Matilla MA, Salmond GP. Bacteriophage varphiMAM1, a viunalikevirus, is a broad-host-range, high-efficiency generalized transducer that infects environmental and clinical isolates of the enterobacterial genera Serratia and Kluyvera. Appl Environ Microbiol. 2014;80:6446–57.

8. Matilla MA, Fang X, Salmond GP. Viunalikeviruses are environmentally common agents of horizontal gene transfer in pathogens and biocontrol bacteria. ISME J. 2014;8:2143–7.
9. von Wintersdorff CJ, Penders J, van Niekerk JM, et al. Dissemination of antimicrobial resistance in microbial ecosystems through horizontal gene transfer. Front Microbiol. 2016;7:173.
10. Wellington EM, Boxall AB, Cross P, et al. The role of the natural environment in the emergence of antibiotic resistance in gram-negative bacteria. Lancet Infect Dis. 2013;13:155–65.
11. Blair JM, Webber MA, Baylay AJ, Ogbolu DO, Piddock LJ. Molecular mechanisms of antibiotic resistance. Nat Rev Microbiol. 2015;13:42–51.
12. Carattoli A. Resistance plasmid families in Enterobacteriaceae. Antimicrob Agents Chemother. 2009;53:2227–38.
13. Carattoli A. Plasmids in gram negatives: molecular typing of resistance plasmids. Int J Med Microbiol. 2011;301:654–8.
14. Hulter N, Ilhan J, Wein T, Kadibalban AS, Hammerschmidt K, Dagan T. An evolutionary perspective on plasmid lifestyle modes. Curr Opin Microbiol. 2017;38:74–80.
15. Gullberg E, Albrecht LM, Karlsson C, Sandegren L, Andersson DI. Selection of a multidrug resistance plasmid by sublethal levels of antibiotics and heavy metals. MBio. 2014;5:e01918–4.
16. Hooton SP, Timms AR, Cummings NJ, Moreton J, Wilson R, Connerton IF. The complete plasmid sequences of Salmonella enterica serovar Typhimurium U288. Plasmid. 2014;76:32–9.
17. Rozwandowicz M, Brouwer MSM, Fischer J, et al. Plasmids carrying antimicrobial resistance genes in Enterobacteriaceae. J Antimicrob Chemother. 2018;73(5):1121–37.
18. Falgenhauer L, Waezsada SE, Gwozdzinski K, et al. Chromosomal locations of mcr-1 and bla CTX-M-15 in fluoroquinolone-resistant *Escherichia coli* ST410. Emerg Infect Dis. 2016;22:1689–91.
19. Campos JC, da Silva MJ, dos Santos PR, et al. Characterization of Tn3000, a transposon responsible for blaNDM-1 dissemination among enterobacteriaceae in Brazil, Nepal, Morocco, and India. Antimicrob Agents Chemother. 2015;59:7387–95.
20. Brolund A, Franzen O, Melefors O, Tegmark-Wisell K, Sandegren L. Plasmidome-analysis of ESBL-producing *Escherichia coli* using conventional typing and high-throughput sequencing. PLoS One. 2013;8:e65793.
21. Kav AB, Sasson G, Jami E, Doron-Faigenboim A, Benhar I, Mizrahi I. Insights into the bovine rumen plasmidome. Proc Natl Acad Sci U S A. 2012;109:5452–7.
22. Shi Y, Zhang H, Tian Z, Yang M, Zhang Y. Characteristics of ARG-carrying plasmidome in the cultivable microbial community from wastewater treatment system under high oxytetracycline concentration. Appl Microbiol Biotechnol. 2018;102:1847–58.
23. Dantas G, Sommer MO. Context matters - the complex interplay between resistome genotypes and resistance phenotypes. Curr Opin Microbiol. 2012;15:577–82.
24. Carattoli A. Plasmids and the spread of resistance. Int J Med Microbiol. 2013;303:298–304.
25. Orlek A, Stoesser N, Anjum MF, et al. Plasmid classification in an era of whole-genome sequencing: application in studies of antibiotic resistance epidemiology. Front Microbiol. 2017;8:182.
26. Carattoli A, Zankari E, Garcia-Fernandez A, et al. In silico detection and typing of plasmids using PlasmidFinder and plasmid multilocus sequence typing. Antimicrob Agents Chemother. 2014;58:3895–903.
27. Fang L, Li X, Li L, Li S, Liao X, Sun J, Liu Y. Co-spread of metal and antibiotic resistance within ST3-IncHI2 plasmids from *E. coli* isolates of food-producing animals. Sci Rep. 2016;6:25312.
28. Dolejska M, Villa L, Minoia M, Guardabassi L, Carattoli A. Complete sequences of IncHI1 plasmids carrying blaCTX-M-1 and qnrS1 in equine *Escherichia coli* provide new insights into plasmid evolution. J Antimicrob Chemother. 2014;69:2388–93.
29. Fernandez-Lopez R, de Toro M, Moncalian G, Garcillan-Barcia MP, de la Cruz F. Comparative genomics of the conjugation region of F-like plasmids: five shades of F. Front Mol Biosci. 2016;3:71.

30. Chandran V, Fronzes R, Duquerroy S, Cronin N, Navaza J, Waksman G. Structure of the outer membrane complex of a type IV secretion system. Nature. 2009;462:1011–5.
31. Fronzes R, Schafer E, Wang L, Saibil HR, Orlova EV, Waksman G. Structure of a type IV secretion system core complex. Science. 2009;323:266–8.
32. Christie PJ, Whitaker N, Gonzalez-Rivera C. Mechanism and structure of the bacterial type IV secretion systems. Biochim Biophys Acta. 2014;1843:1578–91.
33. Khan SA. Plasmid rolling-circle replication: highlights of two decades of research. Plasmid. 2005;53:126–36.
34. Anthony KG, Klimke WA, Manchak J, Frost LS. Comparison of proteins involved in pilus synthesis and mating pair stabilization from the related plasmids F and R100-1: insights into the mechanism of conjugation. J Bacteriol. 1999;181:5149–59.
35. Klimke WA, Frost LS. Genetic analysis of the role of the transfer gene, traN, of the F and R100-1 plasmids in mating pair stabilization during conjugation. J Bacteriol. 1998;180:4036–43.
36. Glover JNM, Chaulk SG, Edwards RA, Arthur D, Lu J, Frost LS. The FinO family of bacterial RNA chaperones. Plasmid. 2015;78:79–87.
37. Yoshioka Y, Ohtsubo H, Ohtsubo E. Repressor gene finO in plasmids R100 and F: constitutive transfer of plasmid F is caused by insertion of IS3 into F finO. J Bacteriol. 1987;169:619–23.
38. Gruber CJ, Lang S, Rajendra VK, Nuk M, Raffl S, Schildbach JF, Zechner EL. Conjugative DNA transfer is enhanced by plasmid R1 partitioning proteins. Front Mol Biosci. 2016;3:32.
39. Lawley TD, Klimke WA, Gubbins MJ, Frost LS. F factor conjugation is a true type IV secretion system. FEMS Microbiol Lett. 2003;224:1–15.
40. Lang S, Zechner EL. General requirements for protein secretion by the F-like conjugation system R1. Plasmid. 2012;67:128–38.
41. Lang S, Kirchberger PC, Gruber CJ, et al. An activation domain of plasmid R1 TraI protein delineates stages of gene transfer initiation. Mol Microbiol. 2011;82:1071–85.
42. Utturkar SM, Klingeman DM, Land ML, Schadt CW, Doktycz MJ, Pelletier DA, Brown SD. Evaluation and validation of de novo and hybrid assembly techniques to derive high-quality genome sequences. Bioinformatics. 2014;30:2709–16.
43. Antipov D, Hartwick N, Shen M, Raiko M, Lapidus A, Pevzner PA. plasmidSPAdes: assembling plasmids from whole genome sequencing data. Bioinformatics. 2016;32:3380–7.
44. Wattam AR, Davis JJ, Assaf R, et al. Improvements to PATRIC, the all-bacterial Bioinformatics Database and Analysis Resource Center. Nucleic Acids Res. 2017;45:D535–42.
45. Zankari E, Hasman H, Cosentino S, et al. Identification of acquired antimicrobial resistance genes. J Antimicrob Chemother. 2012;67:2640–4.
46. Pal C, Asiani K, Arya S, Rensing C, Stekel DJ, Larsson DGJ, Hobman JL. Metal resistance and its association with antibiotic resistance. Adv Microb Physiol. 2017;70:261–313.
47. Meyer F, Bagchi S, Chaterji S, et al. (2017) MG-RAST version 4-lessons learned from a decade of low-budget ultra-high-throughput metagenome analysis. Brief Bioinform. https://doi.org/10.1093/bib/bbx105. [Epub ahead of print]
48. Jia B, Raphenya AR, Alcock B, et al. CARD 2017: expansion and model-centric curation of the comprehensive antibiotic resistance database. Nucleic Acids Res. 2017;45:D566–73.
49. Seemann T. Prokka: rapid prokaryotic genome annotation. Bioinformatics. 2014;30:2068–9.
50. Bae D, Cheng CM, Khan AA. Characterization of extended-spectrum beta-lactamase (ESBL) producing non-typhoidal Salmonella (NTS) from imported food products. Int J Food Microbiol. 2015;214:12–7.
51. Croxen MA, Law RJ, Scholz R, Keeney KM, Wlodarska M, Finlay BB. Recent advances in understanding enteric pathogenic *Escherichia coli*. Clin Microbiol Rev. 2013;26:822–80.

52. Hooton SP, Timms AR, Moreton J, Wilson R, Connerton IF. Complete genome sequence of Salmonella enterica serovar Typhimurium U288. Genome Announc. 2013;1:e00467–13.
53. Ibrahim DR, Dodd CE, Stekel DJ, Ramsden SJ, Hobman JL. Multidrug resistant, extended spectrum beta-lactamase (ESBL)-producing *Escherichia coli* isolated from a dairy farm. FEMS Microbiol Ecol. 2016;92:fiw013.
54. Navon-Venezia S, Kondratyeva K, Carattoli A. Klebsiella pneumoniae: a major worldwide source and shuttle for antibiotic resistance. FEMS Microbiol Rev. 2017;41:252–75.
55. Bradford PA, Kazmierczak KM, Biedenbach DJ, Wise MG, Hackel M, Sahm DF. Correlation of beta-lactamase production and colistin resistance among enterobacteriaceae isolates from a global surveillance program. Antimicrob Agents Chemother. 2016;60:1385–92.
56. Grundmann H, Glasner C, Albiger B, et al. Occurrence of carbapenemase-producing Klebsiella pneumoniae and *Escherichia coli* in the European survey of carbapenemase-producing Enterobacteriaceae (EuSCAPE): a prospective, multinational study. Lancet Infect Dis. 2017;17:153–63.
57. Lam KN, Cheng JJ, Engel K, Neufeld JD, Charles TC. Current and future resources for functional metagenomics. Front Microbiol. 2015;6:1196.
58. Gillings MR, Gaze WH, Pruden A, Smalla K, Tiedje JM, Zhu YG. Using the class 1 integron-integrase gene as a proxy for anthropogenic pollution. ISME J. 2015;9:1269–79.
59. Deng Y, Bao XR, Ji LL, et al. Resistance integrons: class 1, 2 and 3 integrons. Ann Clin Microbiol Antimicrob. 2015;14:45.
60. Barraud O, Ploy MC. Diversity of class 1 integron gene cassette rearrangements selected under antibiotic pressure. J Bacteriol. 2015;197:2171–8.
61. Ghaly TM, Chow L, Asher AJ, Waldron LS, Gillings MR. Evolution of class 1 integrons: mobilization and dispersal via food-borne bacteria. PLoS One. 2017;12:e0179169.
62. Essa AMM, Julian DJ, Kidd SP, Brown NL, Hobman JL. Mercury resistance determinants related to Tn21, Tn1696, and Tn5053 in enterobacteria from the preantibiotic era. Antimicrob Agents Chemother. 2003;47:1115–9.
63. Kholodii G, Mindlin S, Petrova M, Minakhina S. Tn5060 from the Siberian permafrost is most closely related to the ancestor of Tn21 prior to integron acquisition. FEMS Microbiol Lett. 2003;226:251–5.
64. Liebert CA, Hall RM, Summers AO. Transposon Tn21, flagship of the floating genome. Microbiol Mol Biol Rev. 1999;63:507.
65. Gillings M, Boucher Y, Labbate M, Holmes A, Krishnan S, Holley M, Stokes HW. The evolution of class 1 integrons and the rise of antibiotic resistance. J Bacteriol. 2008;190:5095–100.
66. Dawes FE, Kuzevski A, Bettelheim KA, Hornitzky MA, Djordjevic SP, Walker MJ. Distribution of class 1 integrons with IS26-mediated deletions in their 3′-conserved segments in *Escherichia coli* of human and animal origin. PLoS One. 2010;5:e12754.
67. Amos GCA, Ploumakis S, Zhang L, Hawkey PM, Gaze WH, Wellington EMH. The widespread dissemination of integrons throughout bacterial communities in a riverine system. ISME J. 2018;12:681–91.
68. Randall CP, Gupta A, Jackson N, Busse D, O'Neill AJ. Silver resistance in Gram-negative bacteria: a dissection of endogenous and exogenous mechanisms. J Antimicrob Chemother. 2015;70:1037–46.
69. Izumiya H, Sekizuka T, Nakaya H, et al. Whole-genome analysis of Salmonella enterica serovar Typhimurium T000240 reveals the acquisition of a genomic island involved in multidrug resistance via IS1 derivatives on the chromosome. Antimicrob Agents Chemother. 2011;55:623–30.
70. Baker-Austin C, Wright MS, Stepanauskas R, McArthur JV. Co-selection of antibiotic and metal resistance. Trends Microbiol. 2006;14:176–82.
71. Poole K. At the nexus of antibiotics and metals: the impact of Cu and Zn on antibiotic activity and resistance. Trends Microbiol. 2017;25:820–32.

72. Becerra-Castro C, Machado RA, Vaz-Moreira I, Manaia CM. Assessment of copper and zinc salts as selectors of antibiotic resistance in Gram-negative bacteria. Sci Total Environ. 2015;530-531:367–72.
73. Yu Z, Gunn L, Wall P, Fanning S. Antimicrobial resistance and its association with tolerance to heavy metals in agriculture production. Food Microbiol. 2017;64:23–32.
74. Douard G, Praud K, Cloeckaert A, Doublet B. The Salmonella genomic island 1 is specifically mobilized in trans by the IncA/C multidrug resistance plasmid family. PLoS One. 2010;5:e15302.
75. Carraro N, Matteau D, Burrus V, Rodrigue S. Unraveling the regulatory network of IncA/C plasmid mobilization: when genomic islands hijack conjugative elements. Mob Genet Elements. 2015;5:1–5.
76. Carraro N, Rivard N, Burrus V, Ceccarelli D. Mobilizable genomic islands, different strategies for the dissemination of multidrug resistance and other adaptive traits. Mob Genet Elements. 2017;7:1–6.
77. Bertelli C, Laird MR, Williams KP, et al. IslandViewer 4: expanded prediction of genomic islands for larger-scale datasets. Nucleic Acids Res. 2017;45:W30–5.
78. Stern A, Sorek R. The phage-host arms race: shaping the evolution of microbes. BioEssays. 2011;33:43–51.
79. Carter MQ, Pham A, Huynh S, He X. Complete genome sequence of a Shiga toxin-producing Enterobacter cloacae clinical isolate. Genome Announc. 2017;5:e00883–17.
80. Arnold JW, Koudelka GB. The Trojan Horse of the microbiological arms race: phage-encoded toxins as a defence against eukaryotic predators. Environ Microbiol. 2014;16:454–66.
81. Steinberg KM, Levin BR. Grazing protozoa and the evolution of the *Escherichia coli* O157:H7 Shiga toxin-encoding prophage. Proc Biol Sci. 2007;274:1921–9.
82. Zinder ND, Lederberg J. Genetic exchange in Salmonella. J Bacteriol. 1952;64:679–99.
83. Haaber J, Leisner JJ, Cohn MT, et al. Bacterial viruses enable their host to acquire antibiotic resistance genes from neighbouring cells. Nat Commun. 2016;7:13333.
84. Keen EC, Bliskovsky VV, Malagon F, Baker JD, Prince JS, Klaus JS, Adhya SL. Novel "Superspreader" bacteriophages promote horizontal gene transfer by transformation. MBio. 2017;8:e02115–6.
85. Enault F, Briet A, Bouteille L, Roux S, Sullivan MB, Petit MA. Phages rarely encode antibiotic resistance genes: a cautionary tale for virome analyses. ISME J. 2017;11:237–47.
86. Sazinas P, Smith C, Suhaimi A, Hobman JL, Dodd CE, Millard AD. Draft genome sequence of the bacteriophage vB_Eco_slurp01. Genome Announc. 2016;4:e01111–6.
87. Kleinheinz KA, Joensen KG, Larsen MV. Applying the ResFinder and VirulenceFinder web-services for easy identification of acquired antibiotic resistance and E. coli virulence genes in bacteriophage and prophage nucleotide sequences. Bacteriophage. 2014;4:e27943.
88. Shousha A, Awaiwanont N, Sofka D, et al. Bacteriophages isolated from chicken meat and the horizontal transfer of antimicrobial resistance genes. Appl Environ Microbiol. 2015;81:4600–6.
89. Jiang SC, Paul JH. Gene transfer by transduction in the marine environment. Appl Environ Microbiol. 1998;64:2780–7.
90. Kok J, van der Vossen JM, Venema G. Construction of plasmid cloning vectors for lactic streptococci which also replicate in Bacillus subtilis and *Escherichia coli*. Appl Environ Microbiol. 1984;48:726–31.
91. Breitbart M, Salamon P, Andresen B, et al. Genomic analysis of uncultured marine viral communities. Proc Natl Acad Sci U S A. 2002;99:14250–5.
92. Rohwer F, Edwards R. The Phage Proteomic Tree: a genome-based taxonomy for phage. J Bacteriol. 2002;184:4529–35.
93. Duhaime MB, Deng L, Poulos BT, Sullivan MB. Towards quantitative metagenomics of wild viruses and other ultra-low concentration DNA samples: a rigorous assessment and optimization of the linker amplification method. Environ Microbiol. 2012;14:2526–37.

94. Lekunberri I, Villagrasa M, Balcazar JL, Borrego CM. Contribution of bacteriophage and plasmid DNA to the mobilization of antibiotic resistance genes in a river receiving treated wastewater discharges. Sci Total Environ. 2017;601-602:206–9.
95. Lekunberri I, Subirats J, Borrego CM, Balcazar JL. Exploring the contribution of bacteriophages to antibiotic resistance. Environ Pollut. 2017;220:981–4.
96. Duhaime MB, Sullivan MB. Ocean viruses: rigorously evaluating the metagenomic sample-to-sequence pipeline. Virology. 2012;434:181–6.
97. Roux S, Enault F, Hurwitz BL, Sullivan MB. VirSorter: mining viral signal from microbial genomic data. PeerJ. 2015;3:e985.
98. Roux S, Krupovic M, Debroas D, Forterre P, Enault F. Assessment of viral community functional potential from viral metagenomes may be hampered by contamination with cellular sequences. Open Biol. 2013;3:130160.
99. Sun M, Ye M, Jiao W, et al. Changes in tetracycline partitioning and bacteria/phage-comediated ARGs in microplastic-contaminated greenhouse soil facilitated by sophorolipid. J Hazard Mater. 2018;345:131–9.
100. Calero-Caceres W, Mendez J, Martin-Diaz J, Muniesa M. The occurrence of antibiotic resistance genes in a Mediterranean river and their persistence in the riverbed sediment. Environ Pollut. 2017;223:384–94.
101. Yang Y, Shi W, Lu SY, et al. Prevalence of antibiotic resistance genes in bacteriophage DNA fraction from Funan River water in Sichuan, China. Sci Total Environ. 2018;626:835–41.
102. Larranaga O, Brown-Jaque M, Quiros P, Gomez-Gomez C, Blanch AR, Rodriguez-Rubio L, Muniesa M. Phage particles harboring antibiotic resistance genes in fresh-cut vegetables and agricultural soil. Environ Int. 2018;115:133–41.
103. Munk P, Andersen VD, de Knegt L, et al. A sampling and metagenomic sequencing-based methodology for monitoring antimicrobial resistance in swine herds. J Antimicrob Chemother. 2017;72:385–92.
104. Spencer SJ, Tamminen MV, Preheim SP, et al. Massively parallel sequencing of single cells by epicPCR links functional genes with phylogenetic markers. ISME J. 2016;10:427–36.
105. Dos Santos DF, Istvan P, Quirino BF, Kruger RH. Functional metagenomics as a tool for identification of new antibiotic resistance genes from natural environments. Microb Ecol. 2017;73:479–91.
106. Simon C, Daniel R. Construction of small-insert and large-insert metagenomic libraries. Methods Mol Biol. 2017;1539:1–12.
107. Boolchandani M, Patel S, Dantas G. Functional metagenomics to study antibiotic resistance. Methods Mol Biol. 2017;1520:307–29.
108. Parnanen K, Karkman A, Tamminen M, Lyra C, Hultman J, Paulin L, Virta M. Evaluating the mobility potential of antibiotic resistance genes in environmental resistomes without metagenomics. Sci Rep. 2016;6:35790.
109. Lorenz MG, Wackernagel W. Bacterial gene transfer by natural genetic transformation in the environment. Microbiol Rev. 1994;58:563–602.
110. Page WJ, Vontigerstrom M. Optimal conditions for transformation of azotobacter-vinelandii. J Bacteriol. 1979;139:1058–61.
111. Trombe MC. Characterization of a calcium porter of streptococcus-pneumoniae involved in calcium regulation of growth and competence. J Gen Microbiol. 1993;139:433–9.
112. Baur B, Hanselmann K, Schlimme W, Jenni B. Genetic transformation in freshwater: *Escherichia coli* is able to develop natural competence. Appl Environ Microbiol. 1996;62:3673–8.
113. Overballe-Petersen S, Harms K, Orlando LAA, et al. Bacterial natural transformation by highly fragmented and damaged DNA. Proc Natl Acad Sci U S A. 2013;110:19860–5.
114. Seitz P, Blokesch M. Cues and regulatory pathways involved in natural competence and transformation in pathogenic and environmental Gram-negative bacteria. FEMS Microbiol Rev. 2013;37:336–63.
115. de Aldecoa ALI, Zafra O, Gonzalez-Pastor JE. Mechanisms and regulation of extracellular DNA release and its biological roles in microbial communities. Front Microbiol. 2017;8:1390.

116. Zhu YG, Gillings M, Simonet P, Stekel D, Banwart S, Penuelas J. Human dissemination of genes and microorganisms in Earth's Critical Zone. Glob Chang Biol. 2018;24:1488–99.
117. Baker M, Hobman JL, Dodd CER, Ramsden SJ, Stekel DJ. Mathematical modelling of antimicrobial resistance in agricultural waste highlights importance of gene transfer rate. FEMS Microbiol Ecol. 2016;92:fiw040.
118. van Bunnik BAD, Woolhouse MEJ. Modelling the impact of curtailing antibiotic usage in food animals on antibiotic resistance in humans. R Soc Open Sci. 2017;4:161067.
119. Volkova VV, Lanzas C, Lu Z, Grohn YT. Mathematical model of plasmid-mediated resistance to ceftiofur in commensal enteric *Escherichia coli* of cattle. PLoS One. 2012;7:e36738.
120. Haber M, Levin BR, Kramarz P. Antibiotic control of antibiotic resistance in hospitals: a simulation study. BMC Infect Dis. 2010;10:254.
121. Volkova VV, Lu Z, Lanzas C, Scott HM, Grohn YT. Modelling dynamics of plasmid-gene mediated antimicrobial resistance in enteric bacteria using stochastic differential equations. Sci Rep. 2013;3:2463.
122. Herman D, Thomas CM, Stekel DJ. Adaptation for protein synthesis efficiency in a naturally occurring self-regulating operon. PLoS One. 2012;7:e49678.
123. Paulsson J, Ehrenberg M. Trade-off between segregational stability and metabolic burden: a mathematical model of plasmid ColE1 replication control. J Mol Biol. 1998;279:73–88.
124. Paulsson JP, Ehrenberg M. Molecular clocks reduce plasmid loss rates: the R1 case. J Mol Biol. 2000;297:179–92.
125. Lardon LA, Merkey BV, Martins S, Dotsch A, Picioreanu C, Kreft JU, Smets BF. iDynoMiCS: next-generation individual-based modelling of biofilms. Environ Microbiol. 2011;13:2416–34.
126. Merkey BV, Lardon LA, Seoane JM, Kreft JU, Smets BF. Growth dependence of conjugation explains limited plasmid invasion in biofilms: an individual-based modelling study. Environ Microbiol. 2011;13:2435–52.
127. Bellanger X, Guilloteau H, Bonot S, Merlin C. Demonstrating plasmid-based horizontal gene transfer in complex environmental matrices: a practical approach for a critical review. Sci Total Environ. 2014;493:872–82.

Index

H. Nishida, T. Oshima (eds.), *DNA Traffic in the Environment*,
https://doi.org/10.1007/978-981-13-3411-5

MIX
Papier aus verantwortungsvollen Quellen
Paper from responsible sources
FSC® C105338

If you have any concerns about our products,
you can contact us on
ProductSafety@springernature.com

In case Publisher is established outside the EU,
the EU authorized representative is:
Springer Nature Customer Service Center GmbH
Europaplatz 3, 69115 Heidelberg, Germany

Printed by Libri Plureos GmbH
in Hamburg, Germany